Part B of *Planetary astronomy from the Renaissance to the rise of astrophysics* is the sequel to Part A (Tycho Brahe to Newton), and continues the history of celestial mechanics and observational discovery through the eighteenth and nineteenth centuries. Twelve different authors (astronomers, historians of astronomy, celestial mechanists and a statistician) have contributed their expertise in some 17 chapters, each of them intended to be accessible to the interested layman. An initial section (six chapters) deals with stages in the reception of Newton's inverse-square law as exact. In the remainder of the book a large place is given to the development of the mathematical theory of celestial mechanics from Clairaut and Euler to LeVerrier, Newcomb, Hill, and Poincaré – a topic rarely discussed – treated at once synoptically and in some detail. This emphasis is balanced by other chapters on observational discoveries and the rapprochement of observation and theory (for instance, the discovery of Uranus and the asteroids, use of Venus transits to refine solar parallax, introduction of the method of least squares, and the development of planetary and satellite ephemerides). Lists of 'further reading' provide entrée to the literature of the several topics.

THE GENERAL HISTORY OF ASTRONOMY

Volume 2

Planetary astronomy from the Renaissance
to the rise of astrophysics

Part B: The eighteenth and nineteenth centuries

THE GENERAL HISTORY OF ASTRONOMY

General Editor: Michael Hoskin, University of Cambridge

Volume 2

Planetary astronomy from the Renaissance to the rise of astrophysics
Part B: The eighteenth and nineteenth centuries

EDITED BY

RENÉ TATON

Centre Alexandre Koyré, Paris

and

CURTIS WILSON

St John's College, Annapolis, Maryland

CAMBRIDGE
UNIVERSITY PRESS

CAMBRIDGE UNIVERSITY PRESS
Cambridge, New York, Melbourne, Madrid, Cape Town, Singapore, São Paulo, Delhi

Cambridge University Press
The Edinburgh Building, Cambridge CB2 8RU, UK

Published in the United States of America by Cambridge University Press, New York

www.cambridge.org
Information on this title: www.cambridge.org/9780521120098

First published 1995
This digitally printed version 2009

A catalogue record for this publication is available from the British Library

Library of Congress Cataloguing in Publication data

Planetary astronomy from the Renaissance to the rise
of astrophysics.
(The general history of astronomy; v. 2)
"Published under the auspices of the International
Astronomical Union and the International Union for the
History and Philosophy of Science."
Includes bibliographies and index.
Contents: pt. A. Tycho Brahe to Newton – pt. B. The
eighteenth and nineteenth centuries.
1. Astronomy – History.   2. Astrophysics – History.
I. Taton, René. II. Wilson, Curtis. III. Series.
QB15.G38 1984 vol. 2   520'.9s   88-25817
ISBN 0-521-24265-1 (pt. A) [520'.9'03]

ISBN 978-0-521-35168-3 hardback
ISBN 978-0-521-12009-8 paperback

CONTENTS

CONTENTS OF VOLUME 2A

PREFACE

Volume 2 of *The General History of Astronomy* deals with the history of the descriptive and theoretical astronomy of the solar system, from the late sixteenth century to the end of the nineteenth century.

In the European tradition from the time of Plato to the sixteenth century, theoretical astronomy viewed its task as the reduction of the apparent celestial movements to combinations of uniform circular motions. This formulation was still axiomatic for Copernicus; indeed, Ptolemy's violation of the axiom in his *Almagest* was an important stimulus leading Copernicus to undertake his renovation of astronomy. To many astronomers working in the later sixteenth century, after the publication of Copernicus's *De revolutionibus* (1543), the major achievement of this work was that it had freed astronomy from such violations of principle; in contrast, the associated rearrangement of circles that put the Sun at the centre and attributed motion to the Earth was, in the view of many, an absurd error. Copernicus's work, both in its adherence to the long-held axiom of uniform circular motion and in its organization, was thoroughly traditional; and it is thus fitting that Volume 1 of *The General History of Astronomy*, which is devoted to the history of ancient and medieval astronomy, should conclude with it. But, as all the world knows, the Copernican rearrangement contained the seeds of a further transformation.

The story of that transformation begins with the work of the Danish astronomer Tycho Brahe, which is also the starting-point for Volume 2. In 1563 young Tycho, then a student at Leipzig, was shocked to discover that the accepted astronomical tables of the day (Alfonsine based on Ptolemaic theory, and the Prutenic based on Copernican theory) were, both of them, days out in predicting the conjunction of Jupiter and Saturn that took place on 24 August of that year. Tycho's response, over the remaining years of the century, was to give to the accumulation of accurate astronomical observations a priority and importance it had never had before. The store of observations he accumulated in the process became the empirical base for Kepler's justly titled *Astronomia nova* (*New Astronomy*), which made possible a new level of accuracy in the prediction of planetary motions.

Inextricably bound up in Kepler's theory was a new celestial physics, founded on the Copernican vision of the solar system. It was an attempt to account for the observed motions by means of hypothesized quasi-magnetic forces and structures. Kepler did not and could not derive his 'laws' of planetary motion – elliptical orbit and areal rule – from observations alone; on the contrary, his analysis of Tycho's observations was directed by his own highly conjectural celestial physics. Consequently, his contemporaries and successors could not but find the hypothetical foundations dubious. Yet the new predicative accuracy that Kepler achieved was too remarkable to be ignored. Astronomers were ineluctably faced with a challenge: to achieve results as successful as Kepler's, but with more convincing physical foundations. It is not easy to imagine what the history would have been if Kepler's *Astronomia nova* had never appeared to pose this problem. At least we can say: exceedingly different.

The seventeenth century was a period of complex transition in thought, belief, and knowledge of the world; delineation of the aspects of the transition that bear on astronomy is a major concern in Part A of Volume 2. From 1610 onward, telescopic observations of the heavens revealed new facts that had to be incorporated in the system of the world, whether this was conceived to be geocentric or heliocentric. Studies of refraction and parallax showed the Earth to be some 18 times further from the Sun than had previously been believed, and so nearly 6000 times smaller, in relation to the volume of the Sun. New studies of motion and force, of falling bodies and the impacts of bodies, were proposed by Galileo, Descartes, and others, a major motive being to show that the heliocentric system was not incompatible with physical principle. The Keplerian planetary tables were tested repeatedly, and continued to prove superior to earlier tables.

An impressively persuasive replacement for Kepler's celestial physics was at length provided in 1687 by Isaac Newton's *Principia*. Part A of Volume 2 concludes with the story of its emergence, and a summary of the astronomical results that Newton succeeded in deriving from his principle of universal gravitation.

Part B, after a section devoted to the gradual acceptance of the Newtonian doctrine during the first half of the eighteenth century, takes up the history of the efforts, from the 1740s onward, to deduce detailed mathematical consequences from universal gravitation. It is the story of what the eighteenth century called 'physical astronomy', but what Laplace in 1799 renamed '*mécanique céleste*', celestial mechanics. The challenge of deducing the consequences triggered the development of new forms of mathematics. Of notable importance here were the trigonometric series, first introduced by Euler, but with later contributions from d'Alembert, Clairaut, Lagrange, and Laplace; a new mechanics of the rotation of rigid bodies, to which d'Alembert and Euler were the chief initial contributors; and the method of variation of the constants of integration in the solution of differential equations, developed by Euler, Lagrange, and Laplace. The story also involves the introduction of statistical calculations into astronomy: at first, with a less than satisfactory outcome, by Euler; then more successfully by Tobias Mayer, whose example was followed by Laplace. With Laplace and Gauss the use of statistical procedures in bringing multiple data to bear on the determination of astronomical constants became *de rigeur*.

Through the nineteenth century the grand theoretical questions remained those that Laplace had forcefully posed and prematurely claimed to answer: the question of the stability of the solar system, and the question of the adequacy of universal gravitation to account for the observed motions of its constituent bodies. By the end of the century Simon Newcomb and his collaborators would achieve a precision in their prediction of planetary motions measured in seconds or fractions of a second of arc. Yet some inconsistencies remained, among them an anomalous motion of the perihelion of Mercury. The first of these was a statistical artifact; the second would become evidence for a new theory of gravitation, which would supersede that of Newtonian physics.

The scope of Volume 2 does not include general relativity. Nor does it embrace astrophysics, which came into being in the 1850s with the development by Kirchhoff and Bunsen of spectral analysis. Both topics belong to volume 4, *Astrophysics and Twentieth-century Astronomy to 1950*. Similarly excluded from the scope of Volume 2 are the topics of stellar astronomy and cosmology, and astronomical instruments, institutions and education, from the Renaissance to the beginnings of astrophysics: these topics constitute the subject-matter of Volume 3. At certain points the concerns of Volumes 2 and 3 overlap. Tycho's sighting instruments were highly relevant to his observational achievement; and from the 1660s and 1670s onward, the pendulum clock, telescopic sights, and the filar micrometer were similarly relevant to the programmes for the improvement of lunar and planetary tables that were adopted by the newly founded Paris Observatory and Greenwich University. Again, Bradley's discoveries of the aberration of light and nutation were prerequisites for the attainment of seconds-of-arc accuracy in the prediction of planetary and lunar positions. For these and similar topics concerned with the stars and with observational instruments and their institutional context the reader must be referred to Volume 3.

Our aim in Volume 2, as in the other volumes of the series, has been to throw light on the development of astronomy as an inventive human activity. We have sought to view the questions and problems of the astronomers of a given time in the very way in which those astronomers saw them, without regard for what has later come to be accepted as 'correct'. We have not attempted an encyclopaedic completeness of coverage; rather, our goal has been to provide an intelligible account of the major endeavours through which astronomy has evolved.

A word is in order with respect to the division of tasks between the two editors of Volume 2, René Taton and myself. Professor Taton resigned his editorship in 1983. By this time he had drawn up the general plan of the volume, and had engaged authors to write rather more than half of the sections envisaged. The engaging of authors for the remaining sections has been my responsibility, and I have also undertaken some reorganization of materials, reducing the original tripartite scheme to a two-part plan, and introducing into Part B a number of new sections which focus on the application of theory to observation.

Finally, I have the grateful duty of acknowledging the extensive guidance and generous assistance provided a neophyte editor by the General Editor of the series, Michael Hoskin. He has again and again given generously of his time, knowledge, and thought to the solution of the problems and difficulties, whether major or minor, encountered in the assembling and editing of this volume. Our undertaking has been, in every aspect, a joint endeavour.

Annapolis, Maryland   Curtis Wilson

PART V

Early phases in the reception of Newton's theory

14

The vortex theory in competition with Newtonian celestial dynamics

ERIC J. AITON

Evidence of the intimate acquaintance that Isaac Newton (1642–1727) had with René Descartes's *Principia philosophiae* (1644) is to be found in some of his earliest manuscripts. Thus when Newton in his *Waste Book* formulates the principle of inertia and the "endeavour from the centre" of a body moved circularly, both the ideas and the wording clearly echo Descartes's *Principia*.

In a manuscript beginning "De gravitatione et aequipondio fluidorum . . .", variously dated to the late 1660s and to the early 1680s, Newton raised fundamental objections to certain Cartesian principles. He rejected Descartes's identification of space and body (which, he claimed, offered a path to atheism) and he demonstrated the inconsistency of Descartes's relativistic concept of motion. Moreover, he observed that the rotation of vortices, from which Descartes deduced the force of the aether to recede from the centre (and thus the whole of his mechanical philosophy) implied a space distinct from bodies as a reference frame. Against the existence of Descartes's non-resisting aether or celestial matter, he argued that, if the resistance were set aside, so would be the corporeal nature; for the faculties of stimulating perception and moving other bodies were essential to matter.

Whenever it was that these objections were formulated, it must be allowed that many of Newton's statements during the 1660s and 1670s argue his acceptance during these years of the hypothesis of aethereal vortices. In 1675 he sent to Henry Oldenburg for presentation to the Royal Society a manuscript entitled "An Hypothesis explaining the Properties of Light discoursed of in my severall Papers", in which he developed the hypothesis of a universal aether as agent for electrical, magnetic and gravitational forces. He supposed this aethereal medium to be "much of the same constitution with air, but far rarer, subtiler

and more strongly elastic". In this manuscript he also mentioned "the aethers in the vortices of the Sun and planets". Again he alluded to the Cartesian theory of vortices in 1680 in a letter to Thomas Burnet.

But when he took up anew the problem of the planetary motions following the correspondence with Robert Hooke in 1679 and 1680, Newton evidently abandoned the hypothesis of aethereal mechanisms. His objections to the Cartesian principles were brought before the public in a forceful and well-reasoned attack in the first edition of the *Principia* (1687). In particular, Newton adduced experimental evidence against the existence of the aether and claimed to demonstrate that the Cartesian vortices could not explain the motions of the planets in accordance with Kepler's laws.

Newton's experiment, which he remarked had been performed some time before, concerned a comparison of the resistance experienced by a pendulum consisting of a suspended wooden box, first when empty and then when filled with metal. The points of return of the empty box after the first, second and third oscillations having been marked, it was found that the full box required 77 oscillations to return to the first mark, and then the same number to return to each of the others in succession. As the weight of the full box was 78 times that of the empty box, Newton concluded that, if the resistances had been equal, the full box (by virtue of its *vis insita*) should have taken 78 oscillations to return to each of the marks. Consequently, the ratio of the resistance in the case of the empty box to that in the case of the full box was 77 to 78. Taking x to represent the resistance on the external surface of the box and y to represent the resistance on its internal superficies (that is, the internal superficies of the wood, so that the resistance on the internal superficies of the wood and metal of the full box will

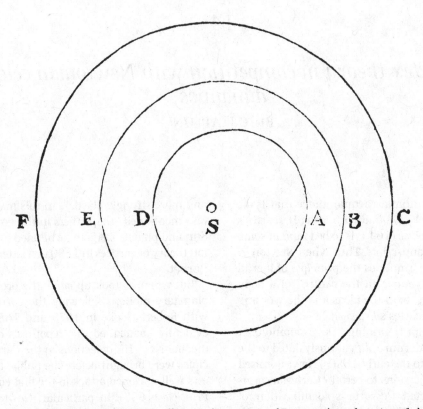

14.1. Newton's diagram in *Principia* (1687) to illustrate his critique of Descartes's explanation of planetary orbits.

be 78*y*), the ratio of the total resistance in the case of the empty box to that in the case of the full box becomes $x + y$ to $x + 78y$. From the equation $(x + y):(x + 78y) = 77:78$, the ratio $x:y$ is found to be 5928:1. The resistance on the internal superficies was therefore less than $\frac{1}{5000}$th part of the resistance on the external surface. This reasoning, Newton remarked, depended on the supposition that the greater resistance in the case of the full box "arises from the action of some subtle fluid upon the included metal". But as the period of the full box was less than that of the empty box, the resistance on the external surface was greater because of the greater speed, and greater again because of the longer distances traversed in oscillating. These considerations showed that "the resistance of the internal parts of the box will be either nil or wholly insensible". Newton directed his demonstration against "the opinion of some, that there is a certain aethereal medium extremely rare and subtle, which freely pervades the pores of all bodies".

Descartes had supposed that the planets floated in the aether moving with the same speeds as the

layers of the rotating vortex in which they were situated. Newton claimed to demonstrate that a solid planet (subject to no forces except that of the aether) could not remain continually in the same orbit unless it was of the same density as the fluid. For if the planet had a greater density, it would have a greater endeavour to recede from the centre (that is, it would have more centrifugal force) and would move further away. Descartes's aether would thus have had to be as dense as the planets themselves.

In a scholium at the end of the section on the circular motion of fluids, Newton concluded that the planets were not carried around in corporeal vortices. For, according to the "Copernican hypothesis", the planets revolve around the Sun in ellipses having the Sun in their common focus, and by radii drawn to the Sun describe areas proportional to the times, whereas "the parts of a vortex can never revolve with such a motion". Suppose that *AD*, *BE*, *CF* (Figure 14.1) represent three orbits, the outermost (for convenience) being taken concentric with the Sun *S*, while *A* and *B* are the aphelia of the others. Newton argued that,

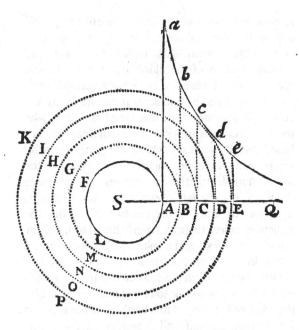

14.2. Newton's diagram in *Principia* (1687) to illustrate his demonstration that planetary motions in a vortex are incompatible with Kepler's third law.

because the space *AC* is narrower than the space *DF*, the fluid of the vortex (and hence the planet) moving in the orbit *BE* would have a greater speed at *B* than at *E*, whereas observations revealed the opposite to be the case.

In order to demonstrate the inconsistency of the explanation of planetary motions in terms of a fluid vortex with Kepler's third law, Newton investigated vortices in an infinite fluid maintained by a rotating cylinder and sphere respectively. He adopted the hypothesis that the resistance or friction between adjacent layers of fluid is proportional to the relative speed and to the area in contact. As Newton followed the same line of reasoning in the two cases, his method may be illustrated by a description of his demonstration for the spherical vortex. First, he supposed the vortex to be divided into solid layers of equal thickness. Let *S* (Figure 14.2) represent the central body, x the radius, dx the thickness and v the speed of a point on the equator of a layer. As each layer moves uniformly, the impressions made by the adjacent superior and inferior layers must be equal and opposite, so that the force must be the same for all layers. According to the hypothesis, the force is proportional to the surface area in contact and the relative speed, here

referred to by Newton as the translations of the layers from one another. Perhaps this terminology was intended to indicate the fact that the relative speed varies with the latitude. Although Newton's demonstration lacks a mathematical treatment of the fluid in different latitudes (perhaps he intended this to be supplied by the reader), such an analysis is quite simple and leads to a result consistent with Newton's conclusions. Taking the relative speed at the equator as a measure of the 'translation', the force on a surface of a layer is proportional to the area (consequently to x^2) and to $-xd(v/x)$. Since the force is the same for all layers, it follows that $-x^3d(v/x) = cdx$, where c is a constant, so that $v = bx + c/x$, where b is another constant. Newton in fact only achieved the particular solution $v = c/x$, from which he deduced that the periodic times of the layers are proportional to x^2.

In the case of the cylindrical vortex, the general solution is $v = bx + c$, but Newton gave only the particular solution $v = c$, and deduced that the periodic times are proportional to x.

In both cases Newton had first supposed that the layers rotated as if they were solid. Then, in the spherical vortex, he reasoned that, if the layers were divided into zones and then the matter in these zones made fluid, the vortex would still rotate in exactly the way he had described. Recognizing the weakness of his argument, however, he added:

But now, as the circular motion, and the centrifugal force thence arising, is greater at the ecliptic than at the poles, there must be some cause operating to retain the several particles in their circles; otherwise the matter that is at the ecliptic will always recede from the centre, and come round about to the poles by the outside of the vortex, and from thence return by the axis to the ecliptic with a perpetual circulation.

Although he thus had to admit that his analysis of vortex motion was at least incomplete, Newton nevertheless claimed that the periodic times could not be brought into line with Kepler's law without hypotheses relating density and friction to distance which he considered to be unreasonable. Again, a number of contiguous vortices, such as Descartes described, would in Newton's view run into each other and so gradually destroy themselves. The formation of a vortex wake behind an obstacle had been observed by Leonardo da Vinci but evidently not by Newton. Having demonstrated to his own satisfaction that the "hypothesis of vortices is utterly irreconcileable with astronomical phenomena", Newton referred the reader to Bks I and III of the *Principia* for an explanation of how "these motions are performed in free space without vortices".

Vortices between the first and third editions of Newton's *Principia*

The first edition of Newton's *Principia* appeared one year after the anonymous publication of Bernard le Bouyer de Fontenelle's *Entretiens sur la pluralité des mondes*. In numerous editions, this popular work introduced the general ideas of the vortex theory to a wide readership. The second edition of Newton's work was published in 1713, one year after the sixth (definitive) edition of Nicolas Malebranche's *Recherche de la vérité*. Malebranche, the leading Cartesian among Newton's contemporaries, completely ignored Newton's *Principia*, even after taking up the study of the differential calculus with the Marquis de l'Hospital, though he described Newton's *Opticks* as an excellent work. When the third and definitive edition of the *Principia* appeared one year before Newton's death, the problem of the Moon's motion (which might be regarded as the Achilles's heel of his system) had not yet been solved.

Four reviews of the first edition of Newton's *Principia* had appeared within a year of its publication. The first was in fact a pre-publication review by Edmond Halley in the *Philosophical Transactions*. This gave a description of the contents, including the remark: "the Cartesian doctrine of the vortices of the celestial matter carrying with them the planets about the Sun, is proved to be altogether impossible". Halley praised Newton throughout but there was perhaps a hint of criticism for Newton's failure to give Johannes Kepler adequate credit in his pointed attribution of the phenomena of celestial motions to Kepler.

Although the other reviews are anonymous, there are some indications that the one in the *Bibliothèque universelle* may have been written by John Locke. This review just listed the contents, though Newton's criticism of vortices was emphasized. Again, the review in the *Acta eruditorum*, now known to have been written by Christoph Pfautz, provided an epitome rather than a critical analysis. In particular, Newton's explanation that

resistance was wholly on the surface of a body and not on the internal superficies, his criticism that vortices were inconsistent with Kepler's laws and his explanation of the planetary motions in free space, without vortices, were all described. Only the review in the *Journal des sçavans* offered critical comment. Its author described the *Principia* as "the most perfect mechanics that we can imagine" but based on "hypotheses which are generally arbitrary" and which can therefore serve only as the foundation of a treatise on mechanics. To complete the work, he advised, "Newton has only to give us a physics as exact as the mechanics". While the reviewer failed to understand the reasoning by which Newton sought to establish universal gravitation in accordance with the inverse-square law (so that he interpreted this as an arbitrary hypothesis), it would be unfair to blame him for regarding the work only as one of mechanics and not physics; for Newton himself had remarked of the words "attraction, impulse or propensity of any sort towards a centre", that he considered those forces not physically but mathematically. This brief review gave little indication of the contents of the *Principia* but it identified the focus to which the criticisms of Newton's opponents were to be directed; namely, his failure to describe the physical cause of the attraction.

None of the reviewers gave the slightest hint of having recognized the weakness of Newton's treatment of fluid vortices. Although his ideas of solidification and interaction of the layers depended on arbitrary assumptions, so that his propositions did not follow from the axioms and laws of motion set out in Bk I, it was only in 1744 that Jean le Rond d'Alembert demonstrated for the first time that the cylindrical and spherical vortices described by Newton could not exist in a permanent state.

The earliest documented criticism of Newton's section on the motion of fluids is probably the marginal note by Gottfried Wilhelm Leibniz (1646–1716) in his copy of the *Principia*, which casts doubt on Newton's hypothesis that the force of resistance between the quasi-solid layers of fluid is proportional to the relative speed of separation. Leibniz's objection stemmed from his distinction between absolute and respective resistance. The first, which he supposed to be the type envisaged in Newton's hypothesis, arose from the rubbing of the particles of fluid against the solid body and was independent of the speed. On the other hand, the respective resistance (not involved in Newton's hypothesis) Leibniz conceived to arise from the impact of fluid on the body and consequently to be proportional to the speed. Although Leibniz offered no further criticism of Newton's treatment of vortices in his marginal notes (which are available to historians in the modern edition of his *Marginalia*), he had in 1689 published his own hypothesis concerning the explanation of the planetary motions by the motion of fluid vortices. This hypothesis, representing the first attempt to reconcile a vortex explanation of the planetary motions with Kepler's laws, will be described in a later section.

Christiaan Huygens (1629–95), who discovered the satellite Titan and the rings of Saturn, accepted the Newtonian system, though with important reservations. In his *Discours sur la cause de la pésanteur*, published in Leiden in 1690, Huygens declared that he had nothing against the *vis centripeta* or gravity of the planets towards the Sun, not only because it was established by experience but also because it could be explained by mechanical principles. The cause of this *vis centripeta*, he suggested, might be similar to that he had proposed for terrestrial gravity; namely the rapid circulation around the central body of a fluid aether in all directions on spherical surfaces, which displaced solid bodies downwards by virtue of its greater centrifugal force. Huygens admitted that he could not offer any explanation of the inverse-square law. But he rejected Newton's universal attraction extending to the smallest particles as an occult quality not explicable by any principle of mechanics. Moreover, he rejected Newton's void on the ground that light could not be transmitted across it.

According to Huygens's interpretation, Newton demonstrated the elliptical orbit by a counterbalancing of the gravity and the centrifugal force. He therefore regarded Newton's theory as a development of that of Giovanni Alfonso Borelli and also, he added, some of the ancients, according to a report of Plutarch. These ideas were set out by Huygens in his *Cosmotheoros*, published posthumously in 1698, where he also described his own modification of the Cartesian theory. Huygens's vortices (surrounding every star) were neither dense nor contiguous but dispersed in space so as not to hinder each other's free rotations. He offered no reason why such vortices, lacking external constraint, should hold together.

When Newton's *Principia* was published in 1687, the influential Cartesian textbook of Jacques Rohault, *Traité de physique* (1671), was well known in England, as on the Continent, for Théophile Bonet's Latin translation (first published in Geneva in 1674) had been published again in Amsterdam and London in 1682. Samuel Clarke, a follower of Newton, in 1697 prepared a new annotated Latin translation of Rohault's work for the use of Cambridge students. At this time he made no criticism of vortices and did not even mention Newton's attraction. In the second edition of 1702, he began to support Newton against Descartes, but the famous refutation of the Cartesian text in the Newtonian footnotes was not fully realized until the 1710 edition.

A new Cartesian work, Abbé Philippe Villemot's *Nouveau système ou nouvelle explication du mouvement des planètes*, published in Lyons in 1707, was praised for its originality by Fontenelle. Malebranche described it as the embryo of an excellent work. He advised Villemot to abandon the opening chapters (which presented a fantastic hypothesis of the formation of the world) and to begin instead with Kepler's third law, which had been received by all astronomers, and then to seek a demonstration of this law from the properties of centrifugal force. Johann I Bernoulli (1667–1748) was less enthusiastic. For he hoped to see in a new edition of Malebranche's *Recherche de la vérité* "the gravity of bodies towards the centre of the earth and the planets towards the sun (which Mr Newton supposes) soundly explained and the phenomena better demonstrated than with Mr Villemot". To Leibniz and Pierre Varignon (1654–1722), the work seemed to have little merit. Leibniz could not find a shadow of a demonstration in it and Varignon commented adversely on Villemot's knowledge of mathematics.

Villemot knew nothing of the calculus and when he came across a copy of Newton's *Principia* shortly before the publication of his own work, he could find nothing in it to cause him to revise his theory. Villemot offered an explanation of Kepler's third law, though he did not mention either the first or second laws. Like all other Cartesians before him, he used the term 'ellipse' only in reference to the lunar variation, while simply remarking of the planetary orbits in general that they were not perfect circles. In his view, the equilibrium of the vortex required the total centrifugal force of each spherical layer to be the same, so that the speed v was proportional to $1/r^{\frac{1}{2}}$, r being the radius of the layer. Consequently, the periodic time t was proportional to $r^{\frac{3}{2}}$. Villemot claimed this to be a "demonstration of the famous problem of Kepler ... of which up to now it has not been possible to render account in any system". Although he recognized that "Ptolemy's theory" required the times needed by a planet to traverse equal arcs in the neighbourhood of the apsides to be proportional to the squares of the distances, he believed that this was only approximately true and that rigorously they "obey Kepler's [third] law exactly".

Villemot also attempted explanations of the precession of the equinoxes, the motions of the planetary and lunar nodes, rotation of the apseline and the rotations of the planets on their axes. The causes of all these phenomena he found in the effects of the celestial matter of the solar and terrestrial vortices. For example, having followed Copernicus in attributing the precession of the equinoxes to the near equality of two rotations resulting in a slow rotation of the Earth's axis relative to the stars, he found the cause of the slight inequality in the difference of the speeds of the aether at the distances of the Earth's centre and its lower extremity from the centre of the solar vortex. The numerical results for precession, rotation of the lunar nodes and the Earth's diurnal motion would have been impressive if his theory had been soundly based, but as we have remarked, its superficiality was clearly evident to Varignon and Leibniz.

Gravity

Gravity was the force, Villemot supposed, "which keeps the heavens in equilibrium, which sustains the planets in their spheres, and in a word which arranges and disposes all the parts of the universe". The acceptance, first by Huygens, then by Villemot and Cartesians generally, of the gravity of the planets towards the Sun demonstrated by Newton, introduced new difficulties for Cartesian explanations of the planetary motions. According to Descartes's explanation, the planets were indeed pushed towards the centre of the vortex but only in the same way as the aether itself; that is, by the external pressure which constrained the whole vortex to circulate. As the planet and surrounding aether circulated with the same speed, neither had more centrifugal force than the other to resist the external pressure, so that both continued to circulate at the same distance from the centre. On the other hand, heavy bodies near the surface of a planet were displaced downwards by the more rapidly moving aether on account of its greater centrifugal force. Thus in Descartes's theory, the causes of deflection of a planet from a straight path and the vertical fall of terrestrial bodies were essentially different. For Newton, of course, they were essentially the same.

Villemot sought to identify the gravity of the planets towards the Sun with terrestrial gravity. But as he supposed, following Descartes, that the planets moved with the same speed as the aether, he deduced that the centrifugal force of the aether could not be the cause of gravity. For this purpose he introduced an outward impulse arising from the hot matter in the centre, so that the aether, impelled upwards, displaced solid bodies downwards. By analogy with rays of light, he deduced the inverse-square law for gravity. While the existence of hot matter at the centre of the solar vortex was obvious, the heat experienced in mines, he declared, provided evidence for the existence of a source of heat in the interior of the Earth. Yet the planets did not fall towards the Sun like heavy bodies towards the Earth. To solve this problem, Villemot introduced a vague force of impact of the planetary vortex on the solar vortex, which served to balance the gravity.

In the sixth edition of the *Recherche de la vérité* (1712), Malebranche modified Villemot's theory of terrestrial gravity while retaining Descartes's explanation of the circulation of the planets. Thus he explained that the gravity of the planets did not come from the centre but from the external compression of the solar vortex. Malebranche gave Villemot's theory of terrestrial gravity some semblance of plausibility by reasoning that the effect of the outward impulse was greater on the small vortices, of which he supposed the aether to be composed, than on solid bodies. Nevertheless the main weakness remained; namely that any theory depending on the action of a fluid on a small body immersed in it without relative motion contradicted the known principles of hydrostatics. For Blaise Pascal had shown that, in a fluid at rest, the pressure was the same in all directions. In the case of the terrestrial vortex, Malebranche supposed, the outward impulse did not arise from the action of hot matter in the centre (as Villemot thought) but from the reaction of the solid Earth on the small elastic vortices. This explanation enabled him to account for the absence of an outward impulse in the solar vortex.

Both Villemot and Malebranche had been led to reject Descartes's theory of terrestrial gravity by the insuperable difficulty they had found in the extreme speed of the aether that Huygens had shown to be necessary. In particular, Malebranche could not see how bodies could escape being carried horizontally with the rapidly moving aether.

The Cartesian theory of terrestrial gravity as the effect of a rapid circulation of the aether was defended by Malebranche's friend Joseph Saurin (1659–1737). First, by a misapplication of mechanical principles, he claimed to show that, contrary to the view of Huygens, the simple circulation about an axis supposed by Descartes could account for the fall of heavy bodies towards the centre of the Earth. Moreover he inferred that a plumb-line on a spherical Earth would not be deflected by the Earth's rotation and he thus rejected Huygens's deduction of the spheroidal form of the Earth from the absence of such a deflection. After this unpromising start, Saurin's second contribution, presented to the Royal Academy of Sciences in Paris in 1709, had considerable merit. Taking Mariotte's experiments on the force of impact of a liquid on a solid surface as a basis, he estimated the feebleness of the force of the aether moving with the speed calculated by Huygens that would be needed to render its effect imperceptible when masked by the air, and so explain why heavy bodies were not carried horizontally by the circulation. Having found the force to be at least three or four million times more feeble than that of air, he speculated that such a non-resisting aether would only be possible if the particles of ordinary matter were small compared with the spaces between, say in the ratio of 1 to 100 000, so that the aether could flow freely between them. Experimental confirmation of this bold idea of the constitution of matter

was first provided by Daniel Bernoulli, who deduced from the Boyle–Mariotte gas law that the particles were infinitesimal compared with the spaces between them.

Saurin had not, of course, rendered the aether incapable of causing gravity. For the mechanism of gravity, in Cartesian physics, was not one of impulsion but of hydrostatic displacement. Solid bodies were not impelled downwards by the force of impact of the aether but were displaced downwards because they had less centrifugal force than the rapidly circulating aether.

This idea of a non-resisting aether that could nevertheless be the cause of gravity might have led to an early reconciliation of the Newtonian and Cartesian systems if it had been generally accepted, though it should be remarked that Saurin himself was firmly attached to the Cartesian system. The idea of a non-resisting aether causing gravity seemed contradictory to Jean Bouillet, who was awarded a prize for an essay on gravity in 1720 by the Royal Academy of Bordeaux. Bouillet rejected Saurin's theory in favour of the ideas of Villemot and Malebranche, which he sought to combine into the explanation of a gravity common to the solar and planetary vortices. First, he supposed the aether to consist of Malebranche's small elastic vortices. Like the celestial matter of Villemot, these could penetrate to the centre of the Earth. In place of Villemot's vague impact of the planetary vortex on the solar vortex, introduced to balance the gravity of the planet and thus prevent it from falling into the Sun, he substituted the centrifugal force of the planetary vortex arising from its motion around the Sun. Then he deduced the inverse-square law from the elasticity of the small vortices. Supposing that the small vortices must have equal quantities of motion, he reasoned that those further from the centre, having a greater speed of translation, must have a smaller speed of rotation about their own centres and hence a smaller elasticity. There was also another reason, he explained, why the elasticity decreased with distance from the centre of the vortex:

The proper motion of the small vortices, or their elasticity, must weaken or diminish to the extent that the layer which contains them finds less resistance from the superior layers, or is less pressed towards the centre of the circulation. But the greater the circle described by the aether, the less it experiences the resistance of the superior layers, the latter having more liberty for filling their motion and consequently less force for pushing those which are inferior to them.

Since each of the two causes weakened the elasticity in proportion to the distance, Bouillet concluded that the elasticity, and hence the effect of gravity, "must diminish according to the squares of the distances"

Leibniz's explanation of planetary motion

According to his own account, Leibniz's "Tentamen de motuum coelestium causis", published in the *Acta eruditorum* in February 1689, was a hasty extemporization of his ideas concerning planetary motion, stimulated by the review of Newton's work in the same journal but written before he had the opportunity to read the *Principia* itself. It is now known that Leibniz's claim to have composed his essay independently of the *Principia* is untrue. For some recently discovered manuscripts show preparatory calculations for the essay that clearly imply a knowledge of Newton's *Principia*. Leibniz attributed to Kepler the idea of the fluid vortex and also (mistakenly) the idea that circular motion engendered a centrifugal force. Descartes, he believed, made use of these ideas without acknowledgement but did not attempt to explain the physical causes of the laws discovered by Kepler, either because he could not reconcile them with his own principles or because he did not believe them true. Evidently Leibniz did not doubt that Descartes had known Kepler's laws.

Leibniz supposed that the vortex carrying the planets rotated in spherical layers. To account for the elliptical orbit he postulated two motions of the planet; a radial motion from layer to layer and a trans-radial motion in which the planet moved with the same speed as the fluid. Kepler's second law, which Leibniz stated explicitly in the exact area form, required the trans-radial speed to vary inversely as the distance, as Kepler had shown in the *Epitome astronomiae Copernicanae*. Consequently, Leibniz described his harmonic vortex or circulation as one in which the speeds of circulation of the layers were inversely proportional to the radii or distances from the centre of circulation. Having supposed the planet M (Figure 14.3) to move in the curve $M_3M_2M_1$ and in equal times to

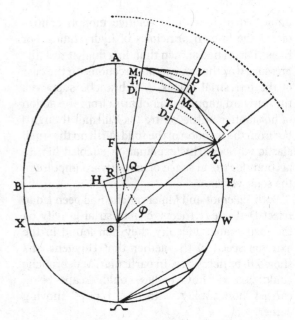

14.3. Diagram by Leibniz to illustrate his theory of planetary motion, published in *Acta eruditorum* in 1689.

describe the elements M_3M_2, M_2M_1 of this curve, he remarked that the motion could be regarded as composed of a circular motion around the Sun $\odot$, whose elements were M_3T_2, M_2T_1, and a rectilinear motion whose elements were T_2M_2, T_1M_1.

Leibniz's work contains at least one major error; namely a calculation of the centrifugal force which gives only half the true value. Following correspondence with Varignon, however, he eliminated this error and published some corrections in 1706. His definitive calculation of centrifugal force was based on the approximation of the circle by a polygon with first-order infinitesimal sides, so that the 'tangent' at a vertex was taken to be the prolongation of the preceding infinitesimal chord and the centrifugal force was expressed by a uniform motion from the circle to the tangent. As Leibniz remarked to Varignon, the method was simpler that did not put the acceleration into the elements. Varignon tried in vain to convince Leibniz that, although the motion in the infinitesimal chord gave correct results, in reality the motion was in the infinitesimal arc.

The original calculation of the gravity did not require correction, probably because (consciously or unconsciously) it had been based on the same principles as the corrected calculation of the centrifugal force. Although he did not make an explicit

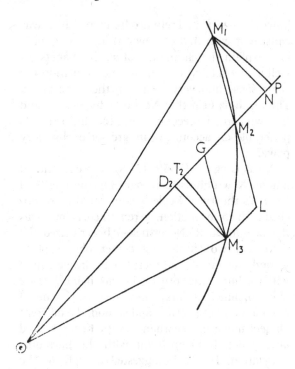

14.4. Interpretation of Leibniz's diagram in accordance
with his definition of tangent.

The radial motion, called by Leibniz *motus para-centricus*, was explained by the combined action of the gravity of the planet towards the Sun and the centrifugal force arising from its circulation with the vortex. Leibniz envisaged two ways of resolving the motion along an infinitesimal element of the curve into components. These are clearly described in a letter to Huygens, where he writes:

the circulation D_1M_2 or D_2M_3 being harmonic, and M_3L parallel to $\odot M_2$, meeting the preceding direction M_1M_2 prolonged in L, then M_1M_2 is equal to M_2L (or to GM_3; the printer has omitted the letter G between T_2 and M_2 marked in my description) and consequently the new direction M_2M_3 is composed as much of the preceding direction M_2L added to the new impression of gravity, that is to say, to LM_3, as of the speed of circulation D_2M_3 of the ambient aether in harmonic progression added to the paracentric speed M_2D_2 already acquired in some progression.

In the case of the first resolution, the planet was supposed to be deflected from its instantaneous inertial path by the action of gravity alone; this would happen in the absence of a vortex.

Consequently Leibniz expressed the effect of the "paracentric solicitation of gravity" by a uniform motion along LM_3. In the case of the second resolution (that of physical reality), the planet was constrained by the vortex to circulate in accordance with Kepler's second law, while the paracentric motion (that is, the motion along the rotating radius vector) was caused by the combined action of gravity and the centrifugal force engendered by the circulation. From the geometry of the figure Leibniz demonstrated that the radial acceleration dd$r$ = centrifugal force – solicitation of gravity.

statement that this was how his calculation was to be understood, Leibniz's demonstrations are consistent with this interpretation. For he expressed the gravity by a uniform motion from the 'tangent' to the curve and the interpretation removes a number of apparent errors, all trivial but otherwise seemingly inexplicable. In the description that follows, the corrections are incorporated and it is assumed that the interpretation according to which the curve is approximated by a polygon with first-order infinitesimal sides is the one that Leibniz held. Such an interpretation is open to the same logical objections as Newton's demonstration of Kepler's second law, in which the curve is similarly approximated.

Leibniz first demonstrated that, in the case of a body moving with harmonic circulation, the centrifugal force varied inversely as the cube of the distance. Let $h\,dt$ be the constant area swept out in equal times dt, equal to twice the area of triangle $M_2M_3\odot$ (Figures 14.3 and 14.4); that is, $D_2M_3 \times \odot M_2$ or $D_2M_3 \times r$, taking $\odot M_2 = r$. Then $D_2M_3 = h\,dt/r$, so that the centrifugal force $2D_2T_2$ (Leibniz's corrected measure) $= (D_2M_3)^2/r = h^2dt^2/r^3$.

From M_1 and M_3, let M_1N and M_3D_2 be normal to $\odot M_2$; as, on account of the harmonic circulation, the triangles $M_1M_2\odot$ and $M_2M_3\odot$ are of equal area, the altitudes M_1N and M_3D_2 also will be equal (because of the common base $\odot M_2$). Taking M_2G equal to LM_3, let M_3G be drawn parallel to M_2L; then the triangles M_1NM_2 and M_3D_2G will be congruent, and M_1M_2 will be equal to GM_3 and NM_2 will be equal to GD_2. Again in line $\odot M_2$ (produced if necessary, as is always understood) let $\odot P$ be taken equal to $\odot M_1$ and $\odot T_2$ equal to $\odot M_3$, then PM_2 will be the difference between the radii $\odot M_1$ and $\odot M_2$, and T_2M_2 the difference between the radii $\odot M_2$ and $\odot M_3$. Now PM_2 equals NM_2 or

$GD_2 + NP$, and T_2M_2 equals $M_2G + GD_2 - D_2T_2$; consequently $PM_2 - T_2M_2$ (the difference of the differences) will be $NP + D_2T_2 - M_2G$, that is (because NP and D_2T_2, the versed sines of two angles and radii having differences less than any assignable magnitude, coincide) twice $D_2T_2 - M_2G$. Now the difference of the radii expresses the radial speed [and] the difference of the differences expresses the element of the radial speed [or the acceleration].

The centrifugal force in the harmonic vortex, $2D_2T_2$, has already been shown to be $h^2\mathrm{d}t^2/r^3$, so that $\mathrm{dd}r = h^2\mathrm{d}t^2/r^3$ – solicitation of gravity. For a body moving in an ellipse in a harmonic vortex, Leibniz demonstrated that $\mathrm{dd}r = h^2\mathrm{d}t^2/r^3 - k\mathrm{d}t^2/r^2$, where $k = h^2/(\text{semilatus rectum})$, from which he deduced the solicitation of gravity to vary inversely as the square of the distance.

The hypothesis of a fluid vortex seemed to Leibniz to be the only possibility remaining after the ideas of planetary intelligences, solid spheres, sympathies and abstruse qualities had been eliminated. Having introduced the harmonic vortex to explain the motion of the planets in accordance with Kepler's second law, he made no attempt to explain the motion of the vortex itself in terms of basic physical principles. Neither did he offer any explanation of gravity in the printed paper – simply alluding to the Cartesian theory (which he attributed to Kepler) – but in another paper published in 1690, he explained gravity as the effect of an outward impulse transmitted through the aether which pushed the planets towards the centre by circumpulsion. By analogy with rays of light, the outward impulse decreases in intensity in proportion to the square of the distance. The outward impulse, he supposed, arose from the circulation of the aether in great circles; this was, in effect, the theory of Huygens. Leibniz distinguished the vortex causing gravity from the harmonic vortex carrying the planets. The two vortices were held to be independent, the first consisting of very tenuous matter revolving in all directions in great circles, the second consisting of coarser matter revolving about an axis with harmonic circulation.

When Huygens pointed out that the harmonic vortex was superfluous, since Newton had demonstrated that gravity alone provided a sufficient physical basis, Leibniz agreed that the hypothesis of Newton explained the elliptical orbit, adding however that he had retained his own (which was consistent with that of Newton) because it could also explain the circulation of all the planets (as well as the satellites of Jupiter and Saturn) in the same sense and very nearly in the same plane. Thus Leibniz held that Newton's hypothesis and his own could be reconciled but that his own was preferable on account of its greater explanatory power.

David Gregory (1659–1708) was exceptional among Newtonians in recognizing the merits of Leibniz's theory. If Kepler's laws could be reconciled with vortex motion, he remarked, Leibniz was the one to do it. Nevertheless he criticized the theory on two counts; first, comets obey Kepler's second law and should therefore be drawn round with the harmonic vortex, whereas their orbits are often inclined at large angles to the plane of circulation of the vortex, and second, the different planets move in accordance with Kepler's third law, which is inconsistent with the harmonic circulation. He even suggested a reply to the second objection which he believed Leibniz would make; namely, that the harmonic circulation was restricted to narrow bands containing the planets, the fluid between circulating in accordance with Kepler's third law. Leibniz indeed accepted this suggestion, although Gregory had warned that it would hardly carry conviction.

In addition to the objection concerning the comets, Newton (as we know from his manuscript notes) made two basic criticisms of Leibniz's theory; first, he claimed that 'centrifugal force', by his third law, was always equal and opposite to 'centripetal force', and second, that Leibniz's mathematical reasoning was unsound, owing to errors in the handling of second-order infinitesimals. The objection concerning 'centrifugal force' assumed that the term could only be interpreted in Newton's definitive sense. Leibniz in fact used it, as Newton had earlier, to denote a force engendered by circular motion. Concerning the other objection, Newton claimed that second-order errors were involved in the assumptions that $NP = D_2T_2$ and $NM_2 = GD_2$. In the case of the second assumption, Newton's assertion would be true if Leibniz's term 'tangent' were interpreted in the Euclidean sense of a tangent at the point M_2 of the curve, but his own understanding of second-order infinitesimals is brought into question by the fact that his statement concerning NP and D_2T_2 is false.

Newton's criticisms formed the basis of the objections of John Keill, published in the *Journal littéraire de la Haye* in 1714. Keill pointed out that NM_2 and GD_2 could be taken equal if the tangent were interpreted as the chord M_1M_2 but that in his view, such an interpretation was not permissible. Newton himself published his objections anonymously in a review of the *Commercium epistolicum*, while Roger Cotes, in his preface to the second edition of the *Principia*, repeated the objection concerning the comets.

Influence of Newtonianism

On the basis of a mistaken inference from a memoir of William Whiston, it has generally been supposed that David Gregory, who was professor of mathematics at Edinburgh between 1683 and 1690, was the first to introduce Newtonianism into courses for students, but a study of manuscripts of Gregory's lectures and a number of extant copies taken by students shows that Newton's theories never formed an integral part of these lectures. The reason was not to be found in any lack of interest on Gregory's part but rather in the immaturity of Scottish students, who entered the universities at about the age of fourteen years, and the utilitarian aims of the curriculum. Gregory began to make a detailed commentary on Newton's *Principia* as soon as he received it in 1687 and in 1690 at least three of his students were being encouraged to "keep Acts" (that is, read out essays) on aspects of Newtonian philosophy. Lectures at Edinburgh began to be based fully on Newton's ideas only towards the turn of the century.

It was on Newton's recommendation that Gregory was appointed Savilian professor of astronomy at Oxford in 1691, where he published his *Astronomiae physicae et geometricae elementa* (1702), which included an exposition of Newton's theories in a form suitable for students.

The first Continental mathematician to interpret Newton's theory in terms of the differential calculus was Varignon. This he did in a series of memoirs contributed to the Royal Academy of Sciences between 1700 and 1710. Having received his first instruction in the differential calculus from Johann Bernoulli in 1692, he was ready by 1698 to apply this calculus to problems of mechanics and planetary motion. Fontenelle brought Varignon's work

to the attention of Leibniz, who encouraged him to undertake further investigations. In his letter to Fontenelle of 3 September 1700, accepting appointment as a foreign member of the Academy, Leibniz raised a number of queries; among them was the question whether G. D. Cassini, doyen of the Paris astronomers, and Philippe de La Hire had abandoned Kepler's ellipse. After confirming that Cassini had replaced the ellipse of Kepler by an oval in which the product of the distances from the foci was constant and explaining La Hire's finding that the planets do not describe any perfectly regular curve, Fontenelle remarked that Varignon, "one of our ablest geometers", had developed a general method for finding the "central forces" which push the planets. He added that, while Newton and Leibniz himself had established these forces by geometry, Varignon used only the differential calculus and never failed to render homage to its inventor, Leibniz.

Although Varignon evidently regarded his work simply as mathematics, Fontenelle, in his reviews for the annual *Histoire* of the Academy, attempted to fit it into a Cartesian framework by finding the cause of the "weights of the planets" in the action of the celestial matter of the solar vortex. Varignon began his investigation by considering rectilinear motion, establishing two general rules defined by the equations $v = dx/dt$ and $f = dv/dt = ddx/dt^2$, where x is the distance from the starting point, v the speed, f the force (acting in the direction of the line) and ddx the distance traversed owing to the increment of speed dv, assumed to be acquired instantaneously at the beginning of the infinitesimal element of time dt. To prove the second rule, Varignon appealed to a generalization of Galileo's law of fall. Taking the distance traversed by a body subjected to a constant force to be proportional to the force and the square of the time, he concluded that $ddx = fdt^2$. Eliminating the time from the two equations, he obtained the relation $fdx = v\,dv$ or $\int f dx = \frac{1}{2}v^2$, which he recognized as Newton's Prop. 39 of Bk I. For this result he claimed independent discovery.

In his second memoir Varignon established a formula for the centripetal force of a body moving in a curve. Let PK (Figure 14.5) be ds, r the distance of P from the centre of force, f the centripetal force and f_1 the component of this force in the direction PK. Then $f/f_1 = PK/PN = ds/(-dr)$, from which it

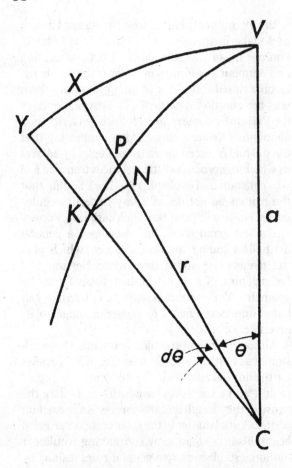

14.5. Varignon's theory of motion in a curve.

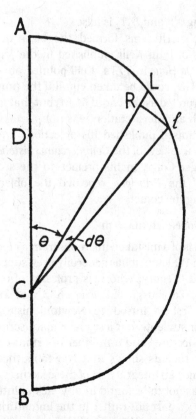

14.6. Varignon's theory of planetary orbits.

follows that $f = -f_1 ds/dr$. By Varignon's second rule of motion, $f_1 = dds/dt^2 = dv/dt$, so that $f = (-dv/dt)(ds/dr) = -v dv/dr$. Varignon failed to recognize that this result was equivalent to Newton's Prop. 40 of Bk I. He described it as "a very simple formula of *central forces*, centrifugal as well as centripetal, which are the principal foundation of the excellent work of Mr Newton".

Although Varignon in 1700 described Kepler's second law as the "most physical", he recognized that some astronomers preferred other laws. At this time he may even have regarded it as an independent hypothesis, for he made no attempt to deduce it from his rules of motion. It could not in fact have been deduced from his formula for centripetal force, for this embodied what was later called the principle of conservation of energy, whereas something equivalent to the principle of conservation of angular momentum would have been needed to establish Kepler's second law. By 1710,

however, Varignon had no doubts about the law, "which Mr Newton has demonstrated ... to be the true one in a space without resistance".

In his third memoir of 1700, Varignon completed the first phase of his investigation of planetary motions, using his formula for centripetal force to demonstrate that, for a planet moving in an ellipse in accordance with Kepler's second law, the centripetal force directed towards a focus must vary inversely as the square of the distance. Suppose that the ellipse (Figure 14.6) with foci C and D represents a planetary orbit, C being the centre of force. Let $CL = r$, $Rl = dz = r d\theta$, where $\theta = \angle ACL$, $Ll = ds$, $AB = a$ and $CD = c$. Also let $b^2 = a^2 - c^2$. Then the equation of the ellipse is $-b\, dr = dz\sqrt{(4ar - 4r^2 - b^2)}$. It follows that $(4ar - 4r^2 - b^2)dz^2 = b^2(dr^2 + dz^2) = b^2 ds^2$ or $(4a - 4r)/r = b^2 ds^2/(r\, dz)^2 = b^2 v^2$, putting $r\, dz = dt$ in accordance with Kepler's second law. On differentiation, this equation becomes $-4a\, dr/r^2 = 2b^2 v\, dv$, so that $f = -v\, dv/dr = 2a/(br)^2$. Thus the centripetal force varies inversely as the square of the distance. As corollaries, Varignon extended his result to the cases of the hyperbola and parabola.

These results, Varignon remarked, were in agreement with the principal propositions concerning planetary motion demonstrated by Newton and Leibniz.

Turning his attention next to the theories of Seth Ward and G. D. Cassini, using firstly their own hypotheses of times and secondly the law of Kepler, Varignon obtained some very complicated expressions for the centripetal forces that would be needed. Evidently there was no hypothesis among those that had been proposed by astronomers to which his general formula for central forces could not be applied.

Having established the principal propositions for a planetary orbit, Varignon set about developing another approach, expressing the centripetal force in terms of the radius of curvature. It was on the basis of this work, communicated in a letter, that Leibniz in 1702 encouraged Varignon to continue his researches, for the fundamental memoirs of 1700 were not published until 1703. Leibniz asked Varignon to consider more than one centre of attraction, mentioning that David Gregory had considered the action of the Sun on a planet and a satellite but not the actions of the principal planets on each other. Although he had not seen Gregory's book, Varignon responded immediately, but his examples were limited to hypothetical cases of little relevance to astronomy. Moreover, they concerned the direct problem; given a curve and a number of fixed centres, to find the forces needed to move a planet with constant speed or speed varying according to a given rule. Leibniz, of course, had in mind the general inverse problem; that is, to determine the orbit, given several forces directed towards moving centres.

Newton had considered the motion of the apseline from the standpoint of the direct problem, assuming the orbit to be a rotating ellipse and finding that the addition of a centripetal force proportional to the cube of the distance would be needed. In a memoir of 1705, Varignon applied the differential calculus to this problem, confirming Newton's result, though without resolving the difficulty of accounting fully for the motion of the lunar apogee. For Varignon, the rotating ellipse was just another "system of astronomy" to which his general theory could be applied. He did not

suggest a physical cause for the extra force, though Fontenelle related the moving apogee to the effect of neighbouring bodies on the basis of the vortex theory.

For the case of a single centripetal force varying inversely as the square of the distance, the inverse problem was solved by Newton in the *Principia*. Having solved the direct problem for ellipse, hyperbola and parabola, he concluded that the inverse-square orbit must be a conic section, because a conic could always be constructed to satisfy the initial conditions of speed and direction of a projected body and two different orbits touching at a point were not possible. A more elegant solution was given in *Principia*, Bk I, Prop. 41, which lacks only the evaluation of the integrals. Both Jakob Hermann and Johann Bernoulli independently translated this proposition into terms of the differential calculus and evaluated the integrals to demonstrate that an inverse-square orbit must be a conic section. Their memoirs were presented to the Paris Academy on the same day in 1710. There followed a controversy in which Bernoulli criticized Hermann's solution for lacking a constant and being somewhat contrived. Hermann was defended by Jacopo Francesco Riccati, who commented on the elegance of the solution and pointed out that the lacking constant only changed the origin, so that no loss of generality was involved. When Varignon saw these solutions, he easily deduced the result of Newton's Prop. 41 from Kepler's second law and his own formula for the centripetal force but he lacked the facility in the integral calculus needed to evaluate the integrals. He had sought instruction in this calculus from Bernoulli but this had been withheld in accordance with a secret agreement between Bernoulli and the Marquis de l'Hospital. The following year Varignon wrote to Leibniz that, although he had considered the problem of fixed multiple centres both in a medium and the void, he had not found anything on moving centres. Neither Varignon nor his contemporaries made any further progress towards the solution of this intractable problem.

Following the publication in 1713 of the second edition of Newton's *Principia*, his system was taught in Leiden by W. J. 'sGravesande, who became professor of mathematics and astronomy in 1717, on his return from a visit to England, where he had met Newton and John Theophilus

Desaguliers. Although he made no original contribution to knowledge, 'sGravesande greatly furthered the cause of Newtonianism, especially through his *Physices elementa mathematica experimentis confirmata sive introductio ad philosophiam Newtonianam* (1720–21), which became one of the most influential popular accounts of Newton's system. This work, however, was ignored by the editors of the *Journal des sçavans*, while the review in the *Mémoires à Trevoux* raised objections against the proscription of hypotheses and the acceptance of the void and attraction in the Newtonian system. These principles of Newtonian philosophy had already become the focus of renewed controversy in the Leibniz–Clarke correspondence, first published in 1717 followed by two further editions in 1720. They continued to be the chief obstacle to the acceptance of the Newtonian system by Cartesians and others who saw in them a return to the occult qualities of the Scholastics. Newton's claim to have deduced the attraction from the phenomena did not convince the critics, for as Fontenelle asked rhetorically in his "Eloge de Newton", were not the Scholastic qualities also causes whose effects could be seen?

Vortices after Newton

In his "Eloge de Newton", Fontenelle contrasted the methods of Descartes and Newton. Descartes, he wrote, "proceeds from what he clearly understands to find the cause of what he sees", whereas "Newton proceeds from what he sees to find the cause, whether it be clear or obscure". A year later, reviewing the first of a series of papers presented to the Royal Academy of Sciences by Joseph Privat de Molières (1677–1742), Fontenelle remarked that, although the Newtonian system had "some very advantageous aspects", the Cartesian system was more "agreeable to the intellect".

Privat de Molières based his defence of vortices on the ideas of Villemot and Malebranche. In his first paper, he sought to establish that the gravity of the planets, varying inversely as the square of the distance, was an effect of the external compression of the vortex. First he divided the vortex into a set of imaginary concentric spherical layers and supposed that the speeds of all the particles in a particular layer were equal. The centrifugal force f of a particle in latitude λ was proportional to $v^2/(r \cos \lambda)$, v being the speed and r the radius of the

layer. Then Privat de Molières supposed that the reaction between the imaginary layers was normal to the surface. Thus each particle was pushed by the layer immediately above with a centripetal force ϕ, which balanced the normal component $f \cos \lambda$ of its centrifugal force. It followed that $\phi = v^2/r$. Moreover, for equilibrium of the whole vortex, the total centripetal force, according to Privat de Molières, had to be the same for each layer. This implied that, if S was the area of the surface of a layer, ϕS was constant and consequently ϕ was proportional to $1/r^2$.

Having established the inverse-square law for gravity, he proceeded to demonstrate Kepler's third law. In the plane of the ecliptic, $\phi = f$, so that $1/r^2 = v^2/r$. Also v is proportional to r/t, where t is the periodic time, so that, eliminating v, it follows that r^3 is proportional to t^2. By this demonstration, Privat de Molières declares, Kepler's law becomes a "principle of mechanics, from which all the celestial motions can be deduced geometrically, as M. Villemot has already attempted to do, and that sustains and confirms the system of Descartes, far from overthrowing it, as has been claimed in our day".

In his second paper, presented in 1729, Privat de Molières showed how an aether composed of Malebranche's elastic globules (expanding and contracting as a result of the inequalities in the external pressure) could revolve in elliptical layers and consequently carry the planets in elliptical orbits. His third paper, presented in 1733, concerned the reconciliation of Kepler's second and third laws in application to the layers of a vortex. At this stage, however, Privat de Molières attempted to assimilate the mathematical results of Newton into his system. As gravity in a spherical vortex had been shown to vary inversely as the square of the distance, he supposed that the same relation would hold very nearly in an elliptical vortex that was nearly spherical. He then concluded that, as Newton had demonstrated the consistency of Kepler's second and third laws with an inverse-square law for gravity, Kepler's laws were consistent also in his elliptical vortex.

Another problem taken up by Privat de Molières was that of the Earth's diurnal motion, which Descartes had attributed ambiguously to the action of the terrestrial vortex and to the persistence of the rotation impressed on the Earth at the time of its creation. The need for a solution had become urgent, for La Hire had remarked, criticizing Villemot's explanation in terms of the action of the terrestrial vortex, that as the inferior layer of the aether moved faster (from west to east) than the superior, the vortex should cause the planet to rotate in the opposite direction (from east to west) to that observed. A more elaborate solution to the problem was given by Jean Jacques d'Ortous de Mairan. On the basis of plausible assumptions (apart from the confusion of mass and weight), Mairan attempted, by a confused argument involving the differences of the impulses of the aether on the superior and inferior hemispheres of the planet and the supposed differences of the weights of these hemispheres in virtue of their different distances from the centre of the vortex, to determine the periods of rotation of the planets on their axes, arriving at remarkably accurate results. For Jupiter, his theory predicted a period of $10\frac{1}{4}$ hours and some minutes, "very nearly as the observations reveal". Even the case of the Moon could be explained by a hypothetical difference in density of the two hemispheres, and he was able to point out that the Newtonian attraction was quite incapable of accounting for the axial rotation of the planets.

In his *Leçons de physique*, published in four volumes between 1734 and 1739 and consisting of lectures delivered to students at the Collège Royal de France, Privat de Molières completed his reconciliation of the systems of Newton and Descartes, assimilating "the void of Newton, or that *non-resisting* space, of which this philosopher has so invincibly established the existence", as well as the "attraction ... from which Newton, without however having been able to discover the mechanical cause, has taken so many good conclusions". Confusion of mass and weight had led him to the conclusion that the aether offered no resistance owing to its lack of gravity towards the Sun. In effect Privat de Molières had adopted the Newtonian system in his treatise, which was the last Cartesian textbook covering the whole field of terrestrial and astronomical physics. His only reservation was that he could never agree that matter was essentially heavy and hard, "absurd hypotheses" which he claimed "Newton does not frame ... but which he adopts".

Privat de Molières had set out in 1728 to defend the Cartesian system and had come to accept the Newtonian system, apart from the attraction considered as a cause. The same trend towards acceptance of the Newtonian system even by Cartesian sympathizers is evident in a number of essays awarded prizes by the Royal Academy of Sciences between 1728 and 1740. In 1728 the prize was awarded to Georg Bernard Bulffinger for an uncompromisingly Cartesian essay in which he offered a modification of Huygens's theory of gravity and a defence of Leibniz's theory. In particular, he showed that Leibniz's reconciliation of Kepler's second and third laws would require the density to be uniform in the harmonic layers of the solar vortex (the layers in which the planets moved) and to vary inversely as the square root of the distance in the intermediate layers (the layers moving in accordance with Kepler's third law).

Another Cartesian essay, on the causes of the elliptical orbit and the rotation of the line of apsides, was adjudged most worthy of the prize in 1730. This was Johann I Bernoulli's "Nouvelles pensées sur le système de M. Descartes". Bernoulli begins his essay with the remark that the reader may be surprised to find him defending the vortices at a time when many philosophers, particularly the English, regard them as pure fancy. Some unpublished correspondence seems to indicate that he defended the vortices simply to improve his chances of winning the prize. His description of the Cartesian system at the beginning, where he seems to suggest that a miraculous arrangement of the vortices by God would be needed to produce the regular effects of the planets, might well lead the reader to wonder whether he is defending it. Yet he declared the Newtonian principles of attraction and void to be unacceptable and there can be little doubt that his criticisms of Newton's propositions were intended seriously. By these propositions Newton had claimed to demonstrate that the hypothesis of vortices was inconsistent with Kepler's third law. In Bernoulli's view, however, Newton's reasoning was "a manifest sophism, being based on two propositions equally false".

Newton's first error, according to Bernoulli, was to suppose the friction between adjacent layers (which he had first regarded as solid) to be dependent on the relative speed and the area. For Guillaume Amontons had shown experimentally that the friction between solid surfaces depended on the normal force between them but was independent of the area in contact. Newton's second error, as seen by Bernoulli, was the neglect of the lever effect of the frictional force. Bernoulli repeated Newton's calculations for the cylindrical and spherical vortices, in his view correctly, estimating the normal force by an integral taking into account the action of all the inferior layers. He deduced periodic times proportional to $x^{\frac{4}{3}}$ and $x^{\frac{5}{3}}$ (where x is the distance from the centre) for the cylindrical and spherical vortices respectively. These results were nearer the $x^{\frac{3}{2}}$ required by Kepler's third law than were those of Newton. From this he surmised that there might be some shape between the cylindrical and spherical that would give periodic times in accordance with Kepler's law. Just a little flattening in the Sun, similar to that observed in Jupiter, he suggested, would suffice.

Bernoulli then considered the explanation of Kepler's third law in terms of a variation in density with distance from the centre of the vortex, finding that the density would need to be proportional to $x^{-\frac{1}{2}}$. This reconciliation of vortex motion and Kepler's law, he remarked, would have had little effect in silencing his opponents if he had not first demonstrated the falsity of Newton's propositions. Evidently Bernoulli did not know that Bulffinger had arrived at exactly the same relation of density with distance using the assumptions of Newton's propositions.

Turning his attention to the problems proposed for the prize, Bernoulli attributed the elliptical orbit to the combination of a circular motion of the planet about the Sun and a radial oscillation about an equilibrium position, taking the planet in turn into layers of greater and smaller density, while the line of apsides rotated, he thought, owing to a lack of synchronism between the circulation and radial oscillation. His treatment of these problems was entirely qualitative and his solutions would have needed the same miraculous arranging of the vortices mentioned at the beginning. Moreover, he did not even attempt to explain why a planet, immersed in fluid layers circulating in accordance with Kepler's third law, should itself rotate about the Sun in accordance with the second.

The prize having been withheld in 1732, a double prize was offered in 1734 for an essay on the physical cause of the inclinations of the planetary orbits to the plane of the solar equator. On this occasion prizes were awarded to Johann I Bernoulli

and his son Daniel, while Pierre Bouguer and Jean Baptiste Duclos received honourable mention. Johann Bernoulli, encouraged perhaps by the appearance of P. L. M. de Maupertuis's *Discours sur les différentes figures des astres* in 1732, undertook to form a new system by combining elements of the Newtonian and Cartesian systems. First, he explained gravity by a central stream of matter from the circumference. Then he postulated an aether consisting mainly of Descartes's first element; that is, he remarked, an aether like Newton's void. Moreover, he abandoned the idea that the planets were carried by the vortex. For if the vortex were to be regarded as an appendage of the Sun, as Bernoulli supposed, the layer in contact with the Sun should have the same speed as the solar rotation. Application of Kepler's third law to the layers of the vortex then gave them speeds of about $\frac{1}{230}$ th part of those of the planets. However, the weak circulation enabled him to explain the direction over a period of time of the planets into orbits in the plane of the ecliptic. The small inclinations of the orbits to this plane he attributed to small drifts arising from the spheroidal form of the planets. The comets spent most of their periodic time where the circulation was extremely weak and were therefore hardly affected by it.

Daniel Bernoulli's replacement of the aethereal vortex by a solar atmosphere itself subject to gravity indicates his preference for the Newtonian system, though he speculated (perhaps as a concession to the judges) that the cause of gravity might be found in a complicated system of independent vortices moving through each other without mutual interference. Having concluded on the basis of a calculation of probability that the concentration of the planetary orbits in a narrow zone could not have occurred by chance, he suggested the effect of the circulating solar atmosphere as the cause. A mathematical analysis of the solar atmosphere in accordance with the gas laws relating density, pressure and temperature, showed a falling off of density in the outer layers almost to zero.

Taking the radius of the Sun to be a and the density, temperature and pressure at a distance x from the centre to be D, T and p respectively, let the density, temperature and pressure at the surface of the Sun be 1. Then by the gas law, $D = p/T$. Supposing the temperature to be inversely proportional to the square of the distance, Bernoulli

Table 14.1

Position in the solar atmosphere	Density
Surface of the Sun	1
Mercury	2200
Venus	3000
Earth	2600
Mars	1300
Jupiter	0.4
Saturn	0.000006

deduced that $D = (x^2/a^2)p$. As the difference in pressure between points of two adjacent layers balances the weight, $-\mathrm{d}p = c(a^2/x^2)D\mathrm{d}x$, where c is a constant. Substituting the value of D in this equation gives $-\mathrm{d}p = cp\mathrm{d}x$ and consequently $p = \mathrm{e}^{-c(x-a)}$. It follows that $D = (x^2/a^2)\mathrm{e}^{-c(x-a)}$ and by differentiation the distance of the layer of maximum density is found to be $x = 2/c$.

For a trial calculation, Bernoulli took Venus to be at the distance of maximum density. Then the distribution of densities was found to be as in Table 14.1. This seemed to suggest that the layer of maximum density was in fact further out, say in the region of Jupiter, so that all the planets would be subject to the action of the solar atmosphere and in consequence of this action over a long period of time reduced "within the narrow limits where they are at present". After an infinite time, Bernoulli believed, the orbits would become perfect circles and united exactly in the plane of the solar equator. As the comets spent most of their time in the region of negligible density, he concluded that they had been almost unaffected by the circulation, so that their orbits should be randomly inclined. In fact, the mean inclination of the orbits of 24 observed comets was found to be 43° 39'.

Duclos's astronomical observations won him membership of the Academy of Sciences in Lyons, where the manuscript of his unpublished essay may be found. His solution of the problem of the inclinations of the planetary orbits was based on the idea that, as the particles of fluid tended to move in great circles but were constrained to move in small circles parallel to the equator, there arose in them an impeded endeavour to move along the meridians. By virtue of this "polar" force, Duclos believed, the aether caused the planet to oscillate about the plane of the equator, like a pendulum, while circulating with the vortex. Duclos remained firmly attached to the Cartesian theory, but in an unpublished history of the vortex theory written in 1737 or 1738, he expressed the view that the best defence of the system of vortices would be the reconciliation of this system with that of Newton along the lines suggested by Privat de Molières. By employing "the good theory of Newton on the central forces without having need of the attraction", this philosopher, he remarked, "profits, without any inconvenience, from the advantages of the English physics and loses nothing of the incontestable superiority of the French physics".

Bouguer's essay took the form of a dialogue between a Newtonian, a Cartesian and a character representing himself. While a modified Cartesianism was allowed to prevail, the Newtonian case was stated quite strongly. To explain the inclination of the planetary orbits, he supposed that originally the various spherical layers of the vortex rotated about different axes, whose inclinations have been reduced in course of time by imperceptible degrees to their present values as revealed by the planets.

The Newtonian system was openly defended by Maupertuis in his *Discours sur les différentes figures des astres*. Although he may have derived some ideas from Bouguer, the main source of inspiration of Maupertuis's work was evidently the inconsistency of the conclusions of Newton and Jacques Cassini concerning the figure of the Earth. This, together with the fact that Cassini wrote papers defending the Cartesian system, may have given rise to the erroneous view that the Cartesian system predicts the Earth to be a prolate spheroid. Cassini's figure of the Earth was, in fact, derived entirely from faulty geodetic measurements. Despite the outcome of the controversy, Fontenelle was not unreasonable when he remarked that the actual measures were to be preferred to those resulting from theories. In his advocacy of the Newtonian system, Maupertuis emphasized its predictive power – in determining the position of comets, for example – and also the fact that impulsion was no more intelligible than attraction.

Jacques Cassini in 1735 and 1736 presented papers to the Royal Academy of Sciences offering Cartesian solutions to two outstanding problems. These papers may have been inspired by dissatisfaction with the explanations of Johann Bernoulli and Privat de Molières, resulting from their combination of Cartesian and Newtonian elements. The first problem concerned the inconsistency of the slow rotation of the Sun with the circulation of the layers of the solar vortex carrying the planets in accordance with Kepler's third law. Cassini solved the problem by surrounding the Sun and planets with atmospheres rotating with the same angular speed as the central bodies. The speed of the atmosphere at its outer limit, he supposed, was equal to that of the layer of aether in contact with it. For the terrestrial vortex, the atmosphere would need to extend to a height of six times the Earth's radius. This was considerably greater than the height formerly attributed to the atmosphere. No difficulty would arise in the case of the solar atmosphere, which would need to extend to about half the distance of Mercury. To test the theory, Cassini proposed a comparison of the theoretical period of revolution of Saturn, assuming the atmosphere to extend to the outside of the ring, with the observational evidence that some favourable occasion might provide.

The second problem to which Cassini gave his attention was the compatibility of Kepler's second and third laws. Privat de Molières had not really attempted a physical solution, having been content simply to accept Newton's mathematics. Cassini considered whether the two laws might result from a lack of precision in the observations. As they predicted a difference of 15′ or 16′ in the position of the Sun in the course of a month round the Earth's aphelion or perihelion, he regarded the reality of the problem as established. Cassini based his solution on the continuity of the aether in the plane of the equator of the vortex.

Suppose that the aether at the mean distance R (Figure 14.7) of the planet turns circularly about the Sun S with the speed required by Kepler's third law. Then Cassini maintains that the aether contained in the space RST moves into the space RSV, while an equal volume enters the space RST. Now suppose that the matter is retained in the ellipse $ARPI$ by some force which prevents it from describing the circle. Then, Cassini explains, the aether that would have filled the small space RVM must extend in the ellipse beyond SM and occupy the small sector MSO, so that the planet, approaching perihelion, traverses a greater distance in the ellipse than it would have traversed in the circle.

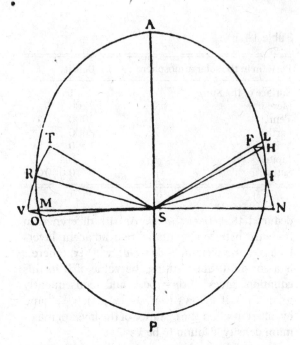

14.7. Jacques Cassini's diagram illustrating his Cartesian theory of planetary orbits.

The area SRO being equal to the area RSV that would be swept out by the line joining planet and Sun in the same time in a circular orbit, it follows that equal areas are described in equal times, as required by Kepler's second law. Cassini's explanation is basically sound, though incomplete. For the aether contained in one sector does not in fact move as a whole into the next sector, though equal volumes of aether leave one and enter the other. Fontenelle was satisfied that Cassini had shown the agreement of the two laws of Kepler in the hypothesis of vortices.

In 1740 the prize for an essay on the tides was divided between four competitors. One of these, the Jesuit Antoine Cavalleri, gave a Cartesian justification of Newton's results. The others, namely Colin Maclaurin, Leonhard Euler and Daniel Bernoulli, based their explanations unequivocally on the Newtonian system, although Euler insisted that the inverse-square law must be the effect of the action of aethereal vortices. Daniel Bernoulli again emphasized the predictive power of universal gravitation, "this incomprehensible and uncontestable principle, that the great Newton has so well established and that we can no longer hold in doubt".

It was their recognition of the explanatory power of the Newtonian system that led the ablest Cartesians to combine Newton's mathematical theory with physical vortices. Then when they lost faith in vortices, remarks such as those of Maupertuis concerning the equal unintelligibility of impulsion and attraction (of which there are hints in Bouguer's essay) could help them to abandon the vortices with a clear conscience and accept universal gravitation as a physical axiom. In the second edition of his prize essay (published in 1748), Bouguer himself, having in the meantime observed the deflection of a plumb-line in the neighbourhood of a mountain in Peru, announced his conversion to Newtonianism, though with a modified attraction law. To cover all cases, including the phenomena explored by the chemists, he suggested a law consisting of two terms, either of which might be dominant, one inversely proportional to the square and the other to the cube of the distance. Alexis-Claude Clairaut, he remarked, had also needed a modification of Newton's attraction to explain the motion of the Moon's apogee; in fact, the addition of a term inversely proportional to the fourth power of the distance. Later in the year 1748, however, Clairaut succeeded in explaining this motion on the basis of the inverse-square law, thus advancing beyond Newton to achieve a confirmatory triumph for the theory of universal gravitation with an inverse-square law in application to the solar system.

Bouguer clearly indicated the common path to progress when he remarked that, for peaceful coexistence between Newtonians and Cartesians, it was sufficient for them to declare,

the ones perhaps without much belief, and the others without much hope, that the word attraction, the same as that of weight, simply describes a fact, while awaiting the discovery of the cause.

A new generation of Continental Newtonian mathematicians, beginning with Clairaut (who helped the Marquise du Châtelet to translate Newton's *Principia* into French), had no difficulty in following Bouguer's advice to confine themselves to the truths of induction or admit only the immediate consequences. When Fontenelle's *Théorie des tourbillons* appeared with thinly veiled anonymity in 1752, this work, like its author (born in 1657), must have seemed to its readers as a survival from a former age. Euler did not even mention vortices in his letters to the Princess of Anhalt-Dessau but expressed the view prevalent among scientists in 1760 when he asserted that the system of universal attraction was established by the most solid reasons. Yet Euler himself maintained the view that gravity had a physical cause, not yet known in detail but certainly arising from the action of a fluid matter filling space.

Further reading

E. J. Aiton, *The Vortex Theory of Planetary Motions* (London and New York, 1972)

P. Brunet, *L'Introduction des théories de Newton en France au XVIIIe siècle* (Paris, 1931)

S. Delorme *et al*, *Fontenelle: sa vie et son oeuvre* (Paris, 1961).

M. A. Hoskin, Mining all within: Clarke's notes to Rohault's *Traité de physique*, *The Thomist*, vol. 24 (1962), 353–63

A. Koyré, La gravitation universelle de Képler à Newton, *Archives internationales d'histoire des sciences*, vol. 4 (1951), 638–53

L. M. Marsak, Cartesianism in Fontenelle and French science, 1686–1752, *Isis*, vol. 50 (1959), 51–60

H. Parenty, Les tourbillons de Descartes et la science moderne, *Mémoires de l'Académie des Sciences, Belles-Lettres et Arts de Clermont–Ferrand*, deuxième série, fasc. 16 (Clermond–Ferrand, 1903)

15

The shape of the Earth

SEYMOUR L. CHAPIN

The great voyages of discovery and exploration in the fifteenth and sixteenth centuries enormously extended man's knowledge of the globe on which he lives. Although all educated men since Antiquity had known that the Earth was spherical, these voyages – and especially the spectacular circumnavigations – confirmed for the ordinary people that man's abode was indeed round, and, in fact, that it was a sphere of very considerable extent. Further, mathematicians and astronomers of the first half of the sixteenth century undertook to determine its size with some precision. Thus, the Frenchman Jean François Fernel, having astronomically established the northern extremity of 1° of the meridian which began in Paris, determined the length of that distance by counting the number of revolutions of a coach's wheel, while a few years later the Netherlander Reiner Gemma Frisius described the method of triangulation for measuring such arcs with greater certainty by means of the terrestrial observation of angles between established sites followed by appropriate trigonometric procedures. It remained for the seventeenth and eighteenth centuries, however, to develop and apply the technique of triangulation as well as to raise the question of the Earth's precise shape and to provide a solution to that question.

An early assault on the first of these tasks was undertaken by Willebrord Snell (1580–1626), a professor at the University of Leiden, who developed the idea of his fellow countryman, Gemma Frisius, to such an extent as to deserve to be called the father of triangulation. Starting his base line with his house in Leiden, he used three spires of two churches as reference sites – solving in the process the so-called recession problem for three points, subsequently named after him – and then established a network of triangles that enabled him to compute the distance between two Dutch towns

about 130 kilometres apart. Although he employed large instruments made by the famous cartographer Willem Janszoon Blaeu – a former assistant to Tycho Brahe who had himself in that capacity undertaken a very accurate arc measure by triangulation – Snell was dissatisfied with the results that he published in 1617. He therefore continued to work toward their perfection, but, because of his early death, these efforts did not become publicly available until more than a century later.

Snell's work was probably inspired in part by an understandable desire to establish precise cartographical details about a newly independent country. It must have owed something also to the fact that the knowledge of a length of a degree was essential to the perfection of sea charts. Certainly Blaeu, for example, had extensive contacts with both merchants and navigators, as evidenced by the fact that he was to become the official cartographer of the Dutch East India Company.

The needs of navigation seem clearly to have been the major incentive for an equivalent English interest in the question of the length of a degree of the Earth's spherical surface. Thus, the English cartographer Edward Wright (1561–1615), whose fame rests largely upon his mathematical explanation of the famous Mercator projection (which yielded maps particularly well suited to navigational needs), in 1599 called for an improvement in the traditional value of 60 miles per degree then employed by English seamen in his work *Certaine Errors in Navigation*.... His suggestion was acted upon by Richard Norwood (1590–1665) who, after having voyaged himself as a young man and then written several books on the mathematics of navigation, decided in 1635 to measure the distance between London and York. He first determined the latitude of each city and then walked from one to the other furnished with a measuring chain and a theodolite, employing the first – along with the measurement of his own steps – to determine the distance covered, and the second to correct for the fact that the route he followed was neither a straight line nor on a flat surface. His value of about 69.5 miles for a degree compares reasonably with what was to become the classic measure of a degree in the second half of the seventeenth century.

That new standard was established in the France of Louis XIV, whose great finance minister, Jean Baptiste Colbert, was sympathetic to proposals made in the early 1660s for the establishment of a state-sponsored centre for scientific studies. Colbert saw that such an institution could be an efficient consultative body for the government, another arm, so to speak, of his whole mercantilistic economic programme. Thus, it would be able to assist in the increase of the country's productivity, not only by passing judgement upon the usefulness of new inventions but by suggesting various ways in which science could serve the nation – as by improving craft practices through scientific theory or playing a role in another of Colbert's broad goals, the exact mapping of his monarch's kingdom.

The latter aim fitted in well with the extensive astronomical programme proposed for the Academy of Sciences that was established in 1666. The proposed programme was put forward by two of the new Academy's original members. One of these was the great Dutch scientist Christiaan Huygens (1629–95), who in 1656 had made Galileo's discovery of the isochronism of the pendulum into the principle underlying an accurate timepiece and who had subsequently developed a marine variant featuring a short pendulum; the latter had been responsible for Colbert's luring him to Paris in 1665 with the offer of an enormous salary. The other was the native Frenchman, Adrien Auzout, who had only recently perfected – if not independently invented – the filar micrometer, and who called for the creation of a royal observatory in Paris and the dispatch of scientific expeditions for the investigation of significant problems or phenomena. Auzout's active support was soon responsible for bringing into the Academy one Jean Picard (1620–82), the closest approximation to a 'professional' astronomer then to be found in France.

It was largely Picard who carried through the revolution in observational astronomy made possible by the filar micrometer, the astronomical pendulum clock, and the application of telescopes to large-scale graduated instruments appropriate for the measurement of small angles. It was with this equipment that he undertook to measure the distance between two locations approximately on the meridian of Paris, to determine the differences in their latitudes, to measure their separation by triangulation, and to deduce from these results the length of a degree of meridian; from this one could easily calculate the size of the Earth, as had been desired by both Huygens and Auzout. This eminently successful arc measure was executed during the summer and autumn academic vacations in 1668, 1669 and 1670. Picard's result provided the basis on which the desired rectification of French cartography could be carried out, and, equally important, a model for all later undertakings of this kind of 'geometrical' geodesy.

The Academy also implemented Auzout's call for scientific expeditions. This occurred first in the search for a solution to a problem that had become crucial with the great oceanic voyages mentioned at the outset, namely the determination of longitude at sea. Since that problem would be solved if the difference in time between a ship's location and its port of departure or some other agreed-upon reference point could be measured, it is not surprising that one of the solutions proposed early on was to have the ship carry along a timepiece keeping the local time of the chosen reference site. This idea was first put forward by Gemma Frisius in 1553, but sufficiently accurate clocks were not available before the appearance of the pendulum-driven clock. We have already mentioned Huygens's development in 1662 of the marine version of this device. The Academy undertook several expeditions between 1668 and 1670 from the Mediterranean to France's possessions on the Atlantic seaboard of North America for the purpose of testing such clocks. Their unhappy outcome was to demonstrate that the utilization of the timekeeper approach would have to await the perfection of spring-driven mechanisms, something that was to require almost another century of development.

Another of Galileo's ideas was destined to bear fruit in the same area. In 1610 he had discovered the first four satellites of Jupiter, and in 1612 he had observed the eclipse of a satellite. He was struck by the possibility of reading the positions of the satellites as the hands of a celestial clock in the determination of longitudinal differences. An expedition to Marseilles, Malta, Cyprus and Tripoli, dispatched under the aegis of Nicolas Claude Fabri de Peiresc (see Chapter 9 of Volume 2A), gave but dubious results: the available tables of the satellites were too inaccurate. But in 1668 the Italian astronomer, Gian Domenico Cassini (Cassini I, 1625–1712), brought out new tables of the motions of Jupiter's satellites which, because they were accurate enough to predict the eclipses of those bodies, finally rendered Galileo's idea practicable for the determination of longitude on *terra firma*. At sea, because of the difficulty of making precise observations through a long telescope on the deck of a swaying ship, the idea proved unusable. But for the cartographical rectification of France, Cassini's ephemerides provided an important tool. Small wonder, then, that Cassini, like Huygens, was invited to enjoy a large salary from the Sun King on the condition of taking up residence in Paris, where he assumed not only membership of the Academy but a leading role in the affairs of the observatory then being constructed.

In July 1671, Picard himself went on an expedition in execution of one part of a programme of astronomical researches that he had presented to the Academy in late 1669. Its purpose was to establish the exact location of Tycho Brahe's observatory at Uraniborg on the island of Hven and its longitudinal separation from Paris, in order to be able effectively to utilize Tycho's star catalogue as well as his planetary and lunar observations, a massive edition of which Picard had begun to prepare for publication in France. Armed with Cassini's predictions of the eclipses of Jupiter's satellites that would be observable in the autumn of 1671, Picard achieved his goal with notable success.

Another aspect of this voyage is more closely related to our chapter's topic: its association with the question of the Earth's possibly non-spherical shape, a question that arose out of concern about the length of a seconds-pendulum (a pendulum adjusted so that its period of swing in one direction was one second of time). In his early work on the pendulum, Huygens had found that the period of a simple pendulum oscillating through small angles is proportional to the square root of its length, and had established a relation between this period and the motion of free fall, so that we can derive from his proportions the modern formula $T = 2\pi\sqrt{l/g}$, where T is the time for a full back-and-forth swing, l is the pendulum's length, and g is the acceleration of gravity. But g does not appear explicitly in his formulas, and he did not imagine it to be variable. Consequently, he proposed to the Royal Society of London at the end of 1661 that the length of the seconds-pendulum – which he stated as 38 Rhineland inches – be accepted as a universal standard of length. In 1666 he made the same proposal to the new Academy in Paris; but by the summer of 1669 he was indicating that the length of the seconds-pendulum, in Parisian units, ought to be 36 inches $8\frac{1}{2}$ lines (there are 12 lines to the Paris inch, as well as to the Rhineland and English inches).

The latter figure appears to be that arrived at by Picard, who had undertaken careful experimentation with pendulums and enthusiastically embraced the suggestion that a seconds device be universally adopted as a standard of length. Near the end of 1670, in fact, he asked a Parisian correspondent of the Royal Society to dispatch the value that he, Picard, had established to that body's secretary with a view to having this adopted as the standard. By that time, however, Huygens had already noted an apparent discrepancy, for when he employed the accepted ratio of 720/695 for the measure of a length in Rhineland inches and its measure in Paris inches, he found that Picard's value corresponded not to the 38 inches he had found, but to 38 inches and $\frac{3}{10}$ of a line (the difference is about 0.6 millimetre). Alerted to this discrepancy, and having learned that the English were finding a longer length for the seconds-pendulum, Picard took advantage of the voyage to Uraniborg to make accurate pendulum observations. His experiments both in Leyden and in Uraniborg appeared to confirm his conviction that the seconds-pendulum was viable as a universal standard of length. But in achieving that result he altered the Rhineland–Paris ratio to 720/696. (A seconds-pendulum in Uraniborg in fact differs in length from a seconds-pendulum in Paris by about 0.63 millimetres, and from a seconds-pendulum in Leyden by about 0.32 millimetres.)

While Picard was in Uraniborg, the Academy

dispatched one of its young "student" members, Jean Richer, to Cayenne (just off the Atlantic coast of South America a little north of the equator) to conduct a series of astronomical observations. Especially important were those designed to take advantage of the close approach of Mars to Earth in 1672 in order to deduce, with the help of corresponding observations made by Cassini in Paris, a new and improved figure for solar parallax (see Chapter 7 of Volume 2A). But what is significant for our present topic is that Richer took with him a pendulum that had been regulated to beat seconds in Paris. In Cayenne he found it necessary to shorten its length by no less than 1.25 lines (2.8 millimetres) in order to make it continue to do so.

This piece of evidence could not immediately be explained. Cassini was inclined to think that Richer had been mistaken or careless. Picard apparently agreed, for he continued to insist that the seconds-pendulum was invariable in length. Indeed, he persevered in that belief in his own subsequent measures in France even when his own assistant suspected a variation, especially in those taken at Bayonne in the extreme south of the country. In 1682, however, the year of Picard's death and a decade after Richer's measurement at Cayenne, another expedition dispatched by the Academy confirmed his results. These new observations were made at Gorée, a small island off Cape Verde on the west coast of Africa, and at Guadaloupe in the West Indies. They were the work of two of the Sun King's engineers for hydrography, Jean Deshayes and a M. Varin, joined by a M. de Glos, a young man trained – as were they – by Cassini at the Royal Observatory.

Today we recognize that the variation with latitude in the duration of the swing of a pendulum of given length must be due to a change in the acceleration of gravity. How was this possible? The answer was soon forthcoming in the *Principia* of Isaac Newton (1642–1727), which appeared in 1687. In Prop. 19 of Bk III Newton set out from the idea that the Earth could originally have been a homogeneous fluid mass, subject to the law of universal gravitation and rotating about an axis; he postulated without demonstration that it would form an ellipsoid of revolution, flattened at the poles. Fluid in two canals leading from centre to surface, one in the equatorial plane and the other coinciding with the polar axis, would be in balance.

By calculation Newton found the centrifugal force at the equator to be to the mean acceleration of gravity as 1:289 (he used 1:290 in the first edition). By a laboriously derived approximation, he argued that the canal along an equatorial radius would exceed the canal along the polar axis by $\frac{1}{289} \times \frac{505}{400} = \frac{1}{229}$th of the polar radius. He claimed (without demonstration and in fact mistakenly) that the ellipticity would necessarily be greater if the Earth were denser toward the centre. In the third edition (1726) he offered the explanation that "the planets are more heated by the Sun's rays toward their equators, and therefore are a little more condensed by that heat than towards their poles".

In Prop. 20 of Bk III Newton undertook to compare effective weights at different latitudes on the surface of the Earth. By the principle of balanced canals, these weights vary inversely as the distances to the centre. It followed that the increase in effective weight varied as the square of the sine of the latitude. From this result he deduced (on the assumption of a homogeneous Earth) that a pendulum beating seconds would have to be shortened by about $\frac{1}{12}$ of an inch when taken from Paris to the Equator (the figure becomes 1.087 twelfths in the second edition of 1713). He likewise supposed that if the Earth were denser toward the centre, the shortening of the seconds pendulum would be greater. And the accumulating evidence suggested that the required shortening was greater.

Newton's treatment of the problem of the shape of the Earth in his *Principia* was the effective beginning of *dynamic* or *physical* geodesy: the attempt to derive the shape of the Earth from assumptions about the forces involved as a rotating, fluid Earth solidified.

Newton's first successor in this endeavour was Huygens, in his *Discours de la cause de la pesanteur* (published in 1690). Huygens did not admit the reciprocal attraction of all particles of matter, but adopted instead a modified Cartesian theory in which gravity was the result of aethereal pressure. He supposed that each particle of a homogeneous fluid mass was impelled only toward the centre of gravity of the mass; the problem of the equilibrium form of the mass under rotation was thus greatly simplified. Huygens, too, arrived at the shape of an oblate spheroid. But, owing to his initial assumption, he found the polar radius to be shorter than the equatorial radius by only $\frac{1}{578}$ th of the latter.

Thus towards the end of the seventeenth century we find Huygens with a quasi-Cartesian theory of gravity, and Newton with a theory of universal attraction, in agreement as to the oblateness of the Earth but not as to the extent thereof. Operations were soon to be under way, however, which would contradict the idea of oblateness itself. As early as 1682, Picard, having completed his work on Tycho's observations and sent it to the royal printing house, was making plans for a voyage to Alexandria in order to establish the bases necessary for a similar study and verification of Ptolemy's observations. Meanwhile Cassini had received backing for a plan that called for a southward extension within France of Picard's earlier arc measure. Unfortunately, neither of these projects came to fulfilment, the first succumbing to Picard's death, the second being cancelled after the demise of Colbert the following year. Colbert's replacement by the Marquis de Louvois as Louis XIV's most influential minister brought changes in the Academy's programme: the publication of Tycho's observations was halted, and the astronomers of the Academy who had by then left on the expedition to extend the meridian measure were called back. Their labours were turned instead to more practical projects nearer to the king's heart.

The work of extending the arc measured by Picard was resumed by Cassini only in 1700, but then in both northerly and southerly directions. These efforts were carried to completion in 1718 by Cassini's son Jacques (Cassini II, 1677–1756). The report of their findings was published in 1720 in his *De la grandeur et de la figure de la terre*.

If the Earth has the shape of an oblate spheroid, the length of one degree of latitude – corresponding to a change of 1° in the altitude of the celestial pole – ought to increase as one moves from the equator to the poles. In the measure undertaken in France, therefore, a northern extension of the same latitudinal amplitude as Picard's arc ought to have been slightly longer than the original arc, which, in turn, ought to have exceeded in length a southern extension. The figures published by Jacques Cassini, however, indicated the opposite result. Having found that the southerly portion of the total arc had the greater length, he and a group of followers maintained that the Earth had the shape of a prolate spheroid – that it was elongated rather than flattened at the poles.

In France at this time opposition to Newton's theory of gravitation was nearly universal. Cassini's result nevertheless posed a problem: how to reconcile it with the flattening effect – undeniable, thanks to Huygens – of the centrifugal forces due to the Earth's rotation? Bernard le Bouyer de Fontenelle, secretary of the Academy of Sciences, charged J. J. d'Ortous de Mairan, a member of the Academy and a defender of Cartesianism, with attempting a theoretical reconciliation of these opposed findings. Mairan's solution, presented in 1720, was arrived at by denying the primitive sphericity of the Earth, which Huygens had assumed as a postulate. If the Earth had originally been a prolate spheroid, then rotation, while reducing its prolateness, could leave it still prolate.

During the 1730s the champion of Newtonianism in France was P. L. Moreau de Maupertuis (1698–1759). In 1732 and early 1733 he undertook to present Newton's law of attraction to the Paris Academy. His *Discours sur les différentes figures des astres* (1732) assumed and defended Newtonian principles. But on the question of the Earth's shape, Maupertuis, like other Continental mathematicians, found Newton's argumentation elliptical and ambiguous. Attempting a derivation of his own based on the same principles, he arrived at a value for the Earth's flattening markedly different from Newton's. The resolution of the question, he concluded, called for a turn from dynamic to geometrical geodesy.

Meanwhile, questions had been raised about the inferences that Jacques Cassini had drawn from his measurements. Joseph-Nicolas Delisle (1688–1768), a professor of mathematics at the Royal College, was the first to raise doubts: the excess in length of the southern degree of latitude over the northern could be nothing more than observational error. Delisle thought the Earth's shape could be derived at least as accurately from variations in lengths of degrees of longitude; in measurements of the latter he believed there were ways of reducing the margin of error that were not available in the measurement of latitudinal variations.

In 1720 Delisle, having earlier tried and failed to interest his fellow Academicians in Newton's lunar theory, had apparently little hope of interesting them in Newton's and his own geodetical ideas; he merely confided them to the king's representative in the Academy. Then, in 1725, he accepted an offer from Peter the Great to found an observatory and an associated school of astronomy in Russia. His stay there, which was to have been for four years, stretched into twenty-two. During this period, he and his students engaged in geodetic and geographical expeditions throughout the country, intended – in a transfer of an established French tradition to the opposite end of the European continent – as a basis for a projected large-scale map of Russia, which remained unrealized.

Meanwhile, the thoughts on geodesy that had occurred to Delisle occurred also to Giovanni Poleni, professor of mathematics at the University of Padua, who in 1724 published a pamphlet challenging the conclusiveness of Cassini's measurements of degrees of latitude, and, like Delisle, advocating measurements of lengths of degrees of longitude. A second edition of the pamphlet appeared in 1729, and gave rise to a lengthy review in 1733 in the *Journal historique de la république des lettres*, published in the Netherlands.

It was a memoir presented to the Academy by Maupertuis on 3 June 1733 that touched off the revival of geometrical geodesy in Paris as a means of determining the Earth's shape. Maupertuis had read Poleni's pamphlet, and made use of it in his memoir. It was now his hope that the Earth's shape could be determined simply as a question of observational fact.

Maupertuis's memoir of June 1733 prompted two of the Academy astronomers, Louis Godin and Charles-Marie de La Condamine (1701–74), to present papers discussing the difficulties in measuring degrees of latitude accurately. The papers of Maupertuis, Godin, and La Condamine led directly to a proposal to send teams of academicians to Peru and Lapland; the large latitudinal difference between these places should give rise to differences in lengths of a degree of latitude or longitude that would settle the question of the Earth's shape.

Jacques Cassini first learned of the Dutch journal review in late 1734, on his return from a year and a half spent measuring the length of a perpendicular to the Paris meridian across France. In a rejoinder published in the Academy's *Mémoires* he argued that, unless errors in timing the eclipses of Jupiter's satellites could be reduced, Poleni's method would lack the accuracy claimed for it.

This may explain why both expeditionary teams proceeded to measure degrees of latitude rather than degrees of longitude.

The Peruvian expedition set out in 1735, and the expedition to Lapland almost a year later. But the Lapland expedition returned in 1737 – seven years before any members of the Peruvian expedition reappeared in France. Maupertuis, who had headed the northern undertaking, published an account of it in 1738. The degree of latitude measured in Lapland was considerably longer than that measured by Picard in France. No wonder that Maupertuis had himself painted bedecked in furs and holding a globe of the Earth that he was squeezing flat at the poles (Figure 15.1).

The result of the Lapland measurement (Figure 15.2) invalidated the results announced after the prolongation of Picard's arc, and thus led to a call for a remeasurement of the meridian of Paris. This task was undertaken in 1739 and 1740 under the auspices of the Academy and the team was led by the Abbé Nicolas-Louis de Lacaille and Cassini III, César-François Cassini de Thury, as he liked to fashion himself. The report of their operations, entitled *Méridienne vérifiée*, appeared under Cassini de Thury's name without mention of Lacaille, although the latter was in fact its author. It reversed the earlier findings and erased any doubts that might have remained. Thus, in 1740, the question was definitely decided in favour of the oblateness of the Earth. The results to be brought back from Peru would corroborate this conclusion.

The return of members of the Peruvian expedition was delayed, partly because of the difficulty of the operations themselves, which took a long time to execute, and partly because of friction that developed among members of the expedition. Godin, who, as the first proponent of the equatorial venture, had been designated leader, did not return to France after completion of the operations, but stayed on until 1751 as a professor of mathematics at the University of San Marcos. He claimed to be writing an account of the voyage, but it never appeared.

La Condamine, another of the astronomers of the expedition, returned to France by the longest and most dangerous route, the Amazon. After reaching the Atlantic, he went to Cayenne where he repeated the observations of Richer that had first raised the problem of the Earth's shape more

15.1. Maupertuis in Lap costume compressing the Earth, with a poem praising his work in geodesy.

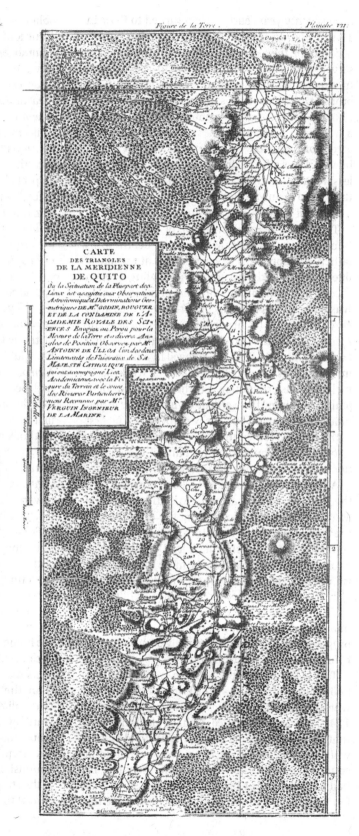

15.2. The triangulation in Lapland used by
Maupertuis and his colleagues.

15.3. The triangulation in Peru used by
Godin, La Condamine and Bouguer.

than seventy years earlier. He returned to Paris in 1745, and in 1751 published a day-by-day account of the actions of the expedition which served as a kind of introduction to the detailed treatment of the geodetic operations that he brought out later that same year (Figure 15.3). Thanks to his pleasant personality and his talent as a writer, La Condamine has received much of the credit for the success of the expedition; but in fact he was a less gifted astronomer than Godin and a less reliable mathematician than the venture's third chief member, Pierre Bouguer (1698–1758) – the most important of the three.

The son of a royal professor of hydrography, Bouguer was a prodigy who, on the death of his father, was appointed to succeed him at the age of fifteen. He quickly became "the leading French theoretical authority on all things nautical", and was named an associate geometrician in the Academy in 1731; to his fellow Academicians he seemed a natural choice to accompany Godin and La Condamine. In Peru he engaged in a number of investigations beyond the geodetical work itself, including determination of the deviation of a plumb bob from the vertical owing to the gravitational attraction of a nearby mountain. The first of the three to return to France (in 1744), he was also the first to publish an account of the expedition, which he did five years later.

Although the expedition to Peru corroborated the results of the Lapland expedition and the *Méridienne vérifiée* in confirming the oblateness of the Earth, it cannot be said to have fixed with certainty the value of the ellipticity. Different calculators obtained different values: Jorge Juan, a Spanish observer who had worked with Godin, found $\frac{1}{266}$; Bouguer decided for $\frac{1}{179}$. Meanwhile, the question of the relation of the empirically determined shape and variation of gravity to Newton's proposed explanation had been given prominence by new theoretical investigations.

The researches of the Scottish mathematician Colin Maclaurin (1698–1746) were presented in a work entitled *A Treatise of Fluxions*, published in 1742. Maclaurin claimed to have demonstrated that if the Earth were a homogeneous fluid mass it would assume the shape of an oblate spheroid in consequence of its diurnal rotation. Actually, he had not proved that the planet *would* assume that form, but had shown for the first time that the

oblate spheroid was *a* form of equilibrium. His value for the ratio of the axes, on the assumption of a homogeneous Earth, was practically the same as Newton's, namely 230 to 229. Since this value disagreed with the ellipticity derived from the measurements in Lapland and in France, Maclaurin proposed to treat the Earth as non-uniform in density, with either less or greater density at the centre. But his investigation of these hypotheses was unsatisfactory in a number of respects, both physical and mathematical; he failed, for instance, to show that any of the hypothesized distributions would be in equilibrium under rotation. One of those aware of the difficulties was the French mathematician, Alexis-Claude Clairaut (1713–65).

Clairaut's epochal *Théorie de la figure de la terre* appeared in 1743. It was the culmination of several years of intensive study of the subject, by a very talented man who had entered the Academy at the extraordinarily young age of eighteen. He had published several papers in the Academy's *Mémoires* on the problem of the Earth's shape, and had been a member of the Academy's Lapland expedition. In his major study of 1743, Clairaut first treated the Earth as a homogeneous fluid; his methods and results were essentially the same as Maclaurin's, his value for the ratio of the axes being 231:230. He then went on to show that, contrary to Newton's claim, the Earth if denser toward the centre need not be more flattened than it would be if homogeneous.

For a heterogeneous stratified ellipsoid of revolution, consisting of ellipsoidal strata of revolution with the denser strata towards the centre, and surrounded by a layer of fluid with its surface everywhere at right angles to the direction of gravity (a "level surface", later called a "geoid"), he showed that the ellipticity would be less than in the homogeneous case, provided that the ellipticity of each sub-surface shell multiplied into the square of its distance from the centre was never greater than the corresponding product at the surface. Generalizing a principle of hydrostatics put forward by Maclaurin, and deriving from it what would later be called the theory of potential, Clairaut argued that the density of the Earth should diminish from centre to surface, with the ellipticity of the strata increasing from centre to surface. The case of the stratified ellipsoid with density decreasing from centre to surface falls between that of the homogeneous ellipsoid, of which the ellipticity is $\frac{1}{230}$, and the case of a spheroid acted upon by a single force directed to the centre (the Huygenian case), in which the ellipticity is about $\frac{1}{578}$. Thus the ellipticity of the stratified ellipsoid – the most plausible model, in Clairaut's view – should fall between these two values, and hence be less than $\frac{1}{230}$.

Newton's principle of balanced canals, Clairaut found, did not hold for the heterogeneous stratified ellipsoid of revolution: the effective gravities at different points on the surface of the ellipsoid did not vary inversely as their distances from the centre. The increase in effective gravity as one moved from equator to pole still proved to be proportional to the square of the sine of the latitude: $g_L - g_E = (g_P - g_E)\sin^2 L$, where g_E, g_P, and g_L are the effective gravities at the equator, pole, and latitude L, respectively. But the total increase compared with the effective gravity at the equator, Clairaut showed, was given by $(g_P - g_E)/g_E = 5\phi/2 - \delta$, where ϕ is the ratio of the centrifugal force at the equator to the effective gravity there, and δ is the ellipticity of the heterogeneous ellipsoid. This is often referred to as "Clairaut's theorem". Since ϕ had been shown to be $\frac{1}{288}$, so that $\frac{5}{2}\phi = \frac{2}{230}$, and measurement of g_L at different latitudes was indicating that $(g_P - g_E)/g_E > \frac{1}{230}$, it followed that, if the Earth were the heterogeneous ellipsoid of revolution of the kind Clairaut favoured, then $\delta < \frac{1}{230}$.

The trouble with this conclusion in 1743 was that the value then current for the ellipticity was the one obtained from comparison of the measurements of the meridian in Lapland and in France, namely $\frac{1}{177}$, which was greater than $\frac{1}{230}$. The measurement of more distant degrees, Clairaut urged, was required to obtain an accurate ratio of the axes; he looked forward to the results of the Peruvian measurement.

Unfortunately, the Peruvian measurement did not lead to general acceptance of an ellipticity less than $\frac{1}{230}$, and so failed to confirm Clairaut's theory. Jean le Rond d'Alembert, for instance, in his *Recherches sur la précession* (1749), obtained from the measurements in Lapland and Peru an ellipticity of $\frac{1}{174}$. By the time he was writing Pt III of his *Recherches sur différens points importans du système du monde* (1756), two further meridian arcs had

been measured, one by Lacaille at the Cape of Good Hope, the other by Rudjer J. Bošković and Christopher Maire in the Papal States of Italy. Comparing these with the earlier measurements, d'Alembert concluded that no value of the ellipticity could reconcile them. And Bošković in his account of the Italian measure ended by remarking that the more one measured degrees, the more uncertain the shape of the Earth became.

Eventually, in the late 1740s, Clairaut did reconcile a value of δ greater than $\frac{1}{230}$ with a value of the fraction $(g_P - g_E)/g_E$ greater than $\frac{1}{230}$, but in doing so he had to give up the hypothesis he had favoured as to the internal constitution of the Earth, namely the assumption that the interior strata were in the form of *ellipsoidal* surfaces of revolution. This led him to abandon as well the supposition that the Earth had originally been in a fluid state, for he believed he had shown that the strata in a heterogeneous, stratified, self-gravitating fluid figure were level surfaces, and further that, in a slowly rotating figure, such level surfaces were, to a close approximation, ellipsoidal. D'Alembert later criticized Clairaut's proofs of these propositions as inconclusive; but the propositions in fact hold if the forces are conservative – which is the case with gravitation.

D'Alembert repeatedly criticized various aspects of Clairaut's work. He objected to Clairaut's belief that the laws of hydrostatics required the denser strata to be nearer the centre, and all the strata to be level surfaces. His own treatment of the problem of the Earth's figure is more general than Clairaut's, in that he did not limit consideration to ellipsoids of revolution. The method that he introduced for estimating the attraction of a spheroid by resolving the body into a sphere and a thin additional shell has proved of value. But as Pierre-Simon Laplace (1749–1827) remarks, d'Alembert's investigations "lack the clarity so necessary in complicated calculations". By contrast, in characterizing Clairaut's *Théorie de la figure de la terre*, Laplace stated that the importance of Clairaut's results and the elegance with which he presents them "puts this work in the class of the most beautiful mathematical productions".

Clairaut's original theory was in fact less in error than the value of the ellipticity Clairaut and d'Alembert had thought it necessary to accept would have implied. This conclusion emerged from analyses undertaken by A.-M. Legendre (1752–1833) and Laplace. In these studies it has been said that "Laplace played leapfrog with Legendre".

The game began in January 1783 when Legendre presented to the Academy a memoir on the attraction of homogeneous spheroids. Here he proved that if the attraction of a solid of revolution is known for every external point on the prolongation of its axis, it is known for every external point. The attraction is given as an integral over the spheroid, and in expanding the integrand as a series, Legendre introduced the 'Legendre polynomials', which serve as coefficients in the series. With the aid of these and on the basis of a suggestion from Laplace, he obtained what was later to be called the potential, from which the attraction was then derivable.

As Legendre was not yet a member of the Academy, his memoir underwent an official review before its eventual publication (in 1785) in the Academy's *Savants étrangers*; the reviewer was Laplace. In March 1783 Legendre was elected a member of the Academy, and so in May 1783 heard Laplace read to the Academy a memoir on the attraction of elliptical spheroids, deepening, as he admitted, his own treatment of the subject.

Laplace's memoir, revised and amplified, became Pt II of his *Théorie du mouvement et de la figure elliptique des planètes*, brought out in 1784 with financial underwriting from Bochart de Saron, honorary member of the Academy. During the 1770s Laplace had written three memoirs on the attraction of solids of revolution, but without using the concept of potential. This concept had been adumbrated in Clairaut's *Théorie* of 1743 and clearly deployed in several of Lagrange's memoirs of the 1770s (see Chapter 21); it was probably from Lagrange's memoirs that Laplace gained his awareness of the concept's value. In his treatise of 1784 he used Legendre's polynomials to calculate the potential of an arbitrary spheroid, not necessarily a solid of revolution, but differing little from a sphere. And here he showed how the departure from sphericity could be developed as a series in what were later to be called spherical functions or spherical harmonics.

Legendre continued his investigations of the attractions of spheroids during the 1780s, publishing three further memoirs in which he demonstrated various properties of the Legendre polynomials and of spherical harmonics.

In the meantime Laplace produced two more memoirs on our subject. The results of these and of his *Théorie* of 1784 were then incorporated, with some revision, in the second volume of his *Traité de mécanique céleste* (1799). From both pendulum experiments and the phenomena of the precession of the equinoxes and nutation of the Earth's axis, Laplace concluded that the flattening of the Earth must be less than $\frac{1}{230}$. The precession and nutation set limits between which this fraction must fall; Laplace in 1799 gave them as $\frac{1}{304}$ and $\frac{1}{578}$ (Nathaniel Bowditch, translator of Laplace's *Mécanique céleste* into English, in taking account of Laplace's later revision of constants, determined that the first of these should be $\frac{1}{279}$).

On the basis of pendulum experiments and the assumption of ellipticity, Laplace found the most probable ellipsoid to have an ellipticity of $\frac{1}{336}$ (Bowditch, correcting errors in the calculation, obtained $\frac{1}{315}$). By "most probable" Laplace here means that the positive and negative errors add to zero, while the sum of their absolute values is a minimum. (The method of least squares was still in the future.)

The total variation in the length of the seconds-pendulum from equator to pole in fact amounts only to 5.24 millimetres. To obtain a reliable value of the Earth's ellipticity from so minute a variation requires that a large number of carefully performed measurements be brought to bear on the formation of a statistical mean. Laplace had only fifteen such measurements at his disposal. Bowditch, who brought out his translation of the second volume of Laplace's *Mécanique céleste* in 1832, had by this date assembled 52 such measurements, and to these he applied the method of least squares; neglecting eight of the measurements as greatly discrepant from the rest, he obtained an ellipticity of $\frac{1}{297}$, precisely the amount adopted internationally some 92 years later.

Laplace also examined the implications of the degrees of the meridian measured at different latitudes; seven of these were available to him in 1799. Determining the ellipticity of an ellipsoid that would reduce the largest error to a minimum, he found this minimum largest error to be 189 metres, and to occur in the degrees of Pennsylvania, the Cape of Good Hope, and Lapland; the ellipticity, on this assumption, was $\frac{1}{277}$. These measures, Laplace believed, could not be so greatly in error; the Earth's shape, therefore, must differ sensibly from the ellipsoidal.

When he calculated the most probable ellipsoid, based on the same seven measures, he found an ellipticity of $\frac{1}{312}$, and the error in the degree of Lapland to be 336 metres. Sure that the error could not be nearly so large – as in fact the remeasurement undertaken by Jons Svanberg in 1801–1803 was to show it to be – he again concluded that the departure from ellipticity must be sizeable. But Bowditch in 1832 later found measurements fitting the elliptic hypothesis more closely. From five measurements in Peru, India, France, England, and Sweden, he obtained an ellipticity of $\frac{1}{310}$.

One of the seven measurements used by Laplace in Vol. 2 of the *Mécanique céleste* was a very recent one, the result of an investigation begun in 1792 and completed in 1799. The establishment of a uniform system of weights and measures had been a goal of the mercantilists from the time of Colbert; and this goal was adopted in 1790 by the revolutionary National Assembly, which directed the Academy of Sciences to prepare a report. An early proposal considered by the Academy's commission on weights and measures was to take the length of the seconds-pendulum at 45° of latitude as the universal standard of length – the idea of Picard, modified as the discovery of latitudinal variation required. It was hoped that this idea would be accepted by the English. It was not, however, and the idea was then set aside in favour of another 'natural' standard: the ten-millionth part of the quarter of the terrestrial meridian. This choice, accompanied by a proposal for the long and expensive operation of measuring an arc of the meridian from Dunkirk to Barcelona, was defended on various grounds, but the chief motivation behind it was to shed new light on the question of the shape of the Earth.

The measurement of the meridian was begun under the direction of two young Academicians, Jean-Baptiste Joseph Delambre and Pierre-François-André Méchain, in 1792. Work was temporarily suspended at the end of the following year, then resumed in 1795 and completed by 1799. In the latter year, the French government invited other states to send representatives to Paris for the purpose of examining the finished work, and the resulting conference may be considered the first international congress of scientists. Its final action

was to adopt an ellipticity of $\frac{1}{334}$ based upon the arc measured in Peru and that newly measured in France and Spain. Delambre and Méchain, who had carried the arduous work of measurement through to completion, published an account of their operations in the three-volume *Base du système métrique*, of which the first volume appeared in 1806. In reviewing the calculations for the meridian while preparing this volume for publication, Delambre concluded that the Earth's ellipticity was given more accurately as $\frac{1}{309}$.

The arc measured by Delambre and Méchain was soon extended southward to the Balearic Islands so that the whole measured arc would centre around a latitude of 45°. This extension was begun by Méchain, and on his death was continued by a new pair of young Academicians, Dominique François Jean Arago and Jean-Baptiste Biot. In 1821 they published an account of their work that in effect constitutes a fourth volume of the *Base du système métrique*.

The geodetic operations involved in the establishment of the metric system did not finally settle the question of the exact shape of the Earth. But by introducing new instruments and techniques, they supplied a model for further measurements of the same kind. The example thus set gave rise to a series of new arc measurements, such as Svanberg's revision of the Lapland measurement.

In Vol. 2 of the *Mécanique céleste* (1799), Laplace could claim with assurance that the Earth's flattening was less than $\frac{1}{230}$, the value implied by the supposition of homogeneity; the result was confirmed by pendulum experiments, the degree measurements, and the precession and nutation. In Vol. 3, which appeared in 1802, he drew support for the same conclusion from the lunar theory. In the lunar tables of Charles Mason (published in 1780) and those of Johann Tobias Bürg (in preparation in 1802, published in 1806), there appeared in the Moon's longitude a term proportional to the sine of the longitude of the lunar node; it had been determined empirically, and Mason put the coefficient at 7".7, while Bürg gave it as 6".8. By gravitational theory, Laplace showed the coefficient to be proportional to the Earth's oblateness; the derivation was free of the uncertainty in derivations of other lunar inequalities, where the convergence of the approximations was slow. By the theory, an oblateness of $\frac{1}{230}$ gave a coefficient of $11''.499$, and $\frac{1}{334}$ gave $5''.552$; thus Bürg's value implied an oblateness of $\frac{1}{305}$.

From the more accurate measurements of latitudinal degrees available in the 1830s, Bowditch, as previously mentioned, concluded a fairly close agreement with Clairaut's original hypothesis of elliptical strata. The primal fluidity of the Earth, proposed by Newton, was once more likely. The prolonged discussion had ended by showing Newton's and Clairaut's ideas to be the apt ones.

Meanwhile, increasing precision was bringing a new problem into focus, the "deflection of the vertical", or departure of the plumb line from perpendicularity to the reference ellipsoid of revolution, owing to anomalies in the distribution of the Earth's mass. This fact would set the problem for the next generation of investigators of the figure of the Earth – the endpoint in one phase of the investigation becoming a point of departure for the next.

Further reading

John Greenberg, Geodesy in Paris in the 1730s and the Paduan connection, *Historical Studies in the Physical Sciences*, vol. 13 (1983), 239–60

Rob Iliffe, Aplatisseur du monde et de Cassini: Maupertuis, Precision Measurement, and the Shape of the Earth in the 1730s, *History of Science*, vol. 31 (1993), 335–73

Tom B. Jones, *The Figure of the Earth* (Lawrence, Kansas, 1967)

Henri Lacombe and Pierre Costabel (eds.), *La Figure de la terre du XIIIe siècle à l'ère spatiale* (Paris, 1988)

Georges Perrier, *Petite histoire de la géodesie* (Paris, 1939)

Isaac Todhunter, *A History of the Mathematical Theories of Attraction and of the Figure of the Earth* (London, 1873; reprinted New York, 1962)

Joella Yoder, *Unrolling Time: Christiaan Huygens and the Mathematization of Nature* (Cambridge, 1988)

16

Clairaut and the motion of the lunar apse: the inverse-square law undergoes a test

CRAIG B. WAFF

In Bk III of his *Principia* (all editions), Isaac Newton (1642–1727) put forward two different arguments for establishing the inverse-square formulation of the law of gravitation. The first was based on the recognized accuracy or near-accuracy of Kepler's third law as applied to the Jovian and (in the second and third editions) Saturnian satellites orbiting about their primary planets, and to the primary planets orbiting about the Sun, together with Cor. 6 of Prop. 4 of Bk I: "If the periodic times are as the three-halves powers of the radii, the centripetal forces will be inversely as the squares of the radii...." But Prop. 4, including this corollary, was demonstrated only for concentric circular orbits – a condition not strictly satisfied by the orbits of the planets. Newton therefore developed his second argument, in which he established a mathematical relation between the motion of the apsides of an orbit and the centripetal force that would produce it. According to his formula for this relation (given in Cor. 1 of Prop. 45 of Bk I), the virtually immobile aphelia of the planets implied a nearly exact inverse-square centripetal force between each of them and the Sun.

The motion of the Moon's apsides, amounting to about 3° per revolution, was more sizeable. Newton had hoped to account for it as a perturbative effect, but a second formula, which he derived in Cor. 2 of Prop. 45, and which mathematically related apsidal motion to a combination of an inverse-square centripetal force and a radial component of a perturbative force, yielded only half the observed motion.

Newton's statements about this calculation in the *Principia* are confusing if not contradictory. As an illustration of Cor. 2 of Prop. 45 (Figure 16.1), he proposed a radial perturbing force that was subtractive (i.e., directed away from the central body), and equal to $\frac{1}{357.45}$ of the inverse-square force directed toward the central body. (By the latter force alone the planet or satellite would revolve in a stationary ellipse.) He attributed this numerical example to no specific astronomical case in the first two editions of the *Principia*, but in the third edition he added the remark: "The apse of the Moon is about twice as swift." In the second edition, however, he inserted in Prop. 3 of Bk III the additional statement that

the action of the Sun, attracting the Moon from the Earth, is nearly as the Moon's distance from the Earth; and therefore (by what we have shown in Cor. II, Proposition 45, Book I) is to the centripetal force of the Moon as 2 to 357.45, or nearly so; that is, as 1 to $178\frac{29}{40}$.

In Prop. 4 of Bk III (second and third editions), in determining the force that the Earth would exert on the Moon if the latter were brought down to the Earth's surface, Newton proceeded as though there were a *subtractive* radial perturbative force due to the Sun, having to the Earth's force a ratio of 1 to $178\frac{29}{40}$.

In Props. 25 and 26 of Bk III, however, Newton identified a perturbative radial force of this value as *additive*, the remaining solar perturbative force being parallel to the line connecting the Sun and the Earth. The latter force had a *subtractive* radial component, and from the analysis of Prop. 26 one can deduce that the *average* radial perturbative force, when both radial components are taken into account, is subtractive, and bears to the centripetal force exercised by the Earth on the Moon the ratio of 1 to 357.45, precisely the value used in the illustration of Cor. 2 of Prop. 45. A careful reading thus indicates that the analysis Newton provides within the *Principia* leads to only half the Moon's apsidal motion. It is no wonder, however, that readers reached differing conclusions as to what

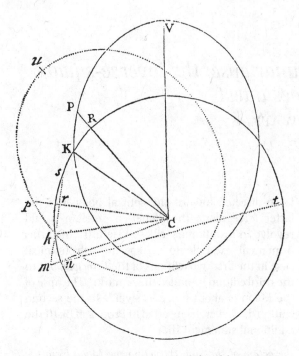

16.1. Newton's figure for Prop. 44 of Bk I of the *Principia*. With the aid of this diagram Newton shows that a body, which by an inverse-square law of force tending to *C* orbits in the stationary ellipse *VPK* about *C*, will orbit in the moving ellipse *upk* if there is added to the inverse-square force an inverse-cube force towards *C*.

In Cor. 2 of Prop. 45, Newton then shows that, in a nearly circular orbit, a force directed away from *C* that is $1/357.45$ times the average inverse-square centripetal force will cause the ellipse to advance 1° 31′ 28″ per revolution. (In fact, the radial component of the Sun's perturbing force on the Moon has just this average value.) In the third edition of the *Principia* Newton added the remark: "The apse of the Moon is about twice as swift". In Props. 3 and 4 of Bk III of the *Principia* (all editions), he surreptitiously assumed the average radial component of the Sun's perturbing force to be twice as great, or $-2/357.45$ times the Earth's force on the Moon, thus appearing to imply that the radial perturbing force accounted for the observed motion of the lunar apse.

Newton had here accomplished or failed to accomplish.

In the 1740s three of the foremost mathematicians of Europe, Leonhard Euler (1707–83), Alexis-Claude Clairaut (1713–65), and Jean le Rond d'Alembert (1717–83), undertook new, analytical derivations of the Moon's motions from the inverse-square law; and all three initially found, for the motion of the lunar apsides, an anomalous result similar to that which can be inferred from the analysis of Prop. 45 of Bk I of the *Principia*: the inverse-square law appeared to yield only about half the observed motion. As a consequence, a considerable discussion and controversy arose, much of it centred round Clairaut, concerning various aspects of the problem: (1) the interpretation of Newton's statements in the *Principia*; (2) the general applicability of the various "remedies" (in particular Clairaut's proposed emendation of the gravitational law) that were proposed in 1747 and 1748 to account for the rest of the apsidal motion; and (3) the accuracy of the calculations that had produced the apparent anomaly.

Responses to the anomaly

Euler

The first of the eighteenth-century mathematicians publicly to call into question the accuracy of the inverse-square law was Euler, in a paper entitled "Recherches sur le mouvement des corps célestes en général", which he presented to the Berlin Academy of Sciences on 8 June 1747. Since Newton's time, Euler pointed out, astronomers had come to the conclusion that the motions of the planetary apsides and nodes were indeed detectable (their detectability had earlier been questioned by Thomas Streete and Nicholaus Mercator); and such motions implied that the planets were subject not only to an inverse-square force directed toward the Sun, but to additional forces as well. Such forces, Euler pointed out, could arise from (1) an irregular distribution of mass in the Sun; (2) sources of force other than the Sun, such as the mutual actions of the planets on one another; or (3) a gravitational law that did not follow exactly the inverse-square ratio.

In his paper, Euler confessed that he was doubtful that the mutual forces among the planets followed the inverse-square law exactly. His doubts were partly due to his commitment to the idea that the action of all forces is by contact; thus in his thinking on the cause of gravity he took up a suggestion of Daniel Bernoulli, which equated gravitation to a pressure in a hydrodynamic aether, the pressure being greater where the aether's velocity was less. In such a model, it seemed unlikely that the law of gravitation, whatever its formula, would extend in mathematical exactness to all distances.

Euler's *a priori* doubts regarding the accuracy of the inverse-square law were strengthened by empirical evidence. Part of the evidence came from his having been unable to account theoretically (on the basis of the law) for all of the anomalies in the motions of the planets Jupiter and Saturn. In addition, Euler pointed out that in a careful study of the inequalities in the Moon's motion, in which he had been engaged since at least 1744, he had found a number of disagreements between observation and the results derived from the inverse-square law. One of these was particularly striking:

Having first supposed that the forces acting on the Moon from both the Earth and the Sun are perfectly proportional reciprocally to the squares of the distances, I have always found the motion of the apogee to be almost two times slower than the observations make it; and although several small terms that I have been obliged to neglect in the calculation may be able to accelerate the motion of the apogee, I have ascertained after several investigations that they would be far from sufficient to make up for this lack, and that it is absolutely necessary that the forces by which the Moon is at present solicited are a little different from the ones I supposed. . . .

After stating several other discrepancies between observation and theoretical calculation, Euler concluded that "all these reasons joined together appear therefore to prove invincibly that the centripetal forces one conceives in the Heavens do not follow exactly the law established by Newton."

Euler expressed a similar opinion in an essay completed at about the same time as the memoir just cited, namely his "Recherches sur la question des inégalités du mouvement de Saturne et de Jupiter", which was his entry in the prize contest of 1748 held by the Paris Academy of Sciences. The problem set for the contest was to formulate a "theory of Saturn and of Jupiter by which the inequalities [of motion] that these planets appear to cause mutually, principally near the time of conjunction, can be explained". Euler's essay, which would eventually win the contest award, arrived at the Paris Academy on 27 August 1747, and at the final pre-vacation meeting of the Academy on 6 September the judges of the contest were selected and were given the submitted essays, including Euler's. Two of the judges – Clairaut and

d'Alembert – had a particular interest in the subject of the contest. As the Marquise du Châtelet (hard at work during this period on a translation of Newton's *Principia*) observed to a correspondent, "Messieurs Clairaut and d'Alembert are after the system of the world, [because] they understandably do not wish to be anticipated by the prize [contest] pieces". In seeking to "crack" what came to be known as the three-body problem, both d'Alembert and Clairaut in the preceding year had undertaken to develop general methods for determining the perturbative effects caused by the mutual actions of the celestial bodies on one another.

Clairaut

Clairaut (Figure 16.2) completed his general solution (by approximation) of the three-body problem toward the end of 1746; this took the form of a memoir placed in a sealed envelope at the Academy on 7 January 1747. In two more sealed envelopes, deposited in March and June, he inserted papers dealing with applications of the theory, in particular to the Moon. He read these three papers to the Academy at meetings in June, July, and August. In a fourth sealed envelope, deposited on 6 September 1747 (the very day on which he received the essays for the prize contest, including Euler's), Clairaut gave the numerical determination of the equations for the Moon's varying radius vector and longitude, and (for the first time) stated his discovery of a discrepancy between the theory and the observed motion of the lunar apogee.

Upon subsequently reading Euler's contest essay, Clairaut was delighted (so he wrote Euler) to find himself in agreement with the doubts that Euler had expressed as to the accuracy of the inverse-square law. Even before reading this essay, however, Clairaut had decided that his discovery of the discrepancy between the observed motion of the lunar apsides and that calculated on the basis of the inverse-square law was sufficiently important to merit discussion in the Paris Academy, whose twice-weekly sessions would resume in November. At the public meeting of 15 November he gave a lecture on the subject, and at two further meetings, on 2 December 1747 and 20 January 1748, he gave readings of the sealed paper he had deposited in September.

Clairaut evidently suspected that his announce-

16.2. Alexis-Claude Clairaut.

ment of the discrepancy, and of the conclusions he had drawn from it, would provoke controversy. He began his public lecture in a good debating style:

In order to justify what I advance here, I am going to expose in all their force the reasons that ordinarily determine [one] in favour of the system of M. Newton; I will then render [an] account of the motives that have impelled me to seek some new proof of this system, of the work that this research required, and of what resulted from it.

An impressive number of phenomena, Clairaut reminded his audience, had been found to agree closely with the results predicted by Newton's law. (1) The general motions of the planets and their satellites were observed to be in near agreement with Kepler's laws, which Newton had shown to be implied by an inverse-square attraction law in the absence of disturbing forces. (2) The motion of the lunar nodes, which Clairaut had himself calculated by a method different from Newton's, appeared "rather consistent" with observation. (3) The theory of the tides "has been verified by the most skilful Mathematicians" (Clairaut no doubt had in

mind the "Newtonian" essays on the tides by Daniel Bernoulli, Colin Maclaurin, and Euler, which shared the prize of the Paris Academy in 1740). (4) And there were "several other questions equally favourable to attraction", which Clairaut left unspecified.

Clairaut had two motives for making a particular study of the motion of the Moon's apogee derivable from gravitational theory. One was the "obscurity" of Newton's research and remarks on the subject. The other was the importance of the apsidal motion in determining the Moon's place. The position of the Moon relative to the apogee of its orbit determined the size of the "equation of centre", which is either added to or subtracted from the mean longitude to obtain the true longitude. Because the equation of centre amounted at maximum to more than 6°, errors due to theoretical misplacement of the apogee could be sizeable.

Like Euler, Clairaut stressed the extreme care he had taken in deriving the apsidal motion:

Seeing therefore the great importance of the determination of the motion of the apogee, I have sought to

derive it from the solution of the general problem . . .;
this operation was more difficult than the solution of
the problem itself, because in determining the orbit of a
planet, some small quantities, which cannot cause any
considerable error during a single revolution, can be
neglected; but these quantities can become of infinite
consequence in the large number of revolutions necessary for knowing the motion of the apogee.

Clairaut then related his astonishment on disco-
vering that his theoretical derivation, despite such
care, generated only half the observed motion,
"that is to say that the period of the apogee [i.e., the
time it takes for the lunar apogee to return to the
same point in the heavens] that follows from the
attraction reciprocally proportional to the squares
of the distances, would be about 18 years, instead
of a little less than 9 which it is in fact."

This result, Clairaut told his audience, swayed
him at first toward abandoning attraction entirely.
Consideration of the impressive quantitative con-
firmations of Newton's law, however, tempered his
reaction. Some means, he felt, needed to be found of
"reconciling the reasons that seem . . . contrary
and favourable to [gravitational] attraction". The
expedient that occurred to him was to suppose that
gravitational attraction does indeed take place in
nature, but by some law other than the inverse-
square. What was needed was a law of gravitational attraction "that will differ very sensibly from the
law of the square at small distances, and diverge
from it so little at large ones that the difference is
undetectable in the observations". As an example,
he suggested a law consisting of two terms, one
having the square of the distance, and the other the
fourth power of the distance, as divisor. Clairaut
argued that the difference in the forces assigned by
the modified law and by Newton's law at the
distance of the Moon might be able to generate the
"missing" half of the Moon's apsidal motion. On
the other hand, at the greater distance of the planet
Mercury from the Sun, any difference in apsidal
motion due to the difference in forces assigned by
the two laws would take centuries to make itself
observationally detectable.

The idea of a non-inverse-square attractive force
was not new. Newton himself in Query 31 of the
later editions of the *Opticks* had suggested that an
attractive force acting sensibly over only rather
small distances might account for the phenomena

of cohesion and chemical reactions. The Oxford
mathematician John Keill, in a paper published in
the *Philosophical Transactions of the Royal Society* in
1708, had argued that attractive forces acting only
over very small distances would decrease with a
triplicate, quadruplicate, or some higher ratio of
the increase of the distances. Clairaut himself in
1739 and 1742 had presented papers to the Paris
Academy of Sciences in which he proposed to
account for the refraction of light and the ascension of liquids in capillary tubes by attractive forces
proportional to some unknown function of the
distance.

Neither was the suggestion of a multi-term
mathematical expression, combining an inverse-
square term with one or more terms involving
higher inverse powers of the distance, unprece-
dented. In a passage noticed by Clairaut in the
Marquise du Châtelet's *Institutions de physique*
(1740), the Marquise reported that

*Some Newtonians . . . have had the idea of explaining all
the Phenomena, as well the celestial as the terrestrial,
by a single and same attractive force, which acts as an
algebraic quantity*

$$\frac{a}{xx} + \frac{b}{x^3} + \dots,$$

*x marking the distance . . .; at some remote distances,
as for example, at those of the Planets, the part of the
attractive force that acts as the cube, is almost nil, and
disturbs only infinitely little the other part of the
attractive force which acts as the square, and on which
depends the ellipticity of the orbits.*

The same idea was also discussed in another book
of which Clairaut had recently made the official
examination for the "Approbation": the Abbé
Pierre Sigorgne's *Institutions Newtoniennes*, which
was published in June 1747. What was novel in
Clairaut's proposal of a multi-term law was only
the suggestion that a higher-order term could have
a sensible effect at the distance of the Moon from
the Earth.

D'Alembert

D'Alembert, like Clairaut, completed his general
solution of the three-body problem toward the end
of 1746. To the Berlin Academy, in December
1746 and January 1747, he sent memoirs on the
determination of planetary perturbations and on

the theory of the Moon, which he subsequently read (possibly in revised form) before the Paris Academy of Sciences in June 1747. In a paper read before the latter academy on 28 February and 6 March 1748, d'Alembert announced publicly that he too had found only half the observed motion of the lunar apsides to follow from the inverse-square law.

His speculations regarding the cause of this discrepancy, unlike those of Euler and Clairaut, remained private. In letters written to Euler and to the Swiss mathematician Gabriel Cramer in 1748 and 1749, d'Alembert expressed his disapproval of Clairaut's attempt to account for both terrestrial and celestial phenomena with a single modified law of gravitation. Newton, on the basis of the inverse-square law, had successfully derived the lunar *variation* and the inequalities of motion to which the lunar nodes and lunar inclination were subject. D'Alembert had confirmed these results, and so concluded that there was no need to question the mathematical form of the gravitation law.

D'Alembert viewed the discrepancy in the lunar apsidal motion as indicating only that the *particular* force acting between the Earth and the centre of gravity of the Moon followed a different law. To suppose that this force followed a law other than the inverse-square, he realized, would weaken the strong correlation between the Moon's centripetal acceleration and terrestrial gravity that Newton had obtained (in Prop. 4 of Bk III of the *Principia*) by assuming this law; nevertheless, he suggested two ways by which a different force relationship between the Earth and the Moon might arise. One way was to suppose, besides the universal inverse-square gravitational force, an additional non-gravitational force acting solely between the Earth and the Moon. In letters written during 1748, d'Alembert inclined to the view that this force, if it existed, was magnetic in origin.

A second way was to suppose that a radially asymmetric distribution of mass in the Moon – an arrangement that could escape observational detection from the Earth owing to the fact that the Moon always keeps virtually the same face towards its parent body – could affect the apsidal motion. To investigate this possibility, d'Alembert imagined the matter of the Moon to be concentrated into two globes connected by a thin rod and rotating around

the rod's midpoint. A calculation for this extreme case showed that in order to produce the "missing" half of the apsidal motion the distance between either globe and the rod's midpoint would have to be about twice the Earth's radius, a result that even d'Alembert conceded to be unrealistic. He soon learned that Euler had undertaken a similar calculation and had reached a similar conclusion.

Interpreting the *Principia*

Meanwhile Clairaut, following his public lecture of 15 November 1747, had been faced with the charge (by his colleague and fellow contest-judge Pierre-Charles Le Monnier) that his announcement of a discrepancy between the inverse-square law and the lunar apsidal motion was based on the research in Euler's prize essay. To rebut this charge, Clairaut on 2 December read the paper he had deposited under seal with the Academy's secretary just prior to receiving the contest essays. In this paper he had made brief mention of Newton's general conclusions in Props. 44 and 45 of Bk I regarding the relation of central force to apsidal motion. This reference led someone to call attention to Newton's "lunar" illustration in Cor.2 of Prop. 45, and as a result Clairaut found himself once more being charged with plagiary, this time in respect of Newton's work.

To answer this new charge, Clairaut prepared yet another paper, which he read to the Academy on 20 and 23 December, and in which he examined, proposition by proposition, all of Newton's work in the *Principia* relevant to the motion of the lunar apogee. Here he sought to show not only *that* Newton had failed to recognize the defect of his law of gravitation, but *why*. To obtain by the calculation of Cor.2 of Prop. 45 the true motion of the lunar apse, it was necessary that the coefficient of the term to be added to the inverse-square force be $1/178\frac{29}{40}$. In Prop. 25 of Bk III, on resolving the Sun's perturbing force into two components, one parallel to the line from Earth to Sun, and the other directed toward the Earth, Newton had demonstrated that the latter component was indeed in its mean quantity $1/178\frac{29}{40}$ of the centripetal force between the Earth and the Moon. Clairaut claimed that Newton had taken this radial component of the Sun's perturbing force, directed *towards* the Earth, as the sole radial component, and had failed to take into account the fact that the component parallel to the line from Earth to Sun could itself be resolved into radial and transverse components, the radial component being *subtractive*. (An additive radial component, we remind the reader, would cause the apsides to *regress*; a subtractive radial component would cause them to *advance*.) For Newton's use of the coefficient $\frac{1}{357.45}$ in Cor. 2 of Prop. 45, Clairaut had an explanation that presupposed that Newton had not known how to evaluate the integral

$$\int_0^{2\pi} \cos\theta \, d\theta,$$

where θ is the elongation between the Moon and the Sun.

Clairaut also had to defend himself against the charge that the deficiency in Newton's calculation of the motion of the apsides had been announced earlier, in the third volume of an edition of the *Principia* edited by the Minim priests Thomas Le Seur and François Jacquier, and published in Geneva in 1742. Here, in an extensive footnote to the scholium following Prop. 35 of Bk III, two different solutions of the problem of determining the motion of the Moon's apogee were supplied, and the second of them, which proceeded by Newton's method of Prop. 45 of Bk I, yielded only half the observed apsidal motion.

The anonymous author of these solutions was the Swiss mathematician Giovanni Ludovico Calandrini (1703–58), who had been professor of mathematics and later of philosophy at the Geneva Academy since 1724. As it happened, Calandrini's long-time colleague Cramer was visiting in Paris in late 1747 and early 1748, and attended the Academy's public meeting on 15 November at which Clairaut announced his discovery of the discrepancy in the motion of the lunar apogee. After being informed by Cramer about this announcement, Calandrini corresponded with both Cramer and Clairaut concerning it. Calandrini did not question the inverse-square law, but considered the method of Newton's Prop. 45 to be defective, in that it failed to take into account the eccentricity of the lunar orbit. Calandrini had been able by a rough calculation using a different method, and taking the Moon's variable orbital eccentricity into account, to obtain a much larger value for the motion of the lunar apse, and he believed that refinement of the calculation (by computing an

accurate *mean* value for the eccentricity) would lead to the observed motion.

Clairaut, who exchanged letters with Calandrini in February and March, did not agree. The orbital eccentricity of the Moon, he insisted, could have only a negligible effect on the motion of the lunar apse. He believed Calandrini's method to be uncertain if not erroneous, and in particular objected to Calandrini's having omitted from consideration the component of the Sun's perturbing force at right angles to the Earth–Moon radius vector. Calandrini, on the other hand, was unable to accept Clairaut's contention that Newton had somehow been able to convince himself that the method of Prop. 45 could account for the entire mean motion of the lunar apogee.

Meanwhile, another debater had entered the fray: Clairaut's colleague in the Paris Academy, Georges-Louis Leclerc, Comte de Buffon (1707–88). On 20 and 24 January Buffon read a paper that, among other things, questioned Calandrini's interpretation of Newton's remarks regarding the motion of the lunar apogee.

It was Buffon's mistaken belief that the proper ratio of perturbing force to centripetal force to be substituted into the formula of Cor. 2 was $1/178\frac{29}{40}$, as calculated by Newton in Prop. 25 of Bk III. He thus shared with Clairaut (although for different reasons) the belief that Newton was convinced of the adequacy of his theory to yield, at least approximately, the entire mean motion of the apogee. Buffon was therefore astonished to find Calandrini claiming that Newton had been well aware that the calculation yielded only half the observational value. Such a conclusion, in Buffon's view, would conflict with the honesty and good faith that had been evident in all of Newton's other publications.

The Buffon–Clairaut polemic

Buffon's main concern in his January paper, however, was not to criticize Calandrini's interpretation – he recognized his own lack of proficiency in mathematical calculations, and proposed to leave the calculational question to others. Rather, he particularly desired to register sharp opposition to Clairaut's conclusion that the inverse-square law of gravitation established by Newton ought to be modified. To base such a conclusion on a discrepancy between theory and observation in but a single phenomenon, as he believed Clairaut had done, was in Buffon's view unreasonable. The correct course would be to "seek the particular reason of this singular phenomenon". Buffon suggested five different possible explanations as to how the missing half of the apogeal motion might be explained, four of them invoking irregular distributions of mass in the Moon, the fifth depending on the Earth's magnetic force. The latter explanation appears to be the one Buffon most favoured.

Was it necessary, Buffon asked, when theoretical calculations of the apogeal motion failed to agree with observation, that Newton's theory be "ruined" at the foundation by changing the law?

M. Clairaut proposes a difficulty contrary to the system of Newton; but it is not a difficulty that should or can become a Principle. It is necessary to seek to resolve it, and not to make from it a Theory all of whose consequences are supported only on a single fact.

According to Buffon, whenever a law is proposed for the variation of a physical quantity, the use of a single term to express this law is mandatory. A physical law must meet two essential requirements: the quantity or quality being measured must vary in a "simple" fashion, and it must be expressed by a single term. This *simplicité* and *unité* is immediately destroyed if a two-term expression is interpreted as implying a complex variation of a single quantity or quality. Two terms had to represent two "qualities" (i.e., two distinct types of force) rather than one.

Buffon was also convinced that Newton's gravitational force could not vary in any ratio other than the inverse-square; this ratio was the only one permitted for a quality or quantity like light or odour propagated in straight lines from a centre.

In the final portion of his paper Buffon posed three questions to Clairaut. (1) If the law of gravitation did not follow the inverse-square ratio, would Newton's "rigorous demonstration" of the identity of terrestrial gravity with the force retaining the Moon in its orbit (in Prop. 4 of Bk III) remain valid? (2) Given Clairaut's newly proposed law, what would be the consequences for comets (such as that of 1680) that came very close to the Sun? Would not the orbital apse of such a cometary orbit undergo a large motion, in contradiction to the empirical verification obtained by Newton and Edmond Halley? (3) Would not the motion of the apsides of the inner planets and of the satellites of

Jupiter and Saturn be affected, contrary to what had been observed?

Clairaut responded in a paper read to the Academy on 17 February 1748. He had given detailed arguments against Calandrini's method, and he resented Buffon's failure to examine them. He denied that Newton had "rigorously demonstrated" the identity of terrestrial gravity with the centripetal force of the Moon. In Prop. 4 of Bk III Newton had considered the lunar orbit as a circle, and had not taken into account the radial component of the Sun's perturbing force. (Clairaut at this point appears unaware of Newton's incorrect use in this proposition, as in the more refined argument of Cor. 7 of Prop. 37 of Bk III, of a subtractive radial component equal to $1/178\frac{29}{40}$ of the Earth's mean force on the Moon.) As for the other questions that Buffon raised, Clairaut argued that the effect of the inverse-fourth-power term on the comet of 1680 would be observationally undetectable, and that the effects on the apsidal motions of the inner planets would be ascertainable only after a long period of time. The satellites of Jupiter and Saturn moved so nearly in concentric circles that their apsidal motions could not be reliably determined.

Turning to Buffon's metaphysical arguments for the correctness of the inverse-square law, Clairaut granted that metaphysics could be helpful in leading through analogy to tentative hypotheses, but urged that "if we allow ourselves to be led by its torch alone, we can lead ourselves astray at any moment". To suppose that, because an inverse-square law holds for the intensity of light, an entirely analogous law must hold for gravitation, presupposes that the mechanisms are similar, and that a body attracts gravitationally by the emission of particles. Such a supposition, Clairaut observed, is quite conjectural; the only reliable way to determine the law of gravitation is by means of empirical facts.

As for Buffon's contention that a physical law must be "simple" and be expressed by a single algebraic term, Clairaut put it down to a resistance against using complex quantities known as *functions* in the study of physico-mathematical problems. "If we are unable to render [these functions] very simple in expressing them, it is the fault of Algebra which, like language, has its imperfections." To claim as Buffon did that each term of Clairaut's two-term law represented a distinct force was presumptuous: "Is it appropriate for us," Clairaut inquired, "to decide whether the Creator has given the attractive virtue to matter by two different decrees, or has endowed it with two forces at the same time by the sole action of his will?"

At this point we must note that, at the meeting of the Academy of 27 January, Pierre Bouguer (1698–1758), a physicist, geodesist, and astronomer, had read a paper "On the institution of the laws of attraction", which he was planning to add to a new edition of his *Entretiens sur la cause d'inclinaison des orbites des planètes*, first published in 1734. Like Buffon, Bouguer claimed that gravitation could be considered as one of several sensible qualities (e.g. light) that are "exercised" in the direction of straight lines emanating from a single point, and that therefore decrease with the square of the distance. But unlike Buffon he believed that laws of attractive forces could differ from the inverse-square, and that the only way of determining these laws was from the phenomena. Some molecules might attract in accordance with the inverse-square, and others in accordance with the inverse-cube, so that the total effect of, say, the Earth on the Moon would be given by

$$\frac{m}{x^2} + \frac{n}{x^3},$$

where m designates the multitude and intensity of the corpuscles acting according to the one law, and n the multitude and intensity of the corpuscles acting according to the other.

In his response to Buffon of 17 February, Clairaut compared Bouguer's hypothesis with his own:

[M. Bouguer's] idea has nothing contrary to my researches, and I am very far from rejecting it: however I do not prefer it to mine, because I find in my hypothesis the advantage of making only a single law for all the phenomena attributed commonly to attraction, and this advantage appears to me superior to the one of the simplicity of analytical expressions.

Clairaut also replied in detail to Buffon's proposal of various asymmetrical distributions of the Moon's mass in order to account for the apsidal motion. Like d'Alembert and Euler, he concluded that the lack of symmetry necessary to yield a sizeable motion of the apsides would be extreme. "If ever I learn that the Moon has been seen otherwise than round", a sceptical Clairaut guaranteed, "I shall

yield myself to this explanation."

Buffon made no immediate public response to Clairaut's paper. Clairaut, consequently, must have assumed that their debate was at an end. He undertook now to unite his several papers on the motion of the lunar apsides into a single composite paper, which he had previously obtained permission to publish in the Academy's *Mémoires* for 1745, then being readied for the press (the collected memoirs for any given year were generally not published till several years later). In preparing the composite paper he ran up against the difficulty of not being able to specify the precise formulation of the proposed additional term in the gravitational law, in the absence of some phenomenon other than the motion of the lunar apsides in which the effect of this term could be determined quantitatively. Among the phenomena in which the effect might reveal itself, capillary action had not yet been sufficiently studied; the figure of the Earth and variation of gravity with latitude depended on the arrangement of the Earth's interior parts, still a matter of considerable controversy; the motion of the apsides of the Jovian and Saturnian satellites were as yet unknown, and those of Mercury and Venus were not precisely enough determined. Clairaut was thus unable to satisfy the desire he had expressed in his sealed paper of 6 September 1747 on the apsidal motion, of demonstrating how a modified law could account for the entire motion of the lunar apsides, while not affecting the agreement that Newton had shown between the inverse-square law and other phenomena attributed to gravitation.

Despite these difficulties, Clairaut still felt that it was simpler to hypothesize a single universal law than to assume two or more different types of attractive force, each following different laws, and perhaps deriving from different physical principles. In Clairaut's opinion, the idea of a modified universal law of gravitation was the most conservative hypothesis that could be made to account for the apsidal discrepancy.

Clairaut's arranging to have his composite paper printed in the *Mémoires* for 1745, together with his omission from this paper of certain paragraphs in the original papers read in 1747, aroused the ire of Buffon, who proceeded to make deletions from his own paper and to arrange to have it published in the same volume of the *Mémoires*. As Clairaut put it in a letter of July 1749 to Cramer, Buffon "subtracted all the things on which I had straightened him up, like the objections derived from comets, the figure of the Moon, etc., and he ... conserved from all that he had objected to me only some metaphysical arguments that could be disputed eternally."

Meanwhile, in the latter part of 1748 Clairaut undertook a review of his theoretical calculation of the Moon's motions. This review led to his discovery in December of that year that a more complete calculation on the basis of the inverse-square law would account for practically all the observed motion of the lunar apsides. The full calculation would require considerable time; to safeguard his priority, Clairaut in late January 1749 arranged for sealed envelopes, containing a description of the method of his new investigation and of the new result for the apsidal motion, to be deposited at the Paris Academy of Sciences and at the Royal Society of London. By the time of the Paris Academy's meeting of 17 May, he had sufficient confidence in his new calculations to make a public announcement of the new discovery that the inverse-square law alone could account for the full motion of the lunar apse. In doing so he asked to be excused from describing his method in detail until he could finish the recalculation, and specifically denied that Buffon's "vague reasons" had had anything to do with the new discovery.

Buffon, however, was convinced that it had been his objections that had persuaded Clairaut to retract the proposal of a modified law, and he insisted, in response to Clairaut's announcement, that the Academy, in printing his own earlier critique, mention the date on which it had been read. He also publicly reproached Clairaut for making changes in the published versions of his memoirs.

Clairaut, on going to the Academy's secretary to arrange for the printing of his retraction, discovered that Buffon not only had inserted in his critique a new metaphysical argument that he had not read at any meeting of the Academy, but also had surreptiously inserted a separate short note containing two additional arguments of the same character. This last step provoked Clairaut to attack his colleague's behaviour and arguments in a new memoir read to the Academy on 21 May and 4 and 11 June. At the last of these sessions Buffon presented a further addition to his original critique,

to which Clairaut responded on 21 June. Buffon continued to insist that a multi-term law involved absurdity; Clairaut believed Buffon's objections to be founded on mistaken algebra and a mistaken view of the relation of algebra to nature.

Clairaut terminated his dispute with Buffon by again asserting that his hypothesis of a modified law had been simply an expedient that he felt could admirably explain a wide variety of both celestial and terrestrial phenomena; he had defended it solely because Buffon had considered it absurd on the basis of metaphysical, mathematical, and physical grounds that he himself considered to have no validity. Buffon had involved him in an unwanted dispute that contributed nothing to the solution of the real problem, which was the use of phenomena to know the true laws of nature.

The resolution of the problem

Clairaut's memoir of January 1749, in which he first described the procedure by which he obtained his new derivation of the apsidal motion, was eventually read to the Academy of Sciences in March 1752, and appeared in the Academy *Mémoires* for 1748, published in 1752. The delay in publication was due to Clairaut's incorporation of the memoir in a larger treatise on the Moon's motion; the latter treatise being Clairaut's award-winning entry in the St Petersburg Academy's prize contest of 1750.

Clairaut's general method is described in Chapter 20 below, and it is there pointed out that, in resolving the equations of motion by approximation, Clairaut substituted for the radius vector r an expression giving the Earth–Moon distance in a rotating ellipse:

$$\frac{k}{r} = 1 - e \cos mv, \qquad (1)$$

where k is the parameter of the rotating ellipse and e its eccentricity, v is the true anomaly (the angle that the radius vector makes with the line of apsides), and m is an undetermined constant differing from 1, so as to allow for motion of the lunar apse. This substitution would find its justification if in the final solution the larger terms could be identified with the terms in (1), and the remaining terms were very small by comparison.

The outcome appeared to justify the substitution: Clairaut obtained for the radius vector the expression

$$\frac{k}{r} = 1 - e \cos mv + 0.007\,098\,8 \cos \left(\frac{2v}{n}\right)$$

$$- 0.009\,497\,05 \cos \left(\frac{2}{n} - m\right) v$$

$$+ 0.000\,183\,61 \cos \left(\frac{2}{n} + m\right) v, \qquad (2)$$

where n is the ratio of the Moon's mean motion to the difference between the mean motions of the Moon and the Sun. The last three terms in (2) are indeed but small oscillations superimposed on the other terms giving the value of k/r. The value of m, on the other hand, proved to be 0.995 803 6, implying that in one sidereal revolution of the Moon the apse moved $(1 - 0.995\,803\,6) \times 360° = 1°\,30'\,38''$ – only about half the observed motion.

Even before he had undertaken to evaluate the coefficients of his equation for the Moon's orbit, Clairaut had indicated a means whereby more exact values of m, n, k, etc., could be obtained. Instead of substituting for k/r the quantity given by (1) above, it was only necessary to substitute

$$\frac{k}{r} = 1 - e \cos mv$$

$$+ \beta \cos \left(\frac{2v}{n}\right)$$

$$+ \gamma \cos \left(\frac{2}{n} - m\right) v$$

$$+ \delta \cos \left(\frac{2}{n} + m\right) v, \qquad (3)$$

which is derived from (2) by putting the indeterminate quantities β, γ, δ in place of the numerical coefficients in (2). Clairaut obtained his anomalous result for the apsidal motion prior to September 1747, but it was not until the latter half of 1748 that he undertook the longer calculation required if the expression (3) were used for k/r. He appears to have been initially convinced that the new calculation would not yield a significantly different value for the motion of the apse. Two events may be singled out as having possibly led him to question this belief.

In a lost letter to Calandrini of early 1748, Clairaut pointed out, in criticism of the Swiss mathematician's procedures for determining the motion of the lunar apse, that Calandrini had failed to take account of the component of the solar perturbing force at right angles to the radius vector. Clairaut made this critique despite his belief at this time that the effect of this component on the motion of the apse could be proved to be negligible. Calandrini responded on 19–20 February, arguing that the criticism did not reflect on his method, but only on his failure to prove the omission legitimate, and pointing out that in a circular orbit the effects of the perturbing force at right angles to the radius vector would, in successive quadrants, cancel each other. Clairaut in his answer of 6 March rejected the supposition of a circular or nearly circular orbit as irrelevant, and seemed to imply that he was still seeking a demonstration that the effect of the perpendicular component was negligible.

Of what would such a demonstration consist? In Clairaut's first determination of m, only the $\cos mv$ term of (2) played a role. If the more refined substitution (3) were used, the last three terms of this expression (of which the last two derived from the perpendicular component of the Sun's perturbing force) would produce some further $\cos mv$ terms. Because of the smallness of the coefficients of the last three terms in (2), Clairaut apparently assumed at first that the new $\cos mv$ terms would be negligible. By the time he penned his second letter to Calandrini on 6 March, he may have begun to wonder whether the sum total of all the coefficients of these new $\cos mv$ terms might affect the value of m more than he had originally suspected. At any rate, in late April he wrote to Euler that he was once more at work on the problem of the Moon, and was making new discoveries regarding gravitational attraction, of which he would inform Euler when he had completed the calculation.

Another stimulus for carrying out the more refined derivation may have come from d'Alembert. In Clairaut's first calculation of both k/r and of the time t as a function of true anomaly v, he had ignored powers of the eccentricity e higher than the first. The square of the eccentricity in fact influenced the equation for k/r very little, but it had a more pronounced effect on the expression for t, owing to the large augmentation that certain terms received by integration. D'Alembert, by his own later account, informed Clairaut of this result in June 1748; and he claimed that the information had been important in persuading Clairaut to carry out the more refined calculation.

In any case, the new calculation was performed.

In the new equations for the orbit and the time, the coefficients of the terms were not very different from the ones determined previously. But, as Clairaut pointed out in his later presentation of the new theory to the Paris Academy,

an important respect in which my new solution differs essentially from the first is the determination of the letter m which gives the motion of the apogee. The term ... by which m is determined is found nearly doubled by the addition that is made of the terms

$$\beta \cos\left(\frac{2v}{n}\right) - \gamma \cos\left(\frac{2}{n} - m\right) v + \text{etc.}$$

to the value of $1 - e \cos mv, \ldots$, and by this means the motion of the apogee is found nearly conforming to the observations, without supposing the Moon impelled toward the Earth by any force other than the one which acts inversely as the square of the distance; and there are grounds for believing that [with further refinement of the calculation] the slight difference that is found between the theory and the observations will vanish entirely.

Thus Clairaut had shown that the inverse-square law of gravitation proposed by Newton could be used to account, in a precise mathematical way, not only for the planets' generally elliptical orbits, but also for the deviation of the Moon's motion from this ideal pattern. The title (here Englished) that he chose for his St Petersburg Academy prize entry reflected his achievement: *Theory of the Moon Deduced from the Single Principle of Attraction Reciprocally Proportional to the Squares of the Distances* (1752).

Euler's praise for Clairaut's achievement was unstinting. In a letter to Clairaut of 29 June 1751 he wrote:

... the more I consider this happy discovery, the more important it seems to me .... For it is very certain that it is only since this discovery that one can regard the law of attraction reciprocally proportional to the squares of the distances as solidly established; and on this depends the entire theory of astronomy.

Consideration of the apsidal motion, according to Euler, offered the safest means of deciding on the sufficiency of the Newtonian theory. And so, in the introduction to the memoir that brought him the prize of 1752 of the Paris Academy, he affirmed that

... because M. Clairaut has made the important discovery that the movement of the apogee of the Moon is perfectly in accord with the Newtonian hypothesis ..., there no longer remains the least doubt about this proportion .... If the calculations that one claims to have drawn from this theory are not found to be in good agreement with the observations, one will always be justified in doubting the correctness of the calculations, rather than the truth of the theory.

Further reading

Craig B. Waff, *Universal Gravitation and the Motion of the Moon's Apogee: The Establishment and Reception of Newton's Inverse-square Law, 1687–1749* (Johns Hopkins University Ph.D. dissertation, 1975; University Microfilms, Ann Arbor, Michigan)

Craig B. Waff, Alexis Clairaut and his proposed modification of Newton's inverse-square law of gravitation, pp. 281–8 in Centré International de Synthèse (ed.), *Avant, avec, après Copernic: La représentation de l'Univers et ses conséquences épistémologiques* (Paris, 1975)

Craig B. Waff, Isaac Newton, the motion of the lunar apogee, and the establishment of the inverse-square law, *Vistas in Astronomy*, vol. 20 (1976), 99–103

17

The precession of the equinoxes from Newton to d'Alembert and Euler

CURTIS WILSON

Isaac Newton was the first to propose a quantifiable cause for the precession of the equinoxes. This slow westward motion of the equinoctial points – the projections from the Earth's centre onto the celestial sphere of the intersections of the terrestrial equator with the ecliptic – had been known since Antiquity, and by Newton's time had come to be evaluated at a steady 50″ of arc per year. From a heliocentric standpoint, it could be viewed as a slow rotation of the Earth's axis about the ecliptic's poles. It was caused, Newton urged, by the gravitational action of the Sun and Moon on the equatorial bulge of the Earth.

Newton's attempt to derive the precession quantitatively, however, was seriously flawed. The first to achieve an essentially correct derivation was Jean le Rond d'Alembert (1717–83), a mathematician and prominent member of the Paris Academy of Sciences from the early 1740s, during the 1750s co-editor with Denis Diderot of the *Encyclopédie*, and in his last thirty years a powerful and somewhat sinister figure in the politics of the Paris and Berlin academies.

D'Alembert's interest in deriving the precession was sparked by the discovery by James Bradley, Astronomer Royal of England, of the nutation. This is a small wobble in the Earth's precessing axis due to the changing orientation of the lunar orbit, which while maintaining a nearly constant inclination to the ecliptic, rotates westward, completing a circle in 18.6 years. Bradley had detected the nutation in the early 1730s, but delayed announcing his discovery till he had traced the phenomenon through a full cycle of the Moon's nodes. His announcement was published in the *Philosophical Transactions of the Royal Society* in January 1748 (Figure 17.1). Copies reached the Continent only in the summer, at which time the Abbé Nicolas-Louis de Lacaille, an astronomer who was in correspon-

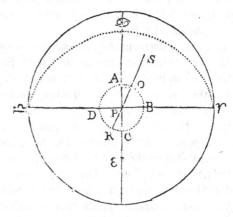

17.1. John Machin's diagram to explain the geometry of nutation (from Bradley's letter announcing nutation published in *Philosophical Transcations* in 1748).

The mean place of the pole of the Earth, as projected onto the celestial sphere, is *P*, which according to Bradley was in the eighteenth century some 29° 29′ from *E*, the pole of the ecliptic's circle. But the true pole moves counterclockwise around *P* in circle *ABCD* with radius 9″ (the size of the circle *ABCD* is greatly exaggerated in the diagram). Later in his letter, to improve the accuracy of the hypothesis, Bradley proposes replacing the circle *ABCD* by an ellipse with major axis *AC* = 18″ and minor axis *DB* = 16″.

The true pole is at *A* when the Moon's ascending node is in the beginning of Aries, and at *C* when the node is in the beginning of Libra; the circle *ABCD* is completed in 18 years, 7 months, which is the period of the revolution of the Moon's nodes.

In relation to the much slower motion of precession – the slow retrogradation of the mean pole *P* in a circle about *E* with a period of nearly 26 000 years – the nutation is a small superimposed wobble.

dence with Bradley, read an extract at a meeting of the Academy of Sciences. D'Alembert immediately attempted a calculation of the nutation, but his result did not agree with Bradley's observations.

From December onward d'Alembert devoted himself full-time to the problem. He appears to have informed no-one of this endeavour save Gabriel Cramer in Geneva; he was afraid (unnecessarily, as the event proved) of being anticipated by the English. For some time he was delayed by his failure to see that the Earth's diurnal rotation was of crucial importance in determining the effect. Finally, on 17 May 1749, he was able to present to the Academy the completed manuscript of his *Recherches sur la précession des equinoxes et sur la nutation de l'axe de la terre, dans le système Newtonien*. It was published in July.

The problem of deriving the precession was set for d'Alembert within a context in which the exactitude of the inverse-square law was under vigorous discussion. Clairaut, having earlier claimed that the inverse-square law could not account for the full motion of the lunar apse, reversed his position in December 1748; but he had not yet revealed how the true motion of the apse was to be derived (see Chapter 16). Through most of 1748 d'Alembert had been at work on the lunar theory, seeking to extract from the inverse-square law the observed anomalies. When he turned in December to the problem of the precession, it was with the same object in view. The solution he finally achieved was a striking confirmation of the theory. As he stated in the preface to his memoir,

The nutation of the terrestrial axis, confirmed by both the observations and the theory, furnishes, it seems to me, the most complete demonstration of the gravitation of the Earth toward the Moon, and hence of the principal planets toward their satellites. Previously this tendency had not appeared manifest except in the ocean tides, a phenomenon perhaps too complicated and too little susceptible to a rigorous calculation to reduce to silence the adversaries of reciprocal gravitation.

In the course of his investigation. d'Alembert carried out a thorough critique of Newton's treatment of the precession; it appears to have been the first such critique to be made.

D'Alembert's critique of Newton's derivation of the precession

Newton's qualitative understanding of the cause of the precession emerged out of his study of the perturbations of the Moon, from the analogy between the retrogression of the Moon's nodes and

the retrogression of the equinoctial points. That the nodes of the Moon's orbit must retrogress on the ecliptic is caused, he showed in Cor. 11 of Prop. 66 of Bk I of the *Principia*, by the action of the Sun. In Cor. 18 of the same proposition he went on to propose that, if the Moon were multiplied so as to become many moons, and these were liquified so as to form a fluid ring, the nodes of the ring would retrogress in the same way as the nodes of the lunar orbit. He hypothesized further (in Hypothesis II of Bk III in the second and third editions) that the result would be the same if the ring became solid. These claims, according to d'Alembert, were correct not precisely but as giving the *average* result.

Newton's next step was to consider a smaller ring, subject to the same perturbative action of the Sun, but with radius equal to the Earth's radius, and period of rotation equal to the Earth's diurnal period. The retrogression of the nodes of this ring, Newton asserted, would be to that of the lunar nodes as the period of the ring to that of the Moon. D'Alembert confirmed this conclusion.

If the imaginary ring be then attached to the underlying Earth, it must share its motion with the latter. But according to what rule? Newton assumed that the total quantity of motion (mv) previously in the ring in virtue of the retrogression of its nodes would be distributed between the ring and underlying sphere in just the way required for the two bodies to rotate together. What Newton should have done, d'Alembert pointed out, was to set the *moment* of the motion lost by the ring equal to the *moment* of the motion gained by the sphere. In other words, he should have assumed conservation of angular – not of linear – momentum. Because of this error, Newton's result for the angular velocity transferred to the Earth was too large by about 16%.

A spherical Earth with an attached equatorial ring is not the same as a spheroidal Earth, and Newton recognized that the Sun's attraction would have a greater moment in the former case than in the latter. In the first edition of the *Principia* he took the ratio of the two moments to be 4:1, in the second edition he showed by a correct deduction that it is 5:2. But he then assumed that the motions engendered would vary as the moments. This, as d'Alembert pointed out, is wrong: the effect of the torques depends not just on the mass to be moved but on its figure or – to use the Eulerian term – its

17.2. Jean le Rond d'Alembert.

moment of inertia.

Another flaw in Newton's derivation was quantitatively more important: the failure to take into account the diurnal rotation of the Earth. "The globe", Newton claimed in Cor. 20 of Prop. 66, "is perfectly indifferent to the receiving of all impressions"; and he evidently meant by this that the finite motion he had deduced for the precession was simply to be added to the diurnal motion. But only infinitesimal rotations add vectorially. D'Alembert in his initial calculation fell into the same error, but then found it implied a precession due to the solar force that is less than half the correct value.

Finally, Newton attempted to derive the ratio of the solar and lunar forces, acting on the equatorial bulge to wheel the Earth about its centre, from data on the heights of the tides; and his result is disastrously wrong. In the first edition of the *Principia* he took the ratio to be $6\frac{1}{3}$ to 1; in the second edition he fiddled with the same data to obtain 4.4815:1.

Daniel Bernoulli, in a memoir that won a prize in the Paris Academy's contest of 1740, sought to determine the ratio not from the heights but from the periods of the tides, and arrived at the value 2.5:1. Bradley, in his announcement of the discovery of the nutation in 1748, suggested determining the ratio from the observed values of the nutation and precession; for the first of these appeared to be due to the action of the Moon alone, while the precession was caused by joint action of the Sun and Moon. D'Alembert was to follow this suggestion, obtaining the ratio 2.35:1. The value accepted today is 2.17:1.

The chief conclusion to be drawn from the foregoing list of errors is that Newton lacked an articulated kinematics and dynamics for extended, rigid bodies. A fully developed mechanics of rigid bodies first appeared in 1765, in Leonhard Euler's great treatise on the subject (*Theoria motus corporum solidorum*). During the 1750s, a number of mathematicians attempted to derive the precession and nutation without relying on d'Alembert's derivation; among them a Benedictine monk, Charles Walmsley (in the *Philosophical Transactions* for 1756), the mathematician Thomas Simpson (in his *Miscellaneous Tracts*, 1757), and Patrick d'Arcy (in the *Mémoires* of the Paris Academy for 1759). These attempts were seriously flawed. Simpson's derivation appeared in slightly modified form in the first edition of J.-J. L. de Lalande's *Astronomie* (1764); d'Alembert, in an essay published in 1768, made a detailed analysis of its errors. All these putative derivations utilized integrations, but none of them formulated differential equations representing accurately the mechanics of a spinning, spheroidal Earth, subject to oblique forces. Such equations first appeared in d'Alembert's *Recherches* of 1749, and then (and in dependence on d'Alembert's work) in Euler's "Découverte d'un nouveau principe de Mécanique", published in the Berlin *Mémoires* for 1750.

D'Alembert's derivation of the precession and nutation

An adequate dynamics of extended, solid bodies must somehow take into account what is happening to all the 'elements' or 'particles' of the body – the forces and accelerations to which they are subject. The mechanics of systems of bodies or mass-elements in rigid or flexible connection was developed quite independently of Newton, starting from the problem of the 'centre of oscillation' of a compound pendulum. The discussion of this and related problems, particularly by Jakob I and Daniel Bernoulli, forms the background for d'Alembert's announcement, in his *Traité de dynamique* of 1743, of the famous "d'Alembert's principle". On this principle turns d'Alembert's derivation of the precession and nutation.

D'Alembert's statement of the principle, both in the *Traité de dynamique* and in his later *Recherches sur la précession*, is less revealing than it could be. It says in effect that the product of each mass-element by its reversed acceleration is to be considered a force, and that the sum of all such forces is in equilibrium with the applied forces. But d'Alembert does not give the true reason why this is so, namely that the forces acting to connect the mass-elements into a rigid body are in equilibrium with each other and therefore equal 0. The actual acceleration $\vec{a}$ of each mass-element is the vector-sum of the acceleration due to the connections, $\vec{a_c}$, and the acceleration imposed by the external forces, $\vec{a_f}$; hence $\vec{a_c} = \vec{a} - \vec{a_f}$. Since the accelerations $\vec{a_c}$ multiplied by their respective masses form a system in equilibrium, so do the accelerations $\vec{a} - \vec{a_f}$ or $\vec{a_f} - \vec{a}$.

According to Pierre-Simon Laplace, d'Alembert was the first to lay down the conditions for translational and rotational equilibrium in the three coordinate directions. In applying these conditions in the *Recherches* of 1749, however, d'Alembert fell into a curious error. Instead of setting the sum of the components of all the forces in each coordinate direction equal to zero, he set the components of the external forces in each direction equal to the corresponding components of the 'forces' represented by the product of each mass-element by its reversed acceleration, summed over the whole body. The three resulting equations for translational equilibrium are wrong, but since d'Alembert never used them, the error did no harm. In expressing the rotational equilibrium d'Alembert similarly set the moments of the external forces about each coordinate axis equal to the corresponding moments of the 'forces' represented by the reversed accelerations of the mass-elements. Because he made a further error in the direction of the externally applied torques, the resulting equations are correct.

Years later, in the first volume of his *Opuscules mathématiques*, d'Alembert stated the conditions of equilibrium correctly, setting the total forces in a given direction, and the total torques about a given axis, equal to zero. If he recognized his original error, he did not acknowledge it. The fact remains that the correctness of his results in the *Recherches* of 1749 depended on compensating errors of sign.

D'Alembert's next task was to derive, from his three equations for rotational equilibrium, differential equations enabling him to determine the motions of the Earth's axis. For this he needed (1) expressions for the external torques in terms of known quantities pertaining to the periods, distances, longitudes and latitudes of the Sun and Moon, and (2) expressions for the moments about three perpendicular axes of the product 'mass × reversed acceleration' for an arbitrary mass-element of the Earth, in terms of the position of the Earth's axis. The latter expressions were to be integrated over all mass-elements, to yield the total moment about each axis.

In deriving expressions for the external torques, d'Alembert first assumed the Earth to be a uniformly dense spheroid with polar radius a and equatorial radius $a(1 + a)$, where a is a very small number. Later he showed that the results still held, with minor modification, if the density varied in such a way that the layers of equal density were bounded by surfaces of revolution nearly spherical in shape. Under these assumptions the only forces contributing to unbalanced torque were those on the equatiorial bulge, bounded externally by the spheroidal surface and internally by the inscribed sphere.

D'Alembert resolved the accelerative force due to the Sun's or Moon's attraction of any mass-element of the equatorial bulge into two components, one directed to the Earth's centre, the other parallel to the line joining the Earth's centre to the perturbing body, Sun or Moon. Only the second of these components, or rather its difference from the parallel accelerative force on a mass-element at the Earth's centre, produced torque on the axis. D'Alembert first summed the moments of these differences for an annulus of the equatorial bulge; then he summed the moments for all the annuli.

In determining the moments of the 'internal forces', as we may call them, d'Alembert used three planes of reference: the ecliptic, in which the Earth's centre of gravity was assumed to lie; a plane passing through the Earth's axis perpendicular to the ecliptic; and a third plane passing through the Earth's centre of gravity and perpendicular to the other two. He specified the position of the Earth's axis by two variables: the angle (ϵ) that the projection of the axis onto the ecliptic made with a fixed line, say the line from the Earth's centre of gravity to the first star of Aries; and the angle (Π) that the axis itself made with the ecliptic plane. To specify the position of an arbitrary mass-element of the Earth, he used the distance b along the Earth's axis from one pole, the radius f at right angles to the axis, and the angle X which this radius formed with the plane through the axis and perpendicular to the ecliptic.

The reversed accelerations of an arbitrary mass-element, in directions perpendicular to each of the three reference planes, were expressed in terms of $d^2\epsilon$, $d^2(\cos \Pi)$, and d^2P ($= d^2X$), where P is the angle through which the Earth rotates in its diurnal motion during the time-interval considered. The sum of the moments about each axis of the products 'mass-element × reversed acceleration' was then obtained by integration.

Once all the required substitutions had been made into the equations for rotational equilibrium, d'Alembert was faced with three differential equations. By combining these equations he obtained an integrable equation for d^2P, and was then left with two differential equations in the variables ϵ and Π. These he resolved using approximations suggested by the known observational facts. As the second-order differentials occurred in relatively small terms, d'Alembert left them out of account, so that only first-order equations remained. Integrating these, he obtained

$$\Pi = \Pi_m - \frac{3Am'(1+\beta)\sin \Pi_m}{2Kk(n'-M)} \cos(n'-M), \quad (1)$$

$$\epsilon = -\frac{3A(2+\beta)\sin \Pi_m}{2Kk} z$$
$$- \frac{3Am'(1+\beta)\cos 2\Pi_m}{2Kk(n'-M)\cos \Pi_m} \sin(n'-M)z. \quad (2)$$

Here Π_m is the mean inclination of the Earth's axis to the ecliptic; $(1 + \beta)$ is the ratio of the Moon's force to the Sun's force in moving the equatorial bulge; z is the mean longitude of the Sun; n' is the rate of motion of the nodes of the lunar orbit, compared to the Earth's motion about the Sun; M is the mean rate of precession per year; m' is the tangent of the inclination of the lunar orbit to the ecliptic; and A, K, k are constants.

Equation (1) is in agreement with Bradley's observational result if we equate $3Am'(1+\beta)\sin \Pi_m/2Kk(n'-M)$ with $9''$. Equation (2) is in agreement with Bradley's observational determinations provided $(3A/2Kk)(2+\beta)\sin \Pi_m$ is identified with M, the mean rate of precession or $50''$ per year. The second term of (2) represents the inequality of the precession, with a maximum value of $16''.8$ (but d'Alembert gave it as $15''$).

Bradley's empirical determinations allowed d'Alembert to obtain a value for the ratio $(2 + \beta)/(1 + \beta)$, and thus for $(1 + \beta)$; he found it to be 2.35. This result in turn permitted a calculation of the mass-ratio of the Moon to the Earth. D'Alembert put it at $\frac{1}{80}$; the present-day value is $\frac{1}{81.3}$. (But this ratio, we should note, is very sensitive to small changes in the values assigned to the mean precession and maximum nutation.)

D'Alembert's equations (1) and (2) have to do with the motions of the Earth's axis of figure. Since this axis is continuously in motion, it can never be identical with the instantaneous axis of rotation, that is, the line within the Earth's body that is instantaneously at rest with respect to the Earth's centre of gravity. From his differential equations d'Alembert could determine the relation between these two axes; he found the tangent of the angle between them to be

$$\frac{[\mathrm{d}\Pi^2 + \cos^2\Pi_m\,\mathrm{d}\epsilon^2]}{k\mathrm{d}z},$$

where the numerator is the angular distance through which the Earth's axis of figure moves, while the Earth is rotating through the angle $k\mathrm{d}z$ about its axis of rotation. The numerator is so small in comparison with the denominator that the angle between the two axes can be regarded as negligible.

In Chap. 11 of his *Recherches* d'Alembert undertook a second derivation, assuming from the start the identity of the axes of figure and rotation. The basic strategy was to combine vectorially the infinitesimal displacements due to the Earth's diurnal rotation with those due to the attractive forces of the Sun and Moon. D'Alembert thus obtained differential equations identical with those he had previously extracted by approximation from the more exact equations of his first derivation. The second derivation thus corroborated the first.

D'Alembert's exposition of his course of reasoning leaves much to be desired. His friend Cramer wrote to him that the work was disorderly and full of typographical errors, and that it contained totally unintelligible diagrams. But what above all mattered to d'Alembert in composing this work was to show that the precession and nutation indeed followed from Newton's law of gravitation. And this, if we set aside two inadvertent and compensating errors, he succeeded in doing.

Euler's work on the precession and mechanics of rigid bodies

D'Alembert's *Recherches sur la précession* appeared in print early in July 1749, and d'Alembert sent off a copy to Euler on 20 July. Not until 3 January 1750 did Euler write to d'Alembert to acknowledge that he had received it, and, albeit with difficulty, had learned enough from it to carry out a derivation of the precession and nutation of his own. A more detailed acknowledgement appears in Euler's letter of 7 March, written just two days before he presented his own derivation to the Berlin Academy:

I applied myself repeatedly and for a long time to the problem of the precession, but I always encountered an obstacle, ... and above all this problem: given a body turning about any axis freely, and acted upon by an oblique force, to find the change caused both in the axis of rotation and in the movement .... With respect to this problem all my investigations had been hitherto unavailing, and I would not have applied myself to it further, if I had not seen that the solution must necessarily be encompassed in your treatise, although I was not able to find it there .... I must confess that I could not follow you in the preliminary propositions you employed, for your way of carrying out the calculation was not yet familiar to me .... But now that I have succeeded better in the investigation of this same subject, having been assisted by some insights in your work by which I was gradually enlightened, I have come to be able to judge your excellent conclusions.

Euler's derivation of the precession was published in 1751 in the Berlin *Mémoires* for 1749, and d'Alembert was astonished and hurt to find that it contained no acknowledgement of his own priority in achieving such a derivation. His complaints constrained Euler to insert, in the Berlin *Mémoires* for 1750 (published in 1752), an *Avertissement* declaring that Euler had written his memoir "only after having read the excellent work of M. d'Alembert on the subject; and that he has not the least pretention to the glory due to him who first resolved this important question."

Euler's derivation of the precession and nutation was a model of clarity. First he determined the moment of inertia of the Earth considered either as a homogeneous sphere or as a sphere with a denser, spherical core. Then, using equatorial coordinates, he computed the torque exerted by a distant attracting body out of the plane of the equator on an Earth considered as spheroidal in shape and either homogeneous or with a denser spheroidal core. To shift from equatorial to ecliptic coordinates, he utilized the propositions of spherical trigonometry, thereby avoiding the geometrical interpretation of differentials by which d'Alembert's two derivations had been bedevilled. Finally, to calculate the angle through which the Earth's axis is shifted by a torque S during the time dt, Euler had recourse to the formula

$$\text{angle} = S dt^2 / 2 I ds,$$

where I is the Earth's moment of inertia (taken to be the same about all axes), ds is the angle described in the Earth's diurnal rotation during dt, and the factor '2' is required because of Euler's way of defining force. This formula Euler promised to explain elsewhere.

In several respects Euler's derivation was less exacting than d'Alembert's first derivation. Euler started from the assumption that the Earth's axis of figure and axis of rotation are identical, claiming without proof that their angle of separation was negligible. He also assumed that at each instant the Earth's axis of rotation is such that the centrifugal forces generated in rotation about it exactly balance, so that the shift in the axis is solely due to the external torques. Finally, he assumed the formula of the preceding paragraph, to be explained "elsewhere". D'Alembert, followed later by Laplace, pointed out that Euler's derivation was identical in its basic strategy with d'Alembert's second derivation; but it is unlikely that Euler was aware of this identity.

Euler's final numerical equations for precession and nutation differ little from d'Alembert's; but they include small terms that d'Alembert neglected, those in the nutation and inequality of the precession due to the Sun's action.

Euler's "elsewhere" proves to be his "Découverte d'un nouveau principe de mécanique", presented to the Berlin Academy in September 1750. Here Euler mounted a frontal attack on the general problem of the free rotation of a rigid body subject to oblique forces. It was a problem he had already envisaged at the time he wrote his *Mechanica* (published in 1736). He had hazarded a tentative hypothesis as to how it might be dealt with in the first volume of his *Scientia navalis*, completed in 1740 but not published till 1749. But what finally set him on the path to a successful solution was a procedure he took over from d'Alembert's *Recherches sur la précession* of 1749.

The main course of argumentation in the "Découverte" begins with a "determination of the movement in general of which a solid body is capable, while its centre of gravity remains at rest". Euler here derives expressions for the components of acceleration of an arbitrary mass-element in the directions of the x, y, and z axes, in terms of the components λ, μ, ν of the instantaneous angular velocity, and the differentials $d\lambda$, $d\mu$, $d\nu$. Paralleling his steps to d'Alembert's, he then finds expressions for the moments of the products 'mass-element × acceleration' with respect to each axis, and integrates over all mass-elements of the body. The latter step is effected for a body of arbitrary mass-distribution by introducing special symbols for the products and moments of inertia. Finally, the moments thus found are set equal to the moments of the external forces. The formula introduced from "elsewhere" in Euler's *Recherches sur la précession* emerges as a corollary.

With the formulas of the "Découverte", Euler was not altogether satisfied. In general, they were difficult to apply. The single coordinate system used was fixed in space; the coordinates of each mass-element dM changed from one interval dt to the next as the body rotated about its centre of gravity. In general, therefore, integrations over time were impossible.

Euler tackled this difficulty in a study entitled "Du mouvement d'un corps solide quelconque lorsqu'il tourne autour d'un axe mobile", presented to the Berlin Academy in October 1751, but published only in 1767. In this work Euler introduced, besides the original coordinate system fixed in space, a second orthogonal system with the same origin but fixed in the body. To express the relation between the two systems, he introduced the now well-known 'Eulerian angles'. The resulting kinematics makes possible the computation of the angle between the axis of figure of a rotating spheroid and the instantaneous axis of rotation, and Euler accordingly calculated the angle between these two axes in the case of the spinning, precessing Earth.

He also investigated the conditions under which a body would continue to rotate about the same axis without application of any externally applied torque. He derived an equation relating the moments and products of inertia about the axes fixed in the body to the ratio of two components of the angular velocity, when the externally applied torques vanish. The equation is cubic; hence there is at least one real root, and therefore at least one axis about which the body can continue to rotate freely.

In 1755 a protege of Euler's, János-András Segner, professor at Halle, proved geometrically that there must be three such axes. Then Euler, in his "Recherches sur la connaissance mécanique des corps", submitted to the Berlin Academy in July 1758, gave a different proof: he showed that the axes through the centre of inertia of a body with respect to which the moments of inertia are either a maximum or a minimum are such that the products of inertia with respect to them vanish. But there must be a maximum and a minimum, unless all the moments of inertia are equal. Hence there are at least two axes about which the body can rotate freely, and all three roots of the cubic equation have to be real, since imaginary roots come in pairs. To these three axes, which are mutually perpendicular, Euler gave the name *principal axes*. Henceforth he would use these axes for the coordinate system fixed in the body.

The triumphant application of these results came in Euler's "Du mouvement de rotation des corps solides autour d'un axe variable", presented to the Berlin Academy in November 1758. Euler found ("with surprise", he says) that he was able to obtain integral solutions to problems he had previously believed "to surpass the powers of the calculus". These included determining the continuation of the motion of a body on which a motion of rotation about an axis other than a principal axis has been imposed. As a result of such successes, Euler then undertook the writing of his extended treatise, *Theoria motus corporum solidorum seu rigi-*

dorum, finished by 1761 but published in 1765.

Epilogue

The fourth volume of d'Alembert's *Opuscules mathématiques*, published in 1768, contains two essays on the mechanics of rigid bodies. D'Alembert's aim here was to show that the results obtained by Euler were obtainable by his own procedures. The Eulerian method "being [he wrote] very complicated, I thought people would not be displeased to see how one can arrive at the same result by a much simpler analysis".

In 1773 Laplace, in the first treatise he submitted to the Paris Academy, in introducing "the general equations of motion of a body of any figure, acted upon by any forces", stated:

M. d'Alembert was the first to give the general solution of this problem, along with the most direct method of arriving at it, in his excellent treatise Sur la précession des équinoxes, *an original work that shines throughout with the genius of invention, and that one can regard as containing the germ of all that has been done since on the mechanics of solid bodies.*

As Laplace proceeded to outline his approach to the mechanics of rigid bodies, however, it emerged as essentially Eulerian. A half century later he was to write: "the equations at which Euler arrives appear to me to be the most simple it is possible to obtain".

Further reading

For a more extended account of the developments here traced, see Curtis Wilson, D'Alembert *versus* Euler on the precession of the equinoxes and the mechanics of rigid bodies, *Archive for History of Exact Sciences*, vol. 37 (1987), 233–73.

18

The solar tables of Lacaille and the lunar tables of Mayer

ERIC G. FORBES and CURTIS WILSON

It is one thing to deduce mathematically the perturbations of the Moon and planets implied by Newton's law of gravitation, another to develop tables yielding precise predictions. The three mathematicians who first derived perturbations analytically (see Chapter 20) – namely Leonhard Euler (1707–83), Alexis-Claude Clairaut (1713–65), and Jean le Rond d'Alembert (1717–83) – all constructed lunar tables; but the errors of these tables ran as high as 5 arc-minutes, roughly half the maximal error of Isaac Newton's modified Horrocksian theory of the Moon (see Chapter 13 of Volume 2A). To do better, skills were required other than those of the mathematical theorist – skills in the making of observations, and in the reduction of observations both new and old. Also required was an understanding of the analytical theory, and a sensitivity to problems involved in fitting it to observational data.

The earliest published astronomical tables incorporating perturbations deduced analytically from the inverse-square law of gravitation appear to have been Euler's lunar tables of 1746 (in his *Opuscula varii argumenti*). But the first such tables to come into general use were the solar tables of Nicolas-Louis de Lacaille (1713–62) and the lunar tables of Johann Tobias Mayer (1723–62). Lacaille's solar tables were published in 1758, and republished in 1764 and 1771 in the first two editions of the *Astronomie* of Joseph-Jérôme Lefrançais de Lalande (1732–1807). From 1761 into the 1790s (but in modified form after 1779), they were the basis for the solar ephemerides of the annual *Connaissance des temps*. Mayer's lunar tables were published in a first version in Göttingen in 1753 and in Lalande's *Astronomie* of 1764. In a revised form they were published in London in 1770, and were the basis of the lunar ephemerides of the *Nautical Almanac* from 1767 to 1813 (but with successive modifications after 1776).

The achievement of Lacaille's and Mayer's tables was not merely a matter of adding theoretically computed perturbations to the purely Keplerian-style tables used earlier. Keplerian orbital elements needed to be redetermined in the light of the perturbations. Also, owing to uncertainties in the planetary and lunar masses and in the solar and lunar parallaxes, the perturbations deduced from theory had themselves to be assessed on an empirical basis. In addition, the discovery by James Bradley of the aberration of light and the nutation of the Earth's axis (officially announced in 1729 and 1748, respectively; see Volume 3) implied that all astronomical observations must be 'corrected' for these effects; and Lacaille and Mayer were the first astronomers to found their tables solely on observations thus reduced.

In pursuing a rapprochement of observation and theory, Lacaille and Mayer were indefatigable. J.-B. J. Delambre in the 1810s, looking back on their achievements from the vantage-point of the Laplacian era, saw them, along with Bradley on whose discoveries their work was premissed, as the giants of eighteenth-century astronomy.

The solar tables of Lacaille

Supported by a scholarship, Lacaille took the three-year course in theology at the Collège de Navarre. In the midst of his theological and literary studies, he discovered Euclid, and developed a secret love of astronomy, which he proceeded to study entirely on his own. A *contre-temps* between Lacaille and the vice-chancellor of the college at the ceremony for the awarding of degrees fortified Lacaille's resolve to devote himself to the mathematical sciences rather than to literature or theology. At some point he had received the title of *abbé*, but he seems never to have served as a priest.

In 1737 Lacaille became an assistant to Jacques Cassini (Cassini II) at the Paris Observatory. In 1739 and 1740 he worked with César-François Cassini de Thury (Cassini III) on the verification of the meridian of the Observatory from Perpignan to Dunkirk. A result of these new measurements was to show that degrees of terrestrial latitude increased in length toward the poles, implying a flattened Earth, in contradiction to the elongated shape that Jacques Cassini had claimed to deduce from earlier measurements. For this work Lacaille was admitted to the Academy of Sciences on 8 May 1741. An account of the meridian measurements was published in 1744 at the expense of Cassini de Thury, whose name alone appeared on the title page, although according to a letter from Lacaille to Bradley of 28 July 1744, the sole author was Lacaille. Earlier, in 1739, Lacaille had been named to the chair of mathematics at the Collège Mazarin, where an observatory was installed for him.

On 25 April 1742 Lacaille read to the Academy a project for a new catalogue of the fixed stars. Improvements in astronomy, he urged, must begin with a star catalogue, since it is from their relation to the fixed stars that the positions of planets and comets are determined. But all previous star catalogues were faulty because previous solar theories, on which the determination of right ascensions of stars must depend, were accurate at best only to about 1 arc-minute, and because star positions had not previously been corrected for aberration and the newly discovered variations in the obliquity of the ecliptic. Lacaille's aim was to develop a star catalogue free of these faults.

To determine right ascensions (RA), Lacaille proposed dividing the sky into a series of zones between parallels of declination, and fixing a telescope in the meridian in each zone in succession, to clock the transits of stars. To obtain the absolute right ascension of some one bright star in the zone, to which the other stars might be referred, he proposed, in the case of zones entered by the Sun, to use the method of John Flamsteed: the differences between the RA of the Sun and that of a bright star near the solstitial colure were determined on two occasions when the Sun was at the same declination; these differences were then added (after account was taken of the star's precession in the intervening time) to give the total difference in RA between the Sun's two positions. Since the solstitial colure is at 6 hours of RA, the RA of the Sun on

each of the two occasions could be calculated; and a simple addition or subtraction would then yield the RA of the star used for comparison.

Where the zone is not entered by the Sun, Lacaille proposed determining the time of meridian transit of the reference star by the method of 'corresponding altitudes': measure the times of passage of the star through a number of different altitudes as it rises in the east, and through the same altitudes as it descends in the west, then average the results to give the time of transit.

Finally, the altitudes of the principal stars were to be determined and corrected for refraction. Lacaille does not enter into the details of how this was to be done; the table of refractions that he later worked out proved to be infected with the errors of division of his instrument.

Lacaille's programme was not taken up by the other astronomers of the Academy, and Lacaille was left to accomplish what he could on his own. The skies in Paris were so frequently overcast as to make singlehanded completion of the programme impracticable; he was able to determine only the positions of the brighter stars of the northern hemisphere. In 1750, however, he won the Academy's endorsement for an expedition to the Cape of Good Hope, and there, between March 1751 and March 1753, he measured the positions of nearly 10 000 stars of the southern hemisphere, while performing other astronomical and geodetical tasks as well, including a determination of solar parallax carried out jointly with Lalande in Berlin, and the first measurement of an arc of a meridian in the southern hemisphere.

How to achieve an accurate solar theory? Most of the known methods for determining both apogee and equation of centre, Lacaille urged, demanded observations more accurate than it is possible to obtain. Methods less direct and less elegant could be surer in practice. Such is the method for the greatest equation of centre proposed by Flamsteed: determine the longitudes of the Sun in spring and autumn when it is approximately 90° from apogee, then from the angle traversed in the interim subtract the angle of mean motion; the difference is just twice the greatest equation of centre. The method assumes an approximate knowledge of the place of the apogee, and an exact knowledge of the year's length, whence the Sun's mean motion is derived.

By this method Lacaille found for the maximum

equation of centre in 1745 two values averaging to 1° 55′ 44″.9, too large by about 7″.8. In Jacques Cassini's *Tables astronomiques* of 1740 the corresponding error was +13″.9; the errors in other tables were larger.

For the solar apogee, Lacaille proposed a method assuming an exact knowledge of the anomalistic year, the time between two successive passages of the Sun through apogee. The apogee and perigee are the only two points of the orbit such that the Sun requires precisely half the anomalistic period to move from one to the other. Hence, given observed longitudes of the Sun very near apogee and very near perigee, and given also the daily motion of the Sun in longitude when at apogee and at perigee, it is possible by trial and error and repeated applications of 'the rule of three' to calculate the position of the apogee. By this method Lacaille found it to be, in mid-1744, at longitude 98° 31′ 35″. According to the tables of Simon Newcomb ("Tables of the Sun", *Astronomical Papers of the American Ephemeris*, Vol. 6), the correct value at this date was 98° 32′ 54″.96, greater by 1′ 20″. No other existing tables came so close (the error in Cassini's tables was −11′ 1″).

Bradley's formulas for the nutation reached Paris in the summer of 1748, and at once Lacaille set out to revise the solar elements, taking the new corrections into account and using a large number of longitudes he had determined from observed differences in RA between the Sun and Procyon, Arcturus, and Vega. At the same time, in order to determine the annual motion of the apogee, and whatever secular changes there might be in the obliquity of the ecliptic and the length of the tropical year, he undertook a study of the solar observations made by Bernhard Walther in Nuremberg around 1500. Astronomers differed as to whether the obliquity of the ecliptic was subject to a secular diminution. Also, Euler in 1746, accepting Ptolemy's equinoxes of AD 139 and 140, and assuming that the Earth's motion was subject to aethereal resistance, had proposed that the length of the year was gradually decreasing. Again, seventeenth-century astronomers such as Thomas Streete had claimed that the apogee was fixed with respect to the stars; others disputed the claim. The construction of solar tables required adjudication of these issues.

In the case of the obliquity, Lacaille found definite evidence of a secular decrease, although his value for this was only about half the correct value. The tropical year he found to be constant to within a few seconds of time (his mean value agrees, to within a fraction of a second, with Newcomb's value for 1750). As for the motion of the apogee, he found it to be 64″.82 per year with respect to the equinox, about 3″ larger than the correct value.

Walther's observations also enabled Lacaille to determine the epoch of the mean longitude of the Sun on 1 January 1500; he found it to be 289° 25′ 36″. Newcomb's value is 1″.69 greater. (The value for the same constant in Cassini's tables errs by +48″.3.) For the equation of centre in Walther's time, Lacaille obtained a value more than 1′ too small, owing primarily to a faulty assumption about the latitude of Walther's observatory; as a result he concluded that the equation of centre was not subject to secular diminution. The establishment of a 17″ to 18″ decrease per century had to wait upon the theoretical work of J. L. Lagrange in the 1780s.

In June 1750 Lacaille presented to the Paris Academy two memoirs on solar theory, reporting his re-determinations of the solar apogee and eccentricity, his investigation of the lunar inequality in the Sun's apparent motion, and a revision in his value for the tropical year. To begin with, Lacaille discusses the problem of precision. He makes it a rule, he states, never to neglect tenths of arc-seconds, whether in the observations or in the calculations. The resulting solar tables, he estimates, are accurate to 12″ or 15″.

Newton had asserted the existence of a detectable lunar inequality in the Sun's apparent motion in Prop. 13 of Bk III of the *Principia* (second and third editions), but did not supply a value for the inequality. Euler in 1744 was the first to incorporate this inequality in solar tables, putting its value at $15″ \sin \theta$, where θ is the elongation of the Moon from the Sun.

Lacaille undertook to establish the existence of the lunar inequality empirically, using eight pairs of observations of the Sun's longitude taken when the Moon was at first quarter, then at last quarter, or vice versa. The difference between the two longitudes, minus the amount of motion predicted by good solar tables that did not take the lunar inequality into account, should be double the maximum lunar inequality, and positive or negative according as the motion is from last to first quarter or vice versa. Lacaille found the differences to be always in the expected direction, but varying over a range from 3″.7 to 35″.9. He concluded that the existence of the lunar inequality, though not its magnitude, was established.

D'Alembert challenged this conclusion in the third part of his *Recherches sur différens points importans du système du monde* (1756), asserting that the scatter in the results was too wide to permit conclusions to be reliably drawn. Lacaille responded in 1757, correcting a calculational error in his earlier memoir and thus somewhat reducing the scatter. D'Alembert returned to the attack in a memoir read on 14 January 1758; Lacaille replied in a section added to his memoir of 1757; d'Alembert replied to the reply in 1762, after Lacaille's death. One of the points at issue was a supposed 40″ variation in the Sun's equation of centre, which the astronomer Pierre-Charles Le Monnier claimed to have observed between September 1746 and April 1747; d'Alembert accepted Le Monnier's observations, while Lacaille challenged the second one as flawed. With hindsight we can conclude that Lacaille was on the right track. The value for the maximum lunar inequality that he initially adopted in 1750 was 12″; the value that he finally adopted in his tables was 8″.5, still too large by about 2″.

In his second memoir of 1750 Lacaille introduced a new method of determining the apogee and equation of centre simultaneously from a triplet of observations, two taken near equinoxes and one near a solstice. The data yield differences in true anomaly and in mean anomaly. By making different assumptions for the eccentricity and place of the apogee, then applying the rule of three to pairs of these assumptions, Lacaille was able to select values for both constants such that they would imply the given differences in true and mean anomaly. He repeated this procedure for eleven triplets formed from nine observations, then took the averages for his final result.

On 29 January 1755 Lacaille presented a supplement to his two memoirs of 1750, in which he once more determined the solar apogee, eccentricity, and epoch of mean longitude, this time from observations of the Sun made at the Cape of Good Hope – observations superior in accuracy, he believed, to those made in Paris.

The entire procedure had to be gone through still one more time, when it became evident that plane-

tary perturbations needed to be taken into account.

It was Euler who first suggested that planetary perturbations were likely to be important in solar theory. Originally he had thought otherwise, claiming in a memoir of 1744 that the astronomical tables of the primary planets, including the Earth, can be constructed on the basis of the exact solution of the two-body problem, just as if they were attracted to the Sun alone. But in studying the inequalities of Jupiter and Saturn in 1747 he came to question this assumption. As he wrote in 1749, when proposing questions for the first prize contest to be sponsored by the St Petersburg Academy of Sciences:

Since it is now established beyond doubt that the motion of Saturn, when it approaches Jupiter, is not a little perturbed, it is asked whether the motion of the Earth is not subject to a similar perturbation from the same cause . . . .

Although the St Petersburg Academy chose the lunar rather than the solar theory for its first contest, Euler continued to urge the question about solar theory, and through his influence it was made the subject of the contests of the Paris Academy for 1754 and 1756. Euler himself entered the contest of 1756; his essay, which won the prize, was the first systematic derivation of the planetary perturbations in the Earth's motion.

To calculate the perturbations, one must have values for the masses of the perturbing planets. The masses of Jupiter and Saturn can be computed from the periods and distances of satellites – Newton had done this, and Euler accepted his results – but Mars, Venus, and Mercury were without known satellites. Euler had recourse to a conjecture. Newton's values for the masses of Saturn, Jupiter, and the Earth (the latter value being in error by about +97%, owing to a faulty value for solar parallax) suggested to him that the densities of the planets vary reciprocally as the square roots of their periods. This relation, together with the solar distances of the planets and their apparent diameters as determined by Le Monnier, yielded values for the masses. From the masses so determined Euler was able to derive a value for the annual motion of the Earth's aphelion, namely 12″.75 with respect to the stars, which agreed very nearly with Cassini's empirically determined value of the same constant; he therefore considered that his conjecture had been satisfactorily corroborated. In the case of Venus his value differs from the modern value by only $+0.7\%$; in the cases of Mars and Mercury his results are much more in error, but in both these cases he found the perturbations to be very small, as they are in fact.

The perturbations in longitude that Euler derived are of the form $A \sin \theta + B \sin 2\theta$, where θ is the difference in mean longitude between the perturbing planet and the Sun. For all the perturbing planets, Euler's values for A and B are surprisingly good; only in one case does the error, as determined from a comparison with Newcomb's tables, rise as high as $1''.4$.

Lacaille, however, did not use Euler's values, but delayed revision of his solar elements until Clairaut had completed his own derivation of these same perturbations. Clairaut, in undertaking the derivation anew, desired to show that his method was of easier application than Euler's; he was also eager to compare the theory with the large number of accurate solar observations that Lacaille had accumulated. Of the planetary perturbations, he dealt only with those due to Jupiter and Venus. Unlike Euler, he took account of the eccentricity of the Earth's orbit, so arriving at formulas with three or four terms each. Unfortunately, the later terms of his formula for the lunar inequality were mistaken, and so was one of the terms in his formula for the Jovian inequality. In the case of both the lunar and Venusian inequalities Clairaut suggested determining the absolute values of the coefficients by a comparison with observations (the formulas were thus proposed as giving only the relative values of the coefficients).

With Clairaut's results in hand, Lacaille proceeded to refine the elements of his solar theory. His calculations were presented in a memoir of 1757. First, he selected observations of the Sun's longitude when the lunar inequality was zero, and comparing these with his provisional solar theory, concluded that the coefficients for the Venusian inequality should be reduced to 82.4% of the values Clairaut had given them (the correct percentage would have been 48.4%). Also, he revised Clairaut's coefficients for the lunar inequality on the basis of the value of the Moon:Earth mass-ratio derived by d'Alembert in his treatise on the precession and nutation, and so arrived at the formula $7''.7 \sin \theta + 3''.5 \sin \theta \sin M$, where θ

Table 18.1

	Lacaille (L)	Newcomb (N)	L − N
Equation of centre	$1°55' 31''.6$	$1°55' 36''.3$	$-4''.7$
Apogee, 1 Jan. 1750	$98°38' 4''$	$98°38' 35''.2$	$-31''.2$
Epoch, 1 Jan. 1750	$280° 0' 43''.4$	$280° 0' 37''.3$	$+6''.1$

represents the difference in mean longitude between the Moon and the Sun, and M the mean anomaly of the Sun.

Next, from 22 empirically determined longitudes of the Sun (each based on a series of observations taken on successive days), Lacaille subtracted out the lunar, Jovian, and Venusian perturbations, so as to obtain the longitudes the Sun would have had if these inequalities were absent. From the 22 longitudes he then formed 34 triplets, and for each triplet determined the equation of centre, apogee, and epoch of mean longitude by the method of his memoir of 1750. His final averages, compared with Newcomb's values for the same constants, are shown in Table 18.1.

Finally, Lacaille compared the resulting theory with 144 observed longitudes of the Sun. Aside from six observations that differed from the theory by more than $15''$ (five of them by more than $27''$), and which Lacaille discarded as defective, he found the average deviation between theory and observation to be $5''.4$. A comparison between Lacaille's tables and Newcomb's for the 22 dates of Lacaille's fundamental observations yields a standard error of $6''.6$; for the tables of Edmond Halley and Jacques Cassini, by contrast, the corresponding figures are $56''.1$ and $32''.2$. Such was the increment in accuracy that Lacaille achieved.

The observational basis of Lacaille's solar theory was published in his *Astronomiae fundamenta* of 1757 (Figure 18.1), "a work of many vigils", as he described it, giving the positions of 400 bright stars and the previously mentioned 144 places of the Sun in the ecliptic. This, and the *Tabulae solares* of 1758 (Figure 18.2) as well, were printed in limited editions at the author's own expense, and distributed to the chief astronomers and academies of science in Europe.

The reception of Lacaille's solar tables, at least by his colleagues in Paris, was mixed. Cassini's son complained that Lacaille should have confined

ASTRONOMIÆ
FUNDAMENTA
NOVISSIMIS
SOLIS ET STELLARUM
OBSERVATIONIBUS
STABILITA
LUTETIÆ IN COLLEGIO MAZARINÆO
ET IN AFRICA AD CAPUT BONÆ SPEI
PERACTIS

A Nicolao – Ludovico DE LA CAILLE, in almâ Studiorum
Univerfitate Parifienfi Mathefeon Profeffore, Regiæ Scientiarum
Academiæ Aftronomo, & earum quæ Petropoli, Berolini,
Holmiæ & Bononiæ florent, Academiarum Socio.

PARISIIS;
E Typographiâ J. J. Stephani COLLOMBAT, Typographi ordinarii Regis.

M. DCC. LVII.

18.1. The title-page of Lacaille's _Astronomiae fundamenta_ of 1757.

himself to correcting the Cassinian solar tables. D'Alembert maintained an attitude of superior scepticism, and Le Monnier, attached to the project of correcting Flamsteed's solar and lunar tables by means of observations, continued to insist that "neither the epochs nor the several other elements of the theory of the Sun are as yet well known to us". Lalande, however, defended the new tables,

and when he became editor of the _Connaissance des temps_ in 1760, adopted them as the basis for the annual ephemerides.

Mayer, on examining Lacaille's solar tables "found that they very nearly agreed with the many and careful observations made by myself from the year 1756 forward with an excellent mural quadrant. Wherefore it did not seem necessary ... to

TABULÆ
SOLARES

Quas è noviſſimis ſuis Obſervationibus deduxit N. L.
DE LA CAILLE, *in almá Studiorum Univerſitate
Pariſienſi Matheſeon Profeſſor , Regiæ Scientiarum
Academiæ Aſtronomus , & earum quæ Petropoli ,
Berolini , Holmiæ , Bononiæ & Gottingæ florent ,
Academiarum Socio.*

PARISIIS,

E Typographia H.L. GUERIN & L.F. DELATOUR.

M. DCC. LVIII.

18.2. The title-page of Lacaille's *Tabulae solares* of 1758.

construct solar tables entirely new, but only ... to correct his tables as far as my observations seemed to require." Mayer left Lacaille's equation of centre unaltered, revised the epoch of mean longitude and the apogee (removing 6″ of error from the former and adding 30″ of error to the latter), and introduced his own tables for the perturbations, which were a good deal less in error than those that Lacaille had derived from Clairaut's formulas. Mayer's version of Lacaille's tables became the basis for the ephemerides of the British *Nautical Almanac* from 1767, and for those of the *Connaissance des temps* from 1779.

The solar theory would not be revised again until the 1780s, when Delambre, inspired by the example of Lacaille, and with the observations of the British Astronomer Royal Nevil Maskelyne at his disposal, undertook to re-do the determination of elements once more from the beginning.

The lunar tables of Mayer

Mayer's investigation of the Moon's motion arose out of his work in cartography, for which more accurate determinations of terrestrial longitudes were greatly in demand.

Mayer, who grew up in Esslingen, showed early talent for drawing and a fascination with practical geometry and mathematical analysis. His first book, published when he turned 18, dealt with the analytic solution of geometrical problems. A second work, entitled *Mathematischer Atlas*, appeared in Augsburg in 1745; in sixty coloured plates it surveyed the range of elementary mathematics, with astronomical, geographical, and other applications.

In 1746 Mayer joined the Homann Cartographic Bureau in Nuremberg. In constructing a map to illustrate the course of a partial lunar eclipse due to occur in August 1748, he discovered that the method of orthographic projection hitherto used in such constructions was inaccurate, because of its neglect of lunar parallax. Also, he found that the best of the existing lunar maps – those of the seventeenth-century observers Giambattista Riccioli and Johann Hevelius – were unreliable.

At the time, lunar eclipses were a principal means for determining differences in terrestrial longitude: one had only to compare the local times at which the same topographical features on the Moon's surface were obscured by (or emerged from) the Earth's shadow, as estimated by observers situated on different meridians of longitude.

One outcome was that Mayer undertook to lay the foundations for an accurate lunar map. Between April 1748 and August 1749, using a telescope fitted with a glass micrometer of his own design, he made a series of measurements of both the Moon's diameter and the positions of various lunar markings, in terms of their relative distances from the centre of the disk and the angle made at the centre with respect to the north–south line.

The task of obtaining an accurate lunar map was complicated by the lunar libration: some 18% of the lunar surface is alternately visible and invisible, owing to (a) the inclination of the Moon's equator to the plane of its orbit, (b) the non-uniformity of its orbital motion contrasted with the uniformity of its axial rotation, and (c) the displacement of the terrestrial observer from the Earth's centre. It was therefore essential to fix a coordinate system in the body of the Moon itself, independent of the position of the terrestrial observer.

For this purpose Mayer used 27 observations of the distance and angle of the crater Manilius from the apparent centre of the lunar disk. From these he derived 27 'equations of condition' involving three unknowns: the inclination of the Moon's equator to the ecliptic, the longitude of the intersection of these two circles, and the 'geographic latitude' of Manilius with respect to the Moon's equator. He next separated the equations into three groups of nine each, and added the equations of each group together to form a single equation; the three resulting equations were then solved for the three unknowns. In separating the original equations into three groups, Mayer was guided by the idea of maximizing the coefficient of a given unknown in each group.

Essentially the same method had been described by Euler in his 1748 prize essay *Recherches sur la question des inégalités du mouvement de Saturne et de Jupiter* (which was published in Paris as a book the following year), as a way of determining empirical corrections to orbital elements and the coefficients of perturbational terms; and since Mayer, in a letter to Euler of 4 July 1751, acknowledged his familiarity with this work, it is likely that he derived the method from Euler's memoir. Euler's application of the method had not been particularly successful, because of the inadequacy of his theory of Saturn – its failure to include certain large perturbational terms first discovered in 1785 by P. S. Laplace. Thus it was Mayer who made the first clearly successful use of the method of equations of condition. His results were published in a memoir of 1750, "Abhandlung über die Umwälzung des Monds um seine Axe".

These results substantiated and refined the theory of G. D. Cassini (Cassini I) on the Moon's libration. Lagrange made use of them in his prize-winning memoir of 1764 on the lunar libration. Lalande in the second edition of his *Astronomie* (1771) reviewed Mayer's procedure as a method generally applicable to the combination of large numbers of observational equations, and Laplace made use of it in his epoch-making "Théorie de Jupiter et de Saturne" (1788). Thereafter this 'method of equations of condition' was widely applied by astronomers to the fitting of planetary and lunar theories to observations, until the superiority of the method of least squares came to be generally recognized (see Chapter 27).

As a means to the determination of differences in geographical longitude, lunar eclipses left much to be desired. They were infrequent, and the edge of the Earth's umbra did not have the sharpness that would permit a very precise fixing of the instant at which a given feature of the lunar surface was eclipsed. The occultation of a star by the Moon, in contrast, could be timed to within a second with the aid of a good telescope, and several such occultations occurred each month. During 1747 and 1748 Mayer observed occultations of the Pleiades and a number of other zodiacal stars. For the timing of immersion and emersion, he utilized not only the directly observed times of these events, but also a number of measured distances between the star and the Moon's limb at times before immersion and after emersion, and applied these multiple data to the determination of an average result. The suddenness with which stars were occulted convinced him that the Moon had no atmosphere.

Mayer's memoirs describing his micrometer, his determination of the lunar libration, his observations of lunar occultations, and his arguments for denying an atmosphere to the Moon, appeared in 1750 in the *Kosmographische Nachrichten und Sammlungen auf das Jahr 1748*. These publications led to his being called in November 1750 to a professorship at the Georg-August Academy in Göttingen. Here he was to have the use of the Göttingen Observatory, then under construction, and after 1755 to be its sole director until his death in 1762.

On 4 July 1751 Mayer initiated a correspondence with Euler. He had tried several times, but in vain, he told Euler, to deduce the lunar inequalities from Newton's theory, until he met with Euler's memoir on the motions of Saturn and Jupiter, which pointed the way. However, unlike Euler, he retained the mean anomalies of the Moon and Sun as the variables from which the inequalities were to be calculated (Euler had used for this purpose the

eccentric anomaly, which falls about midway between the mean anomaly and the true anomaly, and is related to the mean anomaly by Kepler's equation, $E + e \sin E = M$). Mean anomalies are derivable immediately from the time; eccentric anomalies must be obtained from a series expansion or tables constructed for the appropriate eccentricity. Euler in his response pointed out that the perturbations depend on the *true* angular distance between the Moon and Sun, and the attempt to express them in terms of mean anomalies and the differences of mean longitudes inevitably leads to complicated formulas. The point was not lost on Mayer, as we shall see.

In his letter of July 1751, Mayer allowed that the agreement of his theory with observation was less than satisfactory, but expressed disapproval of Clairaut's proposal to emend the Newtonian law of attraction: "there [are] so many circumstances which make the calculation and approximation doubtful, that the error can rather be on the latter side." To obtain more accurate predictions, he proposed using the Saros cycle or period of 223 lunations ($18^y \ 11^d \ 7^h \ 43^m \ 30^s$), by which Halley had earlier sought to provide observational corrections to Newton's lunar theory. Mayer, however, employed his own lunar theory, expressible in ten groups of terms incorporated in the same number of tables. By means of these tables he compared observations separated by the Saros period, and always found his predictions accurate to within 10″ or 20″. "Consequently, if one had a continuing series of observed positions of the Moon throughout 18 years, one could determine the Moon's position very accurately for any future time by means of these tables."

Euler, in replying on 27 July, gave high praise to Mayer's project for predicting the Moon's longitude. He had now succeeded, he informed Mayer, in deducing the whole motion of the lunar apogee from the Newtonian law of attraction. (Clairaut had been the first to succeed in this deduction, as Mayer learned at about this time from newspapers.) Mayer, responding on 15 November, stated that he had not yet determined accurately enough the coefficients of a number of equations, but hoped in repeating the calculation to obtain everything more precisely. However, he urged publication of Euler's and Clairaut's theories, which would make this labour superfluous. In passing he noted that, since the nutation of the Earth's axis is caused by the attraction of the Moon on the spheroidal Earth, there must be reciprocal inequalities in the Moon's motion due to the Earth's shape.

During the second half of 1751 Mayer continued to work on the lunar theory. In a letter to Euler of 6 January 1752 he gave a list of 23 terms for the Moon's inequalities in longitude. "Among these many inequalities there are some which I can completely determine, and some which I must correct, through observations. I have completely discarded still more on account of their smallness." Euler, in his reply of 18 March 1752, compared the terms of Mayer's theory with the corresponding terms in Clairaut's theory (Euler's own theory was not directly comparable, because of his use of the eccentric anomaly as independent variable). Both theories contained considerable error at this stage.

Mayer continued to work on his derivation. In January 1753 he was able to announce to Euler that, although he had still not been able to derive the total observed motion of the Moon's apogee,

I now have the equations of longitude as accurately as I have wished. There are no more than 13 of them, which I have arranged in tables, and through which I shall be able to calculate the longitude of the Moon so correctly, that the error never amounts to 2 minutes.

The equations that Mayer went on to list are, except for a few slight changes in coefficients, the same as those given in his "Novae tabulae motuum solis et lunae", published in Vol. 2 of the *Commentarii* of the Royal Society of Sciences of Göttingen in the spring of 1753. A number of the tables incorporate more than one sinusoidal term. But what chiefly enabled Mayer to reduce the number of tables to thirteen, aside from the neglect of small terms, was a shift in strategy. This was first explained and justified in Mayer's *Theoria lunae juxta systema Newtonianum*, sent (as we shall see) to the British Admiralty in late 1754 but not printed until 1767.

The *Theoria lunae* is a systematic derivation of the lunar inequalities from the inverse-square law. Like Euler, Mayer develops formulas relating mean and true anomaly to eccentric anomaly; unlike Euler, he then eliminates the eccentric anomaly so as to yield true anomaly in terms of mean anomaly.

The resulting formula contained 46 sinusoidal terms. Mayer commented:

Although this formula ... can easily be reduced to tables entirely similar in form to those of the celebrated Clairaut; yet the huge number of inequalities and tables thence derived would make the calculation of the Moon's place so laborious and troublesome that even the most patient calculator would give in to weariness .... I therefore sought a way by which the inequalities comprised in this formula could be reduced in number.

The procedure Mayer now adopted entailed progressive correction of the independent arguments.

(1) First, the Moon's mean longitude was corrected for the secular acceleration discovered by Halley. In his tables of 1753 Mayer gave the resulting increase in longitude per century as 7″; later he revised this figure to 9″.

(2) Next, the Moon's mean longitude and the mean longitude of its apogee and node were corrected for the annual inequalities to which they were subject, depending on the changing distance of the Sun from the Earth during the course of the year. According to Mayer's tables of 1753, their maxima were 11′ 20″, 20′ 36″, and 10′ 18″ respectively. The corrected mean longitude of the Moon less its corrected mean apogee yields the corrected mean anomaly.

(3) The Moon's mean longitude was then corrected for seven inequalities dependent on certain linear combinations of the corrected mean anomaly of the Moon, the mean anomaly of the Sun, and twice the difference in mean longitude of the Moon and Sun. The largest of these in Mayer's tables of 1753 came at maximum to 1′ 48″.

(4) The Moon's mean longitude was then further corrected for two inequalities, dependent on the distances of the Moon's apogee and node, respectively, from the *true* longitude of the Sun. In the tables of 1753 the maximum of the first of these was 3′ 45″; of the second, 44″.

(5) With the mean anomaly of the Moon resulting from all the preceding corrections, Mayer now calculated the Moon's equation of centre, amounting at maximum to about $6\frac{1}{3}$ degrees. The Moon's corrected apogee (step 1) plus the equation of centre (step 4) yielded the "equated longitude of the Moon".

(6) In his tables of 1753 Mayer next computed the 'evection', dependent on twice the corrected distance between the Moon and Sun minus the corrected mean anomaly of the Moon, and amounting at maximum to 1° 20′ 43″. In his later tables he found it possible to combine the calculation of this inequality with that of a similar inequality at stage 2.

(7) Finally, Mayer calculated the Tychonic 'variation', depending on the equated longitude of the Moon minus the true longitude of the Sun. This is a further correction to the Moon's longitude, amounting at maximum to about 37′.

In his *Theoria lunae* Mayer showed that if the terms derived from theory, and depending solely on mean arguments, were transformed so as to depend on the progressively modified arguments employed in the foregoing sequence of steps, then many of the resulting terms – those that he left out of account in his tables – were so small as to be safely neglected. This theoretical justification, however, was not made public till long after the tables of 1753 were published.

In the preface to his tables of 1753, Mayer made a large claim for their accuracy. In comparisons with more than 200 observations of his own and the preceding century, he claimed to have found scarcely ten observations that differed from the predictions of his tables by as much as $\frac{1}{2}′$, and none differing by as much as 2′. By contrast, he noted, the tables of Euler and Clairaut were subject to errors (as their authors conceded) of as much as 4′ or 5′.

D'Alembert, in remarks appended to his *Théorie de la lune* (1754), doubted that Mayer's tables were as accurate as claimed. The manner in which Mayer had derived the formulas underlying his tables was left unclear, and the tables failed to take account of terms that d'Alembert believed to be sizeable.

In nothing Mayer published did he provide a description of his manner of fitting the lunar theory to observations. In the introduction to the tables of 1753 he claims to have spared no labour in his investigation of the mean motions and the secular acceleration of the Moon. Values for these constants must be so chosen as to fit both the less reliable ancient observations and the more accur-

ate modern ones, each with an appropriate degree of precision. In determining the secular acceleration Mayer made use of the Saros cycle, and he also employed two solar eclipses observed by Ibn Yūnus on 13 December AD 977 and 8 June AD 978. These eclipses were especially precious because Ibn Yūnus recorded the altitudes of the Sun on the two occasions, so that the precise times of the eclipses could be calculated. Independently of Mayer, Richard Dunthorne in the *Philosophical Transactions* for 1749 had used these same observations in the same calculation, obtaining 10″ as the change in mean longitude per century, as compared with Mayer's initial value of 7″.

The tables that Mayer published in 1753 turned out to be only a first version. In December 1754, with Euler's encouragement, Mayer entered the competition for one of the prizes offered by the British Parliament: £20 000 for a method giving the longitude at sea to within 30 nautical miles, £10 000 for a method good to within 60 nautical miles. The lunar tables that Mayer submitted to the British Admiralty in late 1754 differed little in form from the published tables of 1753, but contained revised numerical coefficients. Mayer's last revision of the tables was sent to the Admiralty in 1763 by his widow.

We do not know how Mayer computed these revisions. In the preface to his *Theoria lunae*, he states:

In exhibiting here the way in which I investigated the inequalities in the motion of the Moon on the basis of theory, I am not aiming to demonstrate the reliability and truth of my lunar tables. For the theory has this inconvenience, that many of the inequalities cannot be deduced from it accurately, unless one should pursue the calculation – in which I have now exhausted nearly all my patience – much further. My aim is rather to show that at least no argument against the goodness of my tables can be drawn from the theory. This is most evidently gathered from the fact that the inequalities found in the tables, which have been corrected by comparison with many observations, never differ from those that the theory alone supplies by more than $\frac{1}{2}′$. Therefore . . . it is evident that these small errors derive rather from the side of the theoretical calculus than from the tables. This is put beyond all doubt if the numerical results deduced by others from the same *theory – and especially those of the celebrated Euler, Clairaut, and d'Alembert – are compared either with my tables or my theoretical calculation. They sometimes differ by 3' or more; nor do they agree better with each other, except in a few places . . . .*

Later in the same treatise Mayer states that, besides the terms that can be evaluated only on the basis of observation, such as those depending on the orbital eccentricity or on solar parallax, there are terms whose derivation from the theory is so arduous that they are best evaluated observationally; and he adds that he has corrected all terms of the theory by adding or subtracting a few seconds of arc, to bring about better agreement with the heavens. For so massive an operation of correction, it is plausible to assume that Mayer made use of equations of condition – the only then known statistical procedure for determining a large number of unknowns simultaneously, and one which Mayer alone had previously used with success.

In Vol.3 of the *Commentarii* of the Royal Society of Sciences of Göttingen Mayer published a comparison between his tables of 1753 and some 55 lunar eclipses observed telescopically since 1612, and 139 lunar longitudes observed by Bradley from 1743 to 1745 (Figure 18.3). In the latter group, the average deviation between the tables and the observations was 27″, the maximum being 1′ 37″.

When the Göttingen observatory became ready for use in the summer of 1753, Mayer began his own series of observations, determining the Moon's position through occultations of fixed stars – first through occultations of a single fixed star, so that any error in the position assumed for this star would be reflected only in the epoch of mean motion of the Moon, which could later be corrected through observation of a solar eclipse. In this manner Mayer found considerable errors in the then accepted positions for a number of zodiacal stars, and at the same time obtained precise corrections for the numerical coefficients of his lunar theory.

By 1760 Bradley had completed a laborious comparison of Mayer's tables with some 1100 observations made at Greenwich, and so confirmed that the tables were generally reliable to within 1′.25 – an accuracy not quite high enough to justify the maximum Parliamentary award, but

certainly sufficient to merit the lesser bounty of £10000 offered for a method both "useful and practicable" for determining longitude at sea to within 60 nautical miles. A test of the method at sea was postponed, however, because of the British navy's involvement in the Seven Years War, and in the interim John Harrison completed his fourth marine chronometer, which thus came into rivalry with Mayer's lunar method for the longitude.

Although the revised tables that Mayer's widow submitted to the Board of Longitude in August 1763 were tested on a voyage to Barbados, they were not considered in the final recommendation of the Board of Longitude, made on 9 February 1765. Harrison was to receive £10000 for a method that gave the longitude to within 30 nautical miles; Mayer's heirs were to receive a prize "not exceeding £5000". In May 1765, before Parliament made its final allocation of the prizes, *Gentleman's Magazine* published a letter by Clairaut, in which he protested that he and Euler had developed lunar theories more rigorous than Mayer's, and that Mayer's tables derived their accuracy primarily from his skilful discussion of observational data. Parliament, a few days later, awarded only £3000 to Mayer's heirs, along with £300 to a surprised Euler, as the supposed author of the theory on which Mayer's tables were based.

No doubt the accuracy of Mayer's tables *was* primarily due to the skilful fitting of theory to observational data, but Mayer's artfulness in such empirical determinations of numerical coefficients was new and unmatched among his contemporaries. The lunar tables of Johann Tobias Bürg and Johan Karl Burckhardt, which appeared in 1806 and 1812 respectively, would still depend on the empirical determination of perturbational coefficients by the same method of equations of condition that Mayer probably used. Only after 1820 was the theoretical derivation carried out in sufficient detail to begin to yield observationally satisfactory values of these numbers.

Another consequence of Mayer's careful attention to the accord of theory and observation may be mentioned: his refutation of Euler's hypothesis that the Earth's motion about the Sun was detectably accelerating. Euler had first argued for this hypothesis in his *Opuscula* of 1746, and had repeated the same views in a letter to Caspar Wetstein, published in the *Philosophical Transactions* for 1752.

139. LOCA LUNAE EX OBSERVATIONIBUS CEL. D. BRADLEYI. 393

1743.	temp. med. Grenovic. ftyl. vet.	Longitudo Lunae vera obfervata.		Different. calculi ab obferv.
M. d.	h. ' "	S o ' "		' "
Sept. 11.	3.20.39	7. 24. 10. 57		+ 0. 40
12	4. 8.26	8. 6. 43. 28		+ 0. 56
14	5.48.40	9. 1. 22. 43		+ 0. 14
16	7.30.36	9. 26. 1. 38		— 0. 29
17	8.20. 5	10. 8. 33. 15		— 0. 56
18	9. 7.52	10. 21. 18. 12		— 0. 36
20	10.39.10	11. 17. 39. 19		— 0. 15
22	12.11.36	0. 15. 15. 0		+ 0. 24
24	13.49.19	1. 13. 54. 50		— 0. 2
26	15.41.50	2. 13. 18. 12		— 0. 51
28	17.46. 1	3. 12. 57. 28		— 0. 27
Octob. 14	6.11.30	10. 3. 12. 28		+ 0. 13
15	6.59.24	10. 15. 41. 29		— 0. 22
19	9.59.38	0. 8. 54. 14		— 0. 10
21	11.35.57	1. 7. 21. 57		0. 0
24	14.33.17	2. 22. 45. 57		— 0. 7
26	16.41. 9	3. 23. 8. 7		— 0. 11
27	17.40.52	4. 7. 59. 7		+ 0. 20
28	18.35.48	4. 22. 31. 41		+ 0. 49
Nov. 9	3.14.33	9. 16. 6. 0		+ 0. 2
13	6.22.48	11. 5. 38. 34		— 0. 11
17	9.21.24	0. 29. 42. 59		+ 0. 31
18	10.12.50	1. 14. 27. 13		+ 0. 21
20	12.14. 8	2. 15. 15. 59		+ 0. 49
21	13.20.15	3. 1. 3. 15		+ 0. 30
22	14.27. 2	3. 16. 49. 17		+ 0. 15
27	18.56.42	6. 0. 12. 18		+ 0. 54
Dec. 8	2.48. 6	10. 6. 16. 52		+ 0. 7
10	4.18.41	11. 0. 56. 32		— 0. 5
11	5. 1.20	11. 13. 28. 13		— 0. 2
12	5.43.18	11. 26. 14. 1		+ 0. 3
13	6.25.48	0. 9. 20. 15		+ 0. 1
19	12. 1.32	3. 8. 18. 51		+ 0. 24
20	13.10.21	3. 24. 24. 10		+ 0. 32
23	16. 4.27	5. 10. 48. 4		0. 0
26	18.21.40	6. 22. 30. 40		+ 0. 13
27	19. 5.30	7. 5. 31. 27		+ 0. 10

Ddd 2   1744.

18.3. The first of four pages in which, for various dates between September 1743 and April 1745, Mayer gives the longitude of the Moon as observed by Bradley at Greenwich, and the difference between this observed longitude and that calculated by himself. Note that on this page the error never reaches 1' (and on the other pages exceeds 1' only rarely). (From Vol. 3 (1754) of the *Commentarii* of the Royal Society of Sciences of Göttingen.)

Writing to Euler on 22 August 1753 Mayer cited empirical evidence for assuming the tropical year to be essentially constant. Euler was misled by the equinoxes that Ptolemy claimed to have observed in AD 139 and 140: these were surely in error. Ptolemy had constructed his planetary tables before 139, and such construction presupposed solar tables. He presumably borrowed the solar tables of Hipparchus, thus putting the length of the year at $365^d 5^h 55^m$ – so large, Mayer thought, that it was probably deduced from the luni-solar cycle of 76 years, rather than from actual observations.

This year is now too large by approximately $6\frac{1}{2}$ minutes, which in the 300 years which intervened between Hipparchus and Ptolemy, amount to approximately $1\frac{1}{4}$ day .... And the Ptolemaic equinoxes are found to be too late by just about this much. It can be that Ptolemy perceived this error of his solar tables in his observations of the equinoxes, ... but preferred to *discard his observations rather than attempt to revise his system from the ground up.*

Reluctantly Euler gave up his belief in the Earth's secular acceleration. Light, in his view, was a wave-motion, requiring an aether for its propagation; and the aether must offer resistance to planetary motion, causing planets to fall toward the Sun and so to increase their angular motion. (Even if, as the Newtonians claimed, light consisted of corpuscular projectiles, these corpuscles, Euler argued, having inertia would cause resistance to planetary motion.) At least, he wrote to Mayer, the well-confirmed secular acceleration of the Moon could be attributed to aethereal resistance.

Mayer replied (on 25 November 1753) that it might also be due to a long-term inequality. In addition, he informed Euler that ancient observations of equinoxes and eclipses indicated a secular decrease in the Earth's orbital eccentricity, which likewise could be due to a long-term inequality. The correctness of both suggestions would be confirmed in the work of Lagrange and Laplace (see Chapters 21 and 22).

Further reading

Eric G. Forbes, *Tobias Mayer (1723–1762): Pioneer of Enlightened Science in Germany* (Gottingen, 1980)

Eric G. Forbes, *The Euler–Mayer Correspondence (1751–1755)* (New York, 1971)

E. Doublet, *Le Bicentenaire de l'Abbé de la Caille* (Bordeaux, 1914)

Curtis Wilson, Perturbations and solar tables from Lacaille to Delambre: the rapprochement of observation and theory, Part I, *Archive for History of Exact Sciences*, vol. 22 (1980), 53–188

19

Predicting the mid-eighteenth-century return of Halley's Comet

CRAIG B. WAFF

with an appendix on Clairaut's calculation by Curtis Wilson

In his lecture on 25 April 1759 before the public assembly of the Paris Academy of Sciences, the astronomer Joseph-Jérôme Lefrançais de Lalande (1732–1807) opened with the declaration: "The Universe sees this year the most satisfying phenomenon that Astronomy has ever offered us; unique event up to this day, it changes our doubts into certainty, & our hypotheses into demonstrations." He was speaking of the year's outstanding astronomical event – the apparition of a comet fulfilling the bold prediction of Edmond Halley (1656–1742) that one would appear in late 1758 or early 1759 with orbital elements similar to those of comets observed in 1682, 1607, 1531, and possibly before. Halley, who had laboriously calculated the elements of these comets and had recognized their similarity, had speculated that the comets seen in those earlier years were one and the same and that, if so, this celestial object travelled in a highly elliptical retrograde orbit around the Sun with a period of some 75 or 76 years.

Halley's prediction of a return appearance of a comet was without precedent. Moreover, his suggestion that at least some comets might travel in closed elliptical orbits highlighted the possibility that the motions of these celestial objects, like those of the planets, were governed by the gravitational force posited by Isaac Newton. Although by the late 1750s various eighteenth-century mathematicians had already confirmed that Newton's gravitational theory could explain a wide variety of phenomena (see Chapters 16 and 17), Lalande and other contemporary commentators argued that establishing the periodicity of repeated returns of comets, which are not visible to earthbound observers throughout most of their orbits and whose exact nature was still mysterious, would provide a particularly striking vindication of the theory. Locating the expected comet and verifying

that it had orbital elements similar to those of the comets of 1682, 1607, and 1531 thus became a major objective of astronomers and mathematicians in the mid-eighteenth century.

Such a task was no easy matter. Astronomers could most clearly show that a particular comet appearing in the late 1750s was probably identical to those that had appeared in the aforementioned years if they could observe it over a sufficiently long time to permit computation of its orbital elements. They therefore wished to recover the expected comet at the earliest possible moment. The problem for astronomers was deciding when and where in the sky the comet would most likely appear. The theoretical first visibility of the comet would depend on its intrinsic brightness, the distance between it and the Earth, and the magnifying power of an astronomer's telescope. The comet's location in the sky at the time of theoretical first visibility and thereafter would depend on the relative positions of it and the Earth from moment to moment in their respective orbits. Searching the entire sky was out of the question. To select a region in which the comet might possibly appear at any particular time, an astronomer had to make an assumption as to where it was in its orbit at that particular time.

The period of revolution of this supposedly recurring comet was variable by as much as a year, however, as Halley himself recognized in making his prediction. (By contrast, the periods of revolution for each of the planets never varied by more than a few days.) Such a situation obviously rendered uncertain when the comet would reach its closest approach to the Sun (perihelion) or any other specific location in its orbit (including the point of theoretical first visibility). Halley attributed the unevenness of the periods to the possibility that the motion of the comet had been perturbed when it passed near the giant outer planet Jupiter during its repeated journeys to and from the outer regions of the solar system. If so, any accurate prediction of the comet's return to perihelion would require a detailed calculation, based on gravitational theory, of the specific perturbational influences of Jupiter (and possibly other planets) on the comet's motion. Halley himself made no such calculation, but the French mathematician Alexis-Claude Clairaut (1713–65), who had already made significant contributions to the application of gravitational theory to the figure of the Earth and the motion of the Moon (Chapters 15 and 16), would undertake this challenging task together with several colleagues.

Even before Clairaut publicly announced a predicted date of perihelion, however, astronomers and others had devised a variety of means for guiding the search for the expected comet. Some investigators, like Clairaut, would predict a specific date of perihelion. Others confined themselves to suggesting specific times of the year and areas of the sky in which the comet would most likely appear.

Halley's *Synopsis astronomiae cometicae*

In the first edition of his *Principia* (1687), Newton had demonstrated (Prop. 40, Bk III) that comets, like planets, must move in conic sections with foci in the middle of the Sun. He thus foresaw the possibility of a closed elliptical orbit for some comets, but he offered no examples. It was Halley who first undertook the tasks of calculating, by geometrical construction, orbital elements for twenty comets that he considered the best observed, and of examining these elements for similarities that might indicate that a comet had appeared more than once to earthbound observers and that it consequently travelled in a closed elliptical orbit around the Sun.

Although Halley himself had observed several comets in the 1680s, including the one that appeared in 1682, his first intense period of investigation of cometary motions appears to have begun around August 1685, when he visited Newton at Cambridge to determine for the latter the orbits of several comets, including those seen in 1664–65, 1680–81, 1682, and 1683. Even before he undertook detailed calculations of the comet of 1682, Halley perceived features in the motion of this

comet that it shared with those seen in 1531 and 1607. Perhaps the most prominent similarity was the retrograde direction of motion, that is, it moved opposite to the direction (counterclockwise to a hypothetical observer stationed 'north' of the solar system) of the motions of Earth and the other planets around the Sun. On 28 September 1695, Halley informed Newton that he was "more and more confirmed that we have seen that Comett now three times, since ye yeare 1531". Further calculations revealed additional similarities in the locations of perihelia, the inclinations of the orbital planes to the plane of the Earth's orbit, the locations of the nodes (points of intersection of the cometary orbits with the plane of the Earth's orbit), and the distances of the comets from the Sun at the times of perihelion. These similarities, plus the fact that nearly equal periods of time had elapsed between the apparitions of 1531 and 1607, and between those of 1607 and 1682, convinced Halley that one and the same comet had made successive appearances in the inner solar system at a regular interval (about 75 years) – a result that he announced to the Royal Society of London on 3 June 1696.

His appointment as Deputy Comptroller of the Chester Mint (1695–98) and several sea voyages (1698–1701) interrupted Halley's further work on comets over the next years, so that it was not until 1705 that he published a summary of his investigations of cometary motions. The principal feature of Halley's *Synopsis astronomiae cometicae* was a table (Figure 19.1) listing orbital elements that he had calculated for twenty-four well observed comets seen since 1337. Near the end of his essay Halley remarked that he had considered the orbits of all these comets as parabolic. The reason for doing so, he acknowledged, was that such orbits were easier to calculate than hyperbolic or elliptical ones. If all cometary orbits were parabolic in reality, however, astronomers could have no hope of ever seeing a comet return. Halley justified his approach by arguing that the frequency of cometary appearances, plus the fact that none had ever been observed to move with a hyperbolic motion, made it probable that they traversed highly eccentric elliptical orbits, whose orbital elements would differ little from the parabolic ones that he had calculated.

In his *Synopsis*, Halley declared himself con-

Cometarum Omnium hactenus rite Observatorum, Motuum in Orbe Parabolico Elementa Astronomica.

Cometæ Anni.	Nodus Ascend. gr. ′ ″	Inclin. Orbitæ. gr. ′ ″	Perihelion in Orbe. gr. ′	Perihelion in Ecliptica. gr. ′ ″	Latitudo Perihelii. gr. ′ ″	Distantia Perihelii à Sole	Log. dist. Perihelii à Sole.	Temp. æquat. Perihelii Londini. die. h.
1337	♊ 24. 21.	0 32. 11.	♈ 7. 59.	♉ 12. 45. 15	22. 4. 30 B	40666	9. 609236	Junii 2. 6. 25
1472	♑ 11. 46.	20 5. 20.	♈ 15. 33. 30	♉ 15. 40. 20	4. 25. 50 A	54073	9. 734584	Feb. 28. 22. 23
1531	♌ 19. 25.	0 17. 56.	♒ 1. 39.	♒ 0. 48. 15	17. 3. 05 B	56700	9. 753583	Aug. 24. 21. 18¼
1532	♊ 20. 27.	0 32. 36.	♋ 21. 7.	♋ 16. 59. 40	15. 57. 00 B	50910	9. 706830	Oct. 19. 22. 12
1556	♍ 25. 42.	0 32. 6. 30	♍ 6. 50.	♍ 11. 0. 00	31. 10. 20 B	46390	9. 666424	Apr. 21. 20. 3
1577	♒ 25. 52.	0 74. 32. 45	♏ 9. 22.	♏ 7. 53. 00	9. 35. 20 A	18342	9. 263447	Oct. 26. 18. 45
1580	♊ 18. 57.	0 64. 40.	♈ 9. 5. 50	♈ 19. 17. 10	64. 40. 40 A	59628	9. 775450	Nov. 28. 15. 00
1585	♊ 7. 42. 30	0 6. 4.	♈ 8. 51.	♈ 8. 51. 10	2. 55. 25 A	109358	0. 038850	Sept. 27. 19. 20
1590	♌ 15. 30. 40	29. 40. 40	♏ 6. 54. 30	♏ 2. 55. 50	22. 45. 50 A	57661	9. 760882	Jan. 29. 3. 45
1596	♉ 12. 12. 30	55. 12.	♈ 18. 16.	♈ 22. 44. 35	54. 44. 30 B	51293	9. 710058	Julii 31. 19. 55
1607	♉ 20. 21.	17. 2.	♏ 2. 16.	♏ 1. 29. 40	16. 10. 5 B	58680	9. 768490	Oct. 16. 3. 50
1618	♌ 7. 30.	37. 34.	♐ 13. 14.	♐ 6. 10. 00	35. 50. 00 A	37975	9. 579428	Oct. 29. 12. 23
1652	♊ 28. 1.	79. 28.	♈ 26. 18. 40	♈ 10. 41. 35	58. 14. C. B	84750	9. 928140	Nov. 2. 15. 40
1661	♊ 22. 30. 30	32. 35. 50	♈ 25. 58. 40	♈ 27. 37. 30	17. 0. B	44851	9. 651772	Jan. 16. 23. 41
1664	♊ 21. 14.	21. 18. 30	♐ 10. 41. 25	♐ 8. 40. 35	16. 4. 50 A	102575½	0. 011044	Nov. 24. 11. 52
1665	♊ 18. 02.	76. 05.	♈ 11. 54. 05	♈ 24. 6. 35	23. 8. 0 B	10649	9. 027300	Apr. 14. 5. 15½
1672	♊ 27. 30. 30	83. 22. 10	♈ 16. 59. 30	♈ 16. 59. 27	40 B	69739	9. 843476	Feb. 20. 8. 37
1677	♑ 26. 49. 10	79. 03. 15	♌ 17. 37. 5	♌ 16. 21. 0	75. 44. 10 B	28059	9. 448072	Apr. 26. 00. 37½
1680	♒ 2. 2.	60. 56.	♋ 22. 39. 30	♋ 27. 26. 50	8. 11. 10 A	00612½	7. 787100	Dec. 8. 00. 6
1682	♋ 21. 16.	17. 56.	♒ 2. 52. 45	♒ 0. 30. 00	16. 59. 20 B	58328	9. 765877	Sept. 4. 07. 32
1683	♍ 23. 23.	83. 11.	♊ 25. 29. 30	♊ 10. 36. 55	52. 00 B	56020	9. 748343	Julii 3. 2. 50
1684	♈ 28. 15.	65. 48. 40	♋ 28. 52.	♋ 2. 15. 15	25. 35. 20 A	96015	9. 982339	Maii 29. 10. 16
1686	♒ 20. 34. 40	31. 21. 40	♒ 17. 00. 30	♒ 16. 24. 00	31. 17. 35 A	32500	9. 511883	Sept. 6. 14. 33
1698	♋ 27. 44.	11. 46.	♐ 0. 51. 15	♐ 0. 47. 20	0. 38. 10 A	69129	9. 839660	Oct. 8. 18. 57

Hæc Tabula vix indiget explicatione, cum ex titulis satis pateat quid sibi velint Numeri. Distantiæ autem perihelii æstimantur in ejusmodi partibus quales media distantia Terræ à Sole habet centies millenas.

Tabula

19.1. Halley's 1705 table of the orbital elements of twenty-four comets, from *Philosophical Transactions*. Note the similarity of the orbits of the comets of 1531, 1607 and 1682 (which are in fact successive appearances of Halley's Comet).

vinced that the comets seen in 1531, 1607, and 1682 were one and the same. The only seemingly contradictory piece of evidence was the inequality of period (varying between 75 and 76 years), but he argued that the magnitude of this inequality could be explained by gravitational perturbation:

For the Motion of Saturn is so disturbed by the rest of the Planets, especially Jupiter, that the Periodick Time of that Planet is uncertain for some whole Days together. How much more therefore will a Comet be subject to such like Errors, which rises almost Four times higher than Saturn, and whose velocity, tho' encreased but a very little, would be sufficient to change its orbit, from an Elliptical to a Parabolical one.

His belief in the common identity of the three comets was now reinforced by one further piece of evidence: "In the Year 1456, in the Summer time, a comet was seen passing Retrograde between the Earth and the Sun, much after the same Manner: Which, tho' no Body made Observations upon it, yet from its period [75 years before 1531], and the Manner of its Transit, I cannot think differently from those I have just now mentioned." Halley therefore declared that he would "dare venture to foretell" that the comet apparently seen on four previous occasions would return again in 1758, that is, about 76 years after 1682.

Halley promised to discuss the subject of cometary motions in a larger volume "if it shall please God to continue my Life and Health", and he in fact did so in a revised and much expanded version of the *Synopsis*, written sometime before 1719, that eventually appeared in his posthumously published *Tabulae astronomicae* of 1749, soon to be translated into English and French. Halley recalled that after the original version of his essay had appeared, he had discovered in some catalogues of ancient comets new or better records of three earlier retrograde comets – those seen in 1305 around Easter, in 1380 in an unknown month, and in June 1456 – that had preceded at like intervals (75 or 76 years) the three that he had already suspected were the same comet. (The two perihelia of Halley's Comet preceding the perihelia of 1456 were in fact on 25 October 1301 and 10 November 1378.) He also announced that he had subsequently developed a method, which he had applied to the comet of 1682, by which the parameters of a highly eccentric elliptical orbit could

be calculated. Calculating the perihelial dates from the best observed positions of the comets seen in 1531 (by Peter Apian), 1607 (by Johannes Kepler and Longomontanus), and 1682 (by John Flamsteed), Halley claimed that "it is manifest that the two Periods of this Comet are finished in 151 years nearly, and that each alternately, the greater and the less, are completed in about 76 and 75 years". Arguing that the comet's close encounter with Jupiter in 1681 had increased both the comet's period and its inclination during the current revolution, he predicted that its next return would take 76 years or more (after 1682), and thus it ought to appear "about the end of the year 1758, or the beginning of the next".

Halley provided no detailed mathematical analysis to back this prediction, however, and so the prediction remained open to question. Until such an analysis was undertaken – one that would examine in detail particularly the perturbational influences of Jupiter and Saturn on the comet's motion and determine a more precise date for its return to perihelion – mid-eighteenth-century astronomers could make only uncertain guesses when, and where in the sky, the expected comet might first be seen.

One comet or two?

The unevenness of the periods of the supposedly identical comet that had appeared in 1305, 1380, 1456, 1531, 1607, and 1682 – appearances separated by alternating intervals of 75 and 76 years – led at least two individuals to suggest that two comets, rather than one, were involved. In his treatise on the spectacular appearance of the comet of 1744, the Swiss astronomer Jean Philippe Löys de Chéseaux (1718–51) argued that it was "very possible & even rather probable" that the apparently alternating periods could be better explained by the presence of two comets, with identical and constant periods, moving in the same orbit, with one comet near but not quite at its aphelion at the same time that the other was at its perihelion. He considered this hypothesis probable because of doubts that gravitation alone could produce substantial changes in a comet's orbit. Calculating that 151 years 10 days had elapsed between the perihelion passages of the supposedly identical comet seen in 1531 and 1682, Chéseaux suggested that the comet seen in 1607 would return after the same period of time and would consequently reach perihelion on 7 November 1758. If so, the comet would appear very brightly in autumn and would pass near the Earth in mid-September.

Fourteen years later the two-comet theory was proposed again by Thomas Stevenson (d. 1764), a plantation owner on the West Indies island of Barbados and a former Surveyor General of that British colony. An amateur astronomer who had observed a comet that appeared in 1757, Stevenson first proposed a two-comet theory, quite possibly without knowledge of Chéseaux's earlier speculation, in a letter to his friend John Rotheram in England that was later published in several London newspapers and magazines late in October 1758. He also discussed his theory in June 1759 in a letter to James Bradley, the English Astronomer Royal. Like Chéseaux, Stevenson resorted to a two-comet theory because of his inability to understand how the force of gravitation could cause major changes in a comet's orbit. He told Bradley that

I cannot apprehend if once Accelerated, how [the comet] was retarded, & after that Accelerated and retarded Alternately, according to the times of his Appearances recorded in History ... Again nor can I conceive, how the attraction of Jupiter or Saturn whose plan[e]s differ little from the plan[e] of the Ecliptic; could increase the Angle of Inclination of the Comits Orbit, which is upwards of 17°.

In his comment on the inclination, Stevenson was referring to the fact that the angle had been 17° 56' for the comets seen in 1531 and 1682 and 17° 2' for the one seen in 1607.

Unlike Chéseaux, Stevenson relied only on the supposed 1305, 1456, and 1607 apparitions of the comet, in order to determine its period of revolution and so predict a date for its next return to perihelion. Uncertainty regarding the exact dates of perihelion in 1305 and 1456, however, complicated the problem of determining the period. In his letter to Rotheram, Stevenson argued that the comet would not be observed until at least ten days after it had passed perihelion; it would then be at an angular distance of 40° from the Sun and a linear distance from the Earth of $\frac{825}{1000}$ AU. On the basis of this reasoning, he established 1 April as the date of perihelion in 1305, because Easter had occurred on 11 April of that year. For unknown reasons, he

made no attempt to pinpoint a perihelial date in 1456. Calculating that 302 years 198 days had passed between 1 April 1305 and 16 October 1607 (Halley's calculated perihelial date for that year's apparition), Stevenson took half this period (151 years 99 days) and added it to the latter date. By this means he derived a perihelial date of 23 January 1758/59 (OS) or 3 February 1759 (NS) for the forthcoming expected apparition.

After determining the actual perihelial date (13 March) of a comet appearing in 1759 that he identified as the expected one, Stevenson, in an apparent effort to fit the facts, changed the perihelion-to-first-appearance interval in 1305 from 10 to 71 days, and suggested to Bradley that the period of the 1305–1456–1607–1759 comet was increasing by 4 days with each revolution around the Sun (beginning with 151 years 128 days for the revolution between 1305 and 1456). Stevenson considered this regularly occurring small increase per revolution plausible. What he could not comprehend was the variability of a year in period that Halley had postulated in asserting that the comet made its returns to perihelion twice as frequently as Stevenson supposed.

The inequality of periods that concerned Chéseaux and Stevenson was interpreted quite differently by Nils Schenmark, a member of the Swedish Academy of Sciences who apparently had access only to Halley's 1705 publication on comets. Noting that about 76 years had passed between the 1531 and 1607 apparitions, and about 75 years between the 1607 and 1682 apparitions, he suggested in 1755 that the period of the comet might be decreasing in an arithmetic proportion. On the basis of this hypothesis he derived a period of 73 years 228 days for the current revolution and hence a perihelial date of 19 April 1756.

Comet ephemerides

Chéseaux and Stevenson proposed their two-comet theories because of their strongly held belief that a comet's period of revolution around the Sun could not be subject to as large a variation (as much as a year) as Halley had assumed in positing the identity of the comets of 1531, 1607, and 1682. Their assumption of at most only a small amount of variation (a few days rather than a year) of course permitted them (by different methods) to make fairly precise predictions of the date in the late

1750s when the expected comet (last seen in 1607 according to them) ought to return once again to perihelion.

Until a sophisticated theoretical analysis of perturbational effects was performed, those who accepted Halley's view that a comet could have a period with significant variation were unable to make such precise predictions of a perihelial date. Eager nevertheless to observe the expected comet at the earliest possible moment, they took a more cautious and practical approach by constructing (for the benefit of themselves, other astronomers, and the general public) a kind of comet ephemeris – guides to where in the heavens the comet would most likely first appear during a given month (or near a specific date) of the year. The constructors of such ephemerides based them on the known orbit of the Earth (and hence the known pattern of visible background stars at any given time), the orbital elements of the comet that had been determined from observations made at its posited previous apparitions, and various suppositions regarding the perihelial date of the comet at its forthcoming expected apparition.

The first person to construct an ephemeris for the expected comet was Thomas Barker (1722–1809), a grandson of William Whiston living in the English village of Lyndon, near Uppingham, Rutland, and an author mainly of religious works. In a letter to Bradley dated 17 December 1754 and promptly published in the 1755 *Philosophical Transactions*, Barker initially constructed a series of twelve tables that gave the apparent path of the comet with each table supposing the perihelion to occur in a different month of the year, and under those suppositions naming the points in the sky where the comet might then be expected first to appear. From the twelve separate tables he produced a further table (Figure 19.2), summarizing where in the heavens he believed the comet would first be visible for any month of the year. This table proved popular. It was subsequently printed anonymously in 1756 in issues of the *Gentleman's Magazine* and the *Scots Magazine* and in full or abbreviated form in at least two English almanacs for the year 1758.

Contemporary with Barker's effort, another attempt to construct a comet ephemeris was made in 1755 by Dirk Klinkenberg, a Dutch astronomer who had discovered comets in 1743 and 1748. He too constructed twelve tables charting the expected path of the comet, each table (Figure 19.3) assuming a different mid-month perihelial return date.

The next person to construct a comet ephemeris was T. Jamard, an Augustinian of the Abbaye Royale de Sainte-Geneviève in Paris. In a memoir that was favourably reviewed by the Academy of Sciences and published in July 1758, Jamard, like Barker and Klinkenberg, made a range of assumptions regarding the perihelial date. For each he gave such information as the times at and constellations in which the comet would be expected to appear and disappear, its expected route, its probable brightness, how long its tail might be, and the likelihood of discovery.

When a comet was indeed independently discovered by several observers in mid-September 1757 – the first confirmed comet to be seen in seven years – there was naturally speculation as to whether or not it was the one whose return was expected. In England that speculation was mixed with some apprehension among the public, thanks to publications in the preceding two years by the Methodist evangelist John Wesley (1703–91) and the science popularizer Benjamin Martin (1705–82). In November 1755, shortly after a devastating earthquake in Lisbon, Wesley wrote a sermon, later published in London, Bristol, Newcastle, and Dublin, that warned of a forthcoming comet, supposedly on a collision course with Earth, that would cause much havoc unless the populace admitted a belief in the existence of God.

Martin was less inflammatory, but he did suggest in the margins of a popular 1757 broadside (Figure 19.4) that if the expected comet passed through its descending node on 12 May, "we would be in a dangerous Situation, as the denser Part of its blazing Tail would then envelope the Earth, which God forbid". The fear engendered by Wesley and Martin seems to have peaked on 7 October 1757, when, according to a newspaper report, an "uncommon darkness [possibly some freakish weather] occasioned terrible Apprehensions in many People from the Effects the present Comet will have upon the Earth, they being enthusiastically persuaded, that the General Conflagration of the World is near at Hand".

The speculation that the comet appearing in 1757 was the one expected did at first have some

A TABLE shewing where the Comet may be expected to begin to appear any Month

Month		Position	Lat.	Time
January	end	Scarce to be seen Retr. between 30° & 15° ♐	Small increasing S.	7 Weeks after Perihelion
February	begin	15 ♑	Small N. or S.	a Month after Perihelion
February	end	30 & ♑	Small N. decreasing	
March	begin	30 & 0 ♒	Small N. decreasing	
March	end	15 & 0 ♓	Small N.	2 or 3 Weeks after
April	begin	Stat. 10 ♈ & 20 ♓		about Perihelion
April	end	middle ♈	N.	1, 2, or 3 Weeks
May	begin	Dir. begin. ♉		
May	end	begin. ♉		
June	begin	end ♉	N. increasing	2 to 5 Weeks before
June	end	begin. ♊		
July	begin	middle ♊		
July	end	end ♊		
August	begin	Stat. 25 & 30 ♊	Small increasing N.	5 to 8 Weeks before
September		Retr. end ♊	Small S. or N.	2 Months before Perihelion
October		begin. ♊		2 or 3 Months
Novem.	begin	5 ♊ & 20 ♉	Small S.	3 Months before Perihelion
Novem.	mid.	begin. ♉		
Novem.	end.	begin. ♉ end. ♈		
Decem.	begin	begin. ♈	Small S. or N.	11 to 14 Weeks
Decem.	end		very faint	

19.2. Thomas Barker's table in *Philosophical Transactions* for 1755, showing where he expected the returning comet might first become visible for any month of the year.

Als de Comeet den 14den February in het *Perihelium* komt.

Op de Plaat den *Weg* Bb.

Dagen	Langte gr. min.		Breedte gr. min.		Distantie ☽ ☉	Distant. à ☉
Octob. 27 in ♊	20	0	4	35 Z	1 24	2 06
Nov. 6 ——	10	10	3	30	1 00	1 92
—— 16 in ♉	23	45	1	50	0 79	1 78
—— 21 ——	13	0	0	50	0 73	1 71
—— 26 ——	1	50	0	8 N	0 71	1 63
Dec. 1 in ♈	20	0	1	35	0 72	1 57
—— 6 ——	9	20	2	45	0 77	1 48
—— 11	1	0	4	5	0 84	1 41
—— 16 in ♓	23	30	4	25	0 92	1 33
—— 26 ——	13	25	5	35	1 09	1 17
Januar. 5 ——	7	0	6	15	1 27	1 02
—— 15 ——	1	30	6	35	1 42	0 87
—— 30 in ♒	24	30	6	55	1 57	0 67
Febr. 19 ——	14	40	6	0	1 46	0 59
Maart 6 ——	7	5	3	15	1 15	0 71
—— 21 in ♑	27	30	2	0 Z	0 75	0 94
—— 31 ——	11	40	11	5	0 49	1 10
April 5 in ♐	26	25	18	35	0 39	1 17
—— 10 in ♏	29	15	27	20	0 33	1 25
—— 15 in ♎	26	45	30	20	0 37	1 33
—— 20 ——	5	20	27	10	0 47	1 41
—— 25 in ♍	23	15	24	5	0 60	1 48
Mey 5 ——	12	25	19	25	0 92	1 63
—— 15 ——	8	0	16	25	1 25	1 78
—— 25 ——	6	45	14	20	1 75	1 99

19.3. Dirk Klinkenberg's table of expected positions of the comet, assuming a mid-February perihelion date. This is one of a dozen tables drawn up by Klinkenberg for hypothetical perihelion dates spread evenly throughout the year.

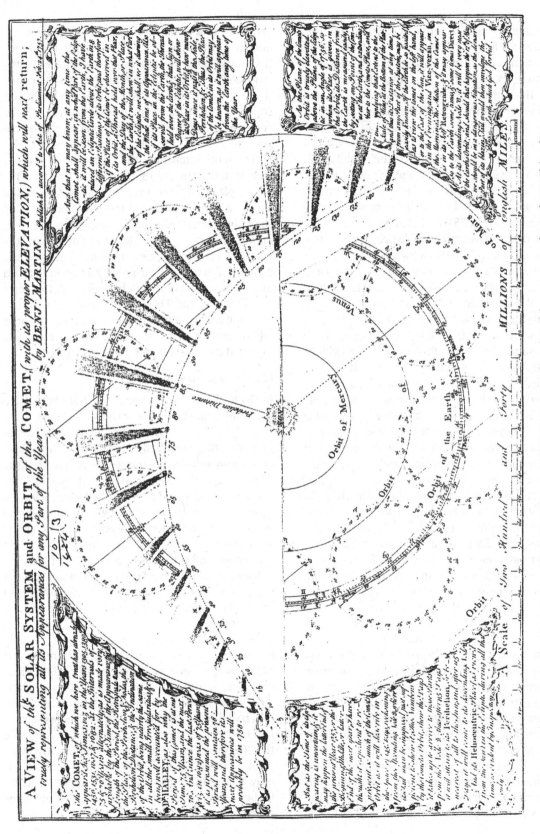

19.4. Benjamin Martin's broadside of 24 February 1757 illustrating the orbit of the comet overlying the Earth's orbit.

supporting evidence. Lalande later noted that this comet, when first seen, had a position conforming to the orbit of the comet of 1682 if one assumed it was then 43 days from reaching its perihelion point. Its subsequent course, however, indicated that it was moving in an entirely different orbit.

The appearance of the comet of 1757 did nevertheless renew public interest in the expected comet and probably inspired the publication shortly thereafter of two new ephemerides. One of these, a twelve-part anonymously written "Account of the COMET's Orbit", appeared serially between October 1757 and October 1758 in the *Boston-Gazette* newspaper, published in the British colony of Massachusetts in North America. For each month, the positions of the ascending and descending nodes, as well as of an intermediate point of the comet's orbit, were given. A circle passing through these three points, the author of the ephemeris noted, "would be, near enough for the present Purpose, the Projection of the Orbit in the Heavens". The author clearly had a good knowledge of astronomy – he explained why he believed that February and September would be, respectively, the most difficult and most favourable times of the year to discover the comet – and was most likely John Winthrop, the Hollis professor of mathematics and natural philosophy at Harvard University and a later observer of the comet appearing in 1759 that was confirmed as Halley's.

Perhaps to give hope to those who were disappointed that the comet of 1757 was not the expected one, Lalande, in the November 1757 issue of the *Mémoires pour l'Histoire des Sciences & Beaux-Arts*, pointed out that in November the Earth would be at the most favourable position for observers to detect the expected comet. It was during this month that the Earth approached the comet's orbit most closely, and if the comet was then in the part of its orbit near the Earth, observers ought easily to see it. The conditions for recovering the comet would become less favourable in the subsequent few months, Lalande explained, both because the distance between the orbits would increase and because around Paris rain often occurred in these months.

An essential factor in knowing where to look for the comet, according to Lalande, was a determination of how far from the Earth the comet could be detected. He pointed out that in 1531 it was

perceived at a distance of $\frac{1}{2}$ AU; in 1607, at $\frac{1}{3}$ AU; and in 1682, again at $\frac{1}{2}$ AU. Lalande concluded that those who had the best view and were the most alert ought to be able to perceive the comet when it came within $\frac{2}{5}$ AU of the Earth. On this supposition he then gave the most likely places the comet would be located if it was first seen, respectively, on 1, 5, 10, 15, 20, and 25 November 1757.

As events happened, the various comet ephemerides that were published in the latter half of the 1750s played no direct role in the recovery of the expected comet. Their value to astronomers and the public, however, was not negligible. By considering the many variables that influenced the potential visibility of the comet, the makers of the ephemerides were able to present a reasonable idea as to where the comet might be first spotted for a given time of year. This service prevented needless, and potentially discouraging, searches at times when, and in areas of the heavens where, the expected comet could not possibly be seen.

Clairaut's perturbational analysis

In his article of November 1757 Lalande recalled that Halley had based his prediction of a late-1758 or early-1759 return to perihelion on the argument that the comet had been sensibly delayed by a close pre-perihelion encounter with Jupiter in June 1681. (Jupiter, Halley had argued, would have accelerated the comet, and so lifted it into a higher orbit with a longer period.) Halley had made no mention of a subsequent post-perihelion encounter with Jupiter, at about the same distance, in November 1683, the result of which, in Lalande's opinion, would have tended to cancel the effect of the pre-perihelion encounter. Lalande noted also that Halley had not remarked on a close encounter of the comet with Venus on 22 September 1682 at a distance of $\frac{1}{7}$ AU. He concluded that nothing definitive could be said about the complicated effects on the comet's motion caused by these close encounters until a rigorous perturbational analysis was performed – an analysis, he announced, that his fellow academician Alexis-Claude Clairaut had undertaken, beginning in June 1757, despite the "frightful" length of the calculations involved. In that calculational effort, Clairaut was aided not only by Lalande but also by Mme Nicole-Reine Étable de Labrière Lepaute, whom Clairaut later called *la savante calculatrice*.

In late autumn of 1758, although his calculation was still incomplete, Clairaut decided he must make an announcement so as not to be forestalled by the comet. At the public meeting of the Academy on 14 November, he gave his preliminary results. The action of Jupiter was such that the period from 1607 to 1682 would be 432 days shorter than the period from 1531 to 1607. Without having fully calculated Saturn's action during the same two periods, Clairaut believed that this action would increase the difference by at least 4 days, making the second period 436 days shorter than the first. The difference as determined by observation, he reported, was 469 days, leaving a discrepancy of 33 days. (As was later pointed out, Clairaut here failed to take into account the shift from Julian to Gregorian calendar, and so the observational difference was 459 days and the discrepancy between observation and calculation 23 days.) The near agreement, Clairaut urged, was as striking a confirmation of the Newtonian system as any of those previously given.

As for the effect of perturbation on the periods from 1607 to 1682 and from 1682 to 1759, Clairaut found that Jupiter's action would lengthen the second of these periods so as to make it 518 days longer than the first. His calculation of Saturn's action was still incomplete, but he believed this would cause the second period to be 100 days longer than the first. The period from 1682 to 1759 would thus be 618 days longer than the preceding period, and the perihelion in 1759 should occur in mid-April. On the basis of the calculation of the perihelion of 1682, Clairaut added: give or take a month.

A brief account of Clairaut's procedures, of the completion and later refinement of his calculation, and of the controversy it gave rise to among his contemporaries, is provided in the appendix to this chapter.

The recovery of the comet

At about the same time that Clairaut began his perturbational analysis of the comet's motion, his academic colleague Joseph-Nicolas Delisle (1688–1768) initiated his own approach to finding the expected comet at the earliest possible moment. Rather than predicting the entire expected course of the comet under various hypotheses of perihelial date, as earlier ephemeris-makers had done, this veteran astronomer limited his calculations to the moment when it ought to become first visible, "because having once found it, one would be able to follow it by observations and calculation during the remainder of the apparition". First visibility, coming earlier for telescopic and later for naked-eye observation, would occur when sunlight reflected from the comet was strong enough to make the latter perceptible from Earth, with the aid of the instrument used.

In order to make the best estimate of how soon before perihelion the comet could be seen, Delisle examined the observations of its apparitions in 1531, 1607, and 1682. Unfortunately, for the 1531 and the 1607 apparitions, no information was available either on the size and figure of the comet at the time of discovery or on how it was first perceived to be a comet. Delisle did learn from Apian's *Astronomicum Caesareum* (1540), however, that the comet of 1531 had been seen 18 days before perihelion and that only six days later it had a tail longer than 15°. The comet of 1607 had been first seen 33 days before perihelion. Three days later, according to Johannes Hevelius's *Cometographia* (1668), it had an exceedingly short tail, its head was not quite round, and though brighter than stars of the first magnitude, it had a pale and weak colour. As for the comet of 1682, Delisle was aware that it had first appeared, whitish and without a tail, 24 days before reaching perihelion. On the basis of this information, he conjectured that at each apparition the comet might have been seen telescopically about a month before perihelion if a search had been undertaken in the place where it was at that time. (The 1531 and 1607 apparitions of course preceded the invention of the telescope.)

On the basis of the Earth's continually changing position throughout the year and assuming a fixed orbit for the comet, Delisle then calculated at 10-day intervals where the comet would be on the supposition that it would be first seen either 35 or 25 days before reaching its perihelion point. He gathered these data points in a table (Figure 19.5) and also plotted them on a celestial map (Figure 19.6). When the points were connected on the latter, they formed two ovals. It was along arcs connecting identical days on these two ovals that Delisle instructed his assistant Charles Messier (1730–1817) to look for the comet.

	Pour trente - cinq jours.		Pour vingt - cinq jours.	
	LONGITUDE.	LATITUDE boréale.	LONGITUDE.	LATITUDE boréale.
Novembre.. 1	♄ 15ᵈ 45'	24ᵈ 14'	⇶ 28ᵈ 5'	17ᵈ 23'
10	23. 50	17. 31	♄ 7. 25	14. 27
20	♒ 1. 5	13. 27	15. 20	12. 13
Décembre... 1	8. 5	10. 34	23. 50	10. 21
10	13. 25	9. 9	♒ 0. 5	9. 18
20	19. 25	7. 55	7. 5	8. 23
Janvier..... 1	26. 5	6. 56	14. 35	7. 34
10	♓ 0. 50	6. 21	20. 15	7. 7
20	6. 20	5. 52	26. 20	6. 40
Février.... 1	12. 55	5. 26	♓ 3. 25	6. 17
10	✗ 17ᵈ 45'	5ᵈ 10'	✗ 8ᵈ 35'	6ᵈ 4'
20	23. 5	4. 57	14. 25	5. 53
Mars...... 1	27. 35	4. 48	19. 35	5. 47
10	♈ 2. 15	4. 41	24. 25	5. 42
20	7. 30	4. 36	♈ 0. 20	5. 39
Avril..... 1	13. 50	4. 34	7. 25	5. 39
10	18. 5	4. 32	12. 25	5. 41
20	23. 15	4. 34	18. 5	5. 47
Mai...... 1	28. 55	4. 36	24. 5	5. 55
10	♉ 3. 50	4. 42	28. 55	6. 6
20	8. 25	4. 51	♉ 4. 50	6. 20
Juin...... 1	14. 27	5. 3	11. 35	6. 41
10	18. 50	5. 14	16. 35	7. 2
20	24. 5	5. 32	22. 25	7. 30
Juillet..... 1	29. 25	5. 57	28. 50	8. 12
10	♊ 3. 55	6. 21	♊ 4. 25	8. 54
20	9. 5	6. 56	10. 50	9. 54
Août..... 1	15. 5	7. 53	19. 25	11. 31
10	20. 25	8. 47	26. 5	13. 9
20	26. 5	10. 12	♋ 4. 50	15. 54
Septembre.. 1	♋ 2. 35	12. 52	18. 35	20. 39
10	8. 25	15. 58	♌ 3. 35	25. 58
20	16. 25	22. 28	♍ 0. 25	33. 59
Octobre... 1	♌ 1. 45	37. 25	♎ 16. 5	37. 19
10	♍ 14. 35	62. 24	♏ 18. 25	31. 36
20	⇶ 25. 5	44. 38	⇶ 11. 35	23. 40

19.5. Joseph Delisle's table of positions of the comet calculated at ten-day intervals, based on the assumptions that it would be seen either thirty-five days (first pair of columns) or twenty-five days (second pair of columns) before perihelion.

After a search lasting more than a year, Messier, guided to some degree by Delisle's map, spotted an object on 21 January 1759 that he and Delisle, during the course of a series of observations that continued through 14 February, began to suspect was the expected comet. Although he claimed that this object was approximately at the position indicated on Delisle's map for a first-appearance of 21 January, Messier clearly extended his search along the arc for this date beyond the confines of the oval. He in fact later speculated that if Delisle had taken looser limits on how soon the comet might be seen

with a telescope before perihelion, he (Messier) might have detected it much earlier. Apparently because he did not wish to make a public announcement regarding the discovery of this object and its probable identification with the expected comet until he knew more about its orbit, Delisle forbade Messier to inform other Parisian astronomers of its existence until after it had rounded the Sun and been recovered again. Delisle apparently felt that the identification of this comet with those seen in 1682, 1607, 1531, and earlier could be definitively established only after it was recovered, in its post-perihelion phase, at a position that clearly indicated its orbit was similar to those determined for its posited earlier apparitions.

Unbeknownst to Delisle and Messier, the same object had already been discovered, 27 days before Messier's first observation, by Johann Georg Palitzsch (1723–88), a German amateur astronomer and prosperous farmer living in Prohlis, a small town near Dresden in Saxony. Whether he made use of any ephemeris to facilitate his discovery remains unclear, but he was certainly aware that a comet was expected. In an account of the discovery published in a Dresden newspaper by his friend Christian Gotthold Hoffman, Palitzsch remarked:

Following my fatiguing habit of wanting to observe as much as possible all that occurred in Nature, and especially the remarkable celestial events, I examined the stars on the 25th of December about six o'clock in the evening with my 8-foot Telescope. The constellation of the Whale [Cetus] presented itself well, and it was also the epoch when the announced Comet ought to approach and show itself. Thus came to pass for me the indescribable pleasure of discovering in the Fish [Pisces], not far from the marvellous star of the Whale [the variable star Mira Ceti], a nebulous star never perceived there before. To tell the truth, it is found between the two stars marked ε and δ on Bayer's Uranometria, or O and N on Doppelmayer's map.

The same observation, renewed on the 26th and the 27th, confirmed my supposition that it was a comet; because from the 25th to the 27th, it had effectively moved from the star O toward the star N.

Palitzsch did not specifically identify this newly discovered comet as the one expected. Hoffman, who observed the object on 27 December, in fact suggested that it might be the same as the one

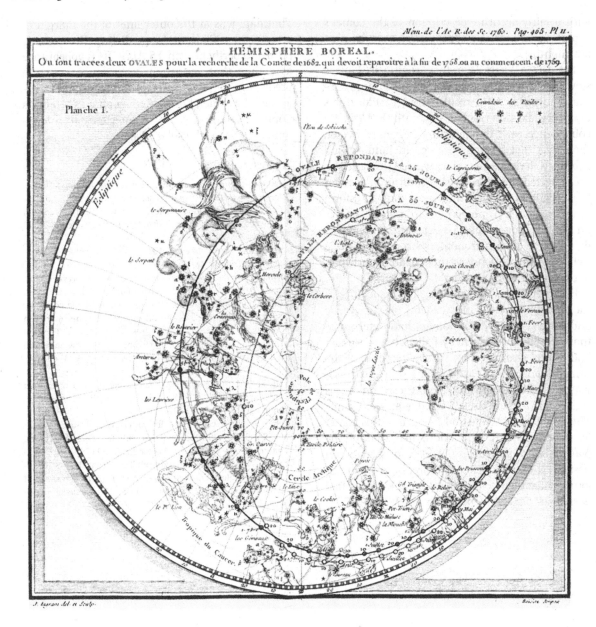

19.6. Delisle's celestial map on which separate lines connecting the thirty-five day and the twenty-five day data points form two oval curves. Charles Messier looked for the comet along arcs connecting identical days on these two ovals.

observed by Thaddeus Hagecius in 1580.

The first person to claim publicly that the object discovered by Palitzsch was the expected comet was the anonymous author of a pamphlet published in Leipzig on 24 January 1759. The author (most likely Gottfried Heinsius, a professor of mathematics at the University of Leipzig), upon learning of the observations of Palitzsch and Hoffman,

immediately suspected that the object was the expected comet. After performing a theoretical analysis (probably based on the orbit of the 1682 comet) of where the comet would be likely to appear, he spotted the object himself on 18 and 19 January at positions predicted by his theory. He reported these new observations, as well as those of Paltizsch and Hoffman, in his pamphlet, together with a fairly accurate description of the comet's expected positions and appearances over the next few months.

The Seven Years War then raging in Europe slowed communications between the countries of that continent. As a result, word of Palitzsch's discovery did not reach Paris until 1 April. By coincidence, early that morning Messier recovered the object that he had observed in January and February at a location that convinced Delisle it was the expected comet. With this confirmation he at last permitted Messier, that same day, to inform other Parisian astronomers regarding the appearance of the long-awaited object. Quite understandably, these other astronomers strongly criticized Delisle for having deprived them of an opportunity to observe the comet earlier.

By 25 April, when Lalande gave his lecture at the public assembly of the Academy of Sciences, he and other Parisian astronomers had determined that the object discovered by Palitzsch and Messier had orbital elements very similar to those of the comets seen in 1531, 1607, and 1682, and thus it was virtually certain that all of these comets, as Halley had hypothesized, were one and the same. They interpreted the periodic return of this comet as indicating that its motion, like that of the planets, was affected by the force of gravitation posited by Newton. That interpretation was reinforced after they determined that the comet had passed its perihelion point on 13 March 1759.

That date was at the outer limit of the margin of error set by Clairaut in predicting the comet's return to perihelion on 15 April. Its reappearance at a time that seemed in accordance with the predictions of Newton, Halley, and Clairaut, transformed the comet, known forever after as "Halley's Comet", from a mysterious interloper to a long-time member of the solar system.

Further reading

Peter Broughton, The first predicted return of Comet Halley, *Journal for the History of Astronomy*, vol. 16 (1985), 123–33

D. W. Hughes, Edmond Halley: why was he interested in comets?, *Journal of the British Interplanetary Society*, vol. 37 (1984), 32–44

S. P. Rigaud, *Some Account of Halley's Astronomiae Cometicae Synopsis* (Oxford, 1835)

Norman J. W. Thrower (ed.), *Standing on the Shoulders of Giants: A Longer View of Newton and Halley* (Berkeley, 1990), especially Halley, Delisle, and the making of the comet by Simon Schaffer (pp. 254–98); Edmond Halley: his interest in comets by David W. Hughes (pp. 324–72); and The first International Halley Watch: guiding the worldwide search for Comet Halley, 1755–1759 by Craig B. Waff (pp. 373–411)

Craig B. Waff, Comet Halley's first expected return: English apprehensions, 1755–58, *Journal for the History of Astronomy*, vol. 17 (1986), 1–37

Craig B. Waff, Tales from the first International Halley Watch (1755–59): 1. Boston waits for and watches the comet, *International Halley Watch Newsletter*, no. 9 (1986), 16–26

Craig B. Waff and Stephen Skinner, Tales from the first International Halley Watch (1755–59): 2. Thomas Stevenson of Barbados and his two-comet theory, *International Halley Watch Newsletter*, no. 10 (1987), 3–9

Appendix: Clairaut's calculation of the comet's return

CURTIS WILSON

Alexis-Claude Clairaut's calculation of a perihelion date for the eighteenth-century return of Halley's Comet was the first large-scale numerical integration ever performed. For the sheer magnitude of the undertaking and the accuracy of the final result, it has been much praised. The calculation itself has received little attention; for a description of it we must turn to Clairaut's own *Théorie du mouvement des comètes* of 1760.

Initially Clairaut believed that it would be necessary only to compute the perturbations of the comet on its near approaches to Jupiter. This hope soon evaporated. Very slight changes in the comet's speed near perihelion produced large changes in the comet's period. Even when the comet was far from both Jupiter and the Sun, Jupiter's action was changing the Sun's place, and as a consequence influencing the time required for the comet to return to perihelion. Moreover, Saturn with about a third of Jupiter's mass could not be ignored. The perturbations induced by both planets must be calculated for entire periods, to determine how they were affecting the times of return.

The main basis of the calculation would be Clairaut's solution of the three-body problem, which he had applied earlier in determining the perturbations of the Moon and the Earth (see Chapters 16, 18, and 20). It consisted of two fundamental equations, one for the radius vector in a perturbed elliptical orbit and the other for the time in the same orbit. The right-hand member of each equation was in two parts, one valid for the unperturbed ellipse, the other giving the added effect of the perturbations. Clairaut used the second part of the second equation in determining the perturbationally induced changes in the times of return of Halley's Comet.

Thus, given the known dates of the perihelia in 1607 and 1682, Clairaut could compute the per-turbational increases or decreases in time for the period 1607–82 and again for the following period, and take the difference; the latter, added to the known length of the period 1607–82, would give the length of the following period and hence the date of perihelion toward the end of the 1750s.

To gauge the accuracy of this computation, Clairaut decided to calculate in an entirely parallel way the date of perihelion in 1682, starting from the dates of the perihelia in 1531 and 1607. The difference between the calculated and actual dates of the perihelion in 1682 would give a measure of the accuracy to be expected in the computation of the perihelion date in the 1750s.

The expression giving the perturbational alteration in the comet's period was an integral; its integrand involved the perturbing forces. To evaluate these forces, it was necessary to know, for any given moment, the positions of both the perturbing planet and the comet. Yet the position of the comet was subject to perturbation. The logical circle was unavoidable: to compute perturbational effects, it was necessary to start from an approximate orbit and motion for the comet.

For this approximate orbit and motion, Clairaut used a single reference ellipse throughout his calculation, namely the ellipse that Halley had determined by assuming a mean period of 75.5 years. Halley had found it to have a semimajor axis of 17.86 AU and an eccentricity of 0.96739. It was this Halleian ellipse, in the slightly altered orientations that Halley found for it in the three successive perihelia of 1531, 1607, and 1682, that Clairaut used when he undertook to approximate the position of the comet in evaluating the integrand of the integral that expressed the perturbationally induced changes in the comet's period.

But Clairaut's integral when applied to the comet could not, in general, be integrated formally.

To be thus integrable, the integrand would have had to be expressed as a function of a single variable, say the eccentric anomaly of the comet. In the case of planets perturbed by planets it had been possible to express the integrand in this way, by using trigonometric series to approximate its value. The goodness of the approximation depended on the rapidity of convergence of the series, which in turn depended on the nearness with which the orbits approximated circles. Such a procedure was inapplicable to the comet because of the large eccentricity of its orbit. Clairaut therefore resorted to 'mechanical quadratures', now called numerical integration. R. J. Bošković, in an essay for the Paris prize contest of 1752, had proposed using mechanical quadratures to compute planetary perturbations (see Chapter 20), but Clairaut was the first actually to launch a large-scale numerical integration.

It was a daunting task. At intervals of 2° of eccentric anomaly, starting at perihelion, it was necessary to compute the distance between perturbing planet and comet – the latter being taken as unperturbed in the orbit Edmond Halley had assigned to it – and then to determine the resulting inverse-square force, with its radial and transverse components. The successive values of the integrand could then be calculated. After that the quadratures could be carried out, essentially by adding up the equally spaced ordinates, but with some use of a correction formula, that had been derived independently by James Gregory and Isaac Newton, to take account of the curvature of the curve atop the ordinates.

With a view to avoiding numerical integration where possible, Clairaut employed two additional processes. When the comet was in the superior half of its orbit, and so farthest from the Sun and from Jupiter and Saturn (the only perturbing planets considered), the main perturbative effect was the action of the perturbing planet on the Sun; and this action could be determined geometrically, to a good approximation, by supposing the comet to be moving in an ellipse about the common centre of gravity of the Sun and the perturbing planet. In some circumstances numerical integration could be avoided by using a 'parabolic line' (a power series in the eccentric anomaly) to approximate the perturbing forces. Clairaut sought shortcuts because he was short of time, and feared the comet

would return before the calculation had been completed (it in fact did so).

Given the pressure of time and the magnitude of the task, Clairaut needed collaborators. The astronomer Joseph-Jérôme Lefrançais de Lalande offered to calculate the distances between the comet and the perturbing planets, the resulting forces, their components, and the products into which they entered. Clairaut was to check all these results for consistency (mainly by constructing graphs to determine whether the ordinates varied in a smooth fashion), and then perform the quadratures. Lalande was assisted by Mme Nicole-Reine Étable de Labrière Lepaute, wife of a noted clockmaker. Clairaut wrote an appreciation of her assistance for inclusion in his initial report, but a jealous ladyfriend insisted that it be suppressed, Lalande commenting years afterward that Clairaut was a *savant judicieux mais faible.*

In his reminiscences Lalande also stated that for six months the three of them calculated from morning till night, sometimes even at table, and from this strenuous work he acquired a malady that changed his temperament for the rest of his life.

For the calculations Clairaut divided the cometary orbit into three parts: from 0° to 90° of eccentric anomaly, the first quadrant of the ellipse from perihelion, requiring more than 7 years to be traversed; from 90° to 270° of eccentric anomaly, the superior half of the orbit, taking more than 60 years to be traversed; and from 270° to 360°, the final quadrant of the ellipse, taking once again about 7 years. The finer subdivision for the quadratures proceeded generally in 2° steps, but in the close approaches to Jupiter before and after the perihelion of 1682 Clairaut used 1° steps.

In calculating the Saturnian perturbations, Clairaut separated the action of Saturn on the Sun from its action on the comet throughout the cometary orbit, and left out of account the eccentricity of Saturn's orbit and its inclination to the cometary orbit through all three revolutions of the comet, except for a short interval near the perihelion of 1682, when the comet made a near approach to Saturn.

As reported in the body of this chapter, Clairaut announced preliminary results, before his calculation was complete, at the public meeting of the Academy of Sciences on 14 November 1758. His prediction was for a perihelion in mid-April 1759, give or take a month. After the recovery of the comet on 31 March, Lalande, N.-L. de Lacaille, and others fitted a parabolic orbit to observations from before and after the conjunction with the Sun, and found that the perihelion had occurred on 13 March. Clairaut's prediction was therefore a month off, at the limit of the range of error he had allowed for it.

A controversy now erupted within the Paris Academy of Sciences as to the accuracy of this prediction. The dispute was bitter, and during the summer led to a spate of letters in the various journals. Its chief instigators, the astronomer Pierre-Charles Le Monnier (1715–99) and the mathematician Jean le Rond d'Alembert (1717–83), maintained that Clairaut's error was large, while the partisans of Clairaut maintained that it was small. Clairaut's followers wanted to compare the error with whole periods of the comet, 75 or 151 years; Le Monnier and d'Alembert insisted that it be compared with the *difference* between two successive periods, since this alone, among the results Clairaut had derived purely from theory, admitted of empirical verification.

In the case of the periods from 1607 to 1682 and from 1682 to 1759, the difference in length had proved to be 585 days, and so 33 days would constitute an error of $(33/585) \times 100\% = 5.6\%$. Such an estimate of error might have been acceptable to Clairaut, but d'Alembert was bent on arguing that in some of Clairaut's calculations the error had been much larger. We shall not enter into the dubious paths d'Alembert followed to extract such a conclusion; the only honest way of checking Clairaut's calculations would have been to repeat them. D'Alembert's posture in this controversy arose from jealousy and a contempt for "number-crunching". He called Clairaut's calculation "more laborious than deep", and he had fundamental doubts about the capacity of theory (whether in Newtonian gravitation, hydrodynamics, or the physics of the vibrating string) to apply with precise accuracy to the world.

Clairaut's attitude was very different, and he constantly sought to improve the precision of predictions from Newtonian theory. Following his preliminary report of November 1758, he completed and revised the cometary calculation for his *Théorie du mouvement des comètes*, which was sub-

mitted to the Academy in August 1759 and published the following April. In 1761, in order to compete in the contest sponsored by the St Petersburg Academy of Science for 1762, he further refined his calculation; he won one-half of the prize (the other half was won by Johann Albrecht Euler, with an essay on the near-approach of Halley's Comet to the Earth in 1760), and his prize essay was published in 1762. In these two revisions Clairaut claimed to reduce the error of his prediction, first to 23 days, and then to 19 days. From a careful summing up of the component numbers entering into the two sums, they are found to be 26.1 and 23.4 days, respectively.

To evaluate the error of the component computations, it would be necessary to re-do them in an entirely parallel way; no such computation has been or is likely to be made. An approximate evaluation can be attempted, however, on the basis of an integration by computer completed by Tao Kiang in 1971; the general result may be summarized as follows (the evaluations of error are to be taken as accurate only to within a day or two).

In Clairaut's calculation as a whole, there were ten main parts, two pertaining to the period 1531–1607, and four to each of the periods 1607–82 and 1682–1759. In five of the parts Clairaut's errors did not exceed 4 days each; in two of the Saturnian parts the discrepancies were much larger, but here what counted was the difference between the two results, and this was in error by only 4 days. Another of the Saturnian parts was off by 8 days. Finally, in two of the Jovian parts of the calculation for the period 1682–1759, the errors were -15 and $+33$ days, fortunately opposite in sign. The goodness of Clairaut's final result, we have to conclude, rested in part on a fortuitous cancellation of opposing errors.

Clairaut was well aware of possible and probable sources of error in his calculation. He nevertheless believed his result to be a verification of the Newtonian system "as striking as any of those previously given". In his final response to d'Alembert (in the *Journal des sçavans* for June 1762) he dismissed the question of a precise measurement of the error of his calculations as unimportant. The important point, he insisted, was that he had silenced the critics of the Newtonian theory of comets. What mattered, it seems, was the general "feel" of the verification, the sense that it augured well for the future of the calculation of cometary perturbations. And since d'Alembert did not have the patience for the calculation of cometary perturbations, Clairaut urged him to leave in peace those who did.

Later developments have vindicated Clairaut's attitude, but not the complete sufficiency of the Newtonian theory to account for cometary returns. Perturbations due to planets other than Jupiter and Saturn caused the return in 1759 to be 8 days earlier than it would otherwise have been. In addition there was a non-gravitational acceleration. Clairaut had considered the possibility of an acceleration due to aethereal resistance, but had dismissed it, marshalling the evidence against it in the last part of his prize essay of 1762.

A second periodic comet was discovered by J.-L. Pons in 1818. It proved to have a period of 3.3 years, and when J. F. Encke computed the perturbations he found a non-gravitational decrease of about 2.5 hours per period, which he attributed to aethereal resistance. As a result celestial mechanicians computing the return of Halley's Comet in 1835 were looking for a similar effect; what they found, however, was a perihelial date later than that calculated from the perturbations, and thus an effect not explicable by resistance. In the 1980s, direct numerical integration from the differential equations has shown that the non-gravitational acceleration of Halley's Comet is constant in each return, and amounts to an increase of about 4.1 days per period. This (like the non-gravitational acceleration of Comet Encke) has been explained by the hypothesis that comets are rotating icy conglomerates. On their near approaches to the Sun they are strongly irradiated, with the result that there is a jet effect due to out-gassing; a time-delay between maximum reception of solar radiation and maximum out-gassing, given the angular rotation of the comet, accounts for the fact that the comet is either accelerated or decelerated in the direction of its orbital motion, and thus pushed into a higher or lower orbit.

Further reading

Alexis-Claude Clairaut, *Théorie du mouvement des comètes* (Paris, 1760)

Alexis-Claude Clairaut, *Recherches sur la comète des années 1531, 1607, et 1759* (St Petersburg, 1762)

T. Kiang, The past orbit of Halley's Comet, *Memoirs of the Royal Astronomical Society*, vol. 76 (1972), 27–66

C. Wilson, Clairaut's calculation of the eighteenth-century return of Halley's Comet, *Journal for the History of Astronomy*, vol. 24 (1993), 1–15

D. K. Yeomans, Investigating the motion of Comet Halley over several millennia, in *Comet Halley: Investigations, Results, Interpretations*, ed. J. W. Mason (New York, 1990) vol. 2, pp. 225–31

D. K. Yeomans, and T. Kiang, The long-term motion of comet Halley, *Monthly Notices of the Royal Astronomical Society*, vol. 197 (1981), 633–46

PART VI

Celestial mechanics during the eighteenth century

20

The problem of perturbation analytically treated: Euler, Clairaut, d'Alembert

CURTIS WILSON

In 1840 William Whewell in his *Philosophy of the Inductive Sciences* proclaimed that astronomy

is not only the queen of the sciences, but, in a stricter sense of the term, the only perfect science; – the only branch of human knowledge in which particulars are completely subjugated to generals, effects to causes ... and we have in this case an example of a science in that elevated state of flourishing maturity, in which all that remains is to determine with the extreme of accuracy the consequences of its rules by the profoundest combinations of mathematics, the magnitude of its data by the minutest scrupulousness of observation ....

This praise of astronomy's state in the 1830s sounds excessive today – the work of Henri Poincaré and Albert Einstein has shown it to be simplistic – but it harbours an historical truth. During the century preceding the appearance of Whewell's book, there had emerged by stages, in a series of innovations and refinements, an extensive body of analytical theory concerning the three-body problem – the problem of the perturbations produced in the motions of one body about another, when a third body is present and all three attract in accordance with Newton's law. With Laplace the honing of the theory reached such a point as to make possible the accurate prediction of lunar and most planetary positions to within a few arcseconds. In intricacy and sophistication, the mathematical machinery marshalled and brought to bear in this development surpassed that in any earlier mathematical theory applying to natural phenomena. *Astronomie physique*, or as Laplace rechristened it in 1796, *mécanique céleste*, was indeed, in the early decades of the nineteenth century, the undisputed queen of the sciences.

The theoretical development was almost exclusively the work of five mathematicians: Leonhard Euler (1707–83), Alexis-Claude Clairaut (1713–

65), Jean le Rond d'Alembert (1717–83), Joseph Louis Lagrange (1736–1813), and Pierre-Simon Laplace (1749–1827). Laplace's achievement would depend crucially on Lagrange's accomplishment; Lagrange's accomplishment would depend in turn on earlier advances made by Euler, Clairaut, and d'Alembert, and these are the subject of the present chapter.

Why, it may be asked, did the development of analytical celestial mechanics not get under way till the 1740s? For it would appear that, much earlier, there was motive enough for the development of such a theory. A means of accurately determining longitude at sea was urgently needed; the British Parliament in 1714 had instituted a handsome prize for the invention of such a procedure – £20 000 if it gave the longitude to within $\frac{1}{2}$°, half as much if to within 1°. A lunar theory accurate to 2' of arc would suffice for the second prize.

That no lunar theory of this accuracy was constructed during the first half of the eighteenth century must be put down to ignorance of how it was to be done. Isaac Newton, as we have seen (Chapter 13 in Volume 2A), for a time hoped to obtain a lunar theory good to 2' by adding on several inequalities to the Horrocksian theory, some of them (such as the annual equations in the motions of the apse and nodes) inferred from the gravitational law, others from observation alone. The result was disappointing, the errors being much higher than 2'. Edmond Halley then proposed comparing the theory with observations throughout a Saros cycle (a period of 18 years after which solar and lunar eclipses return in the same sequence), on the supposition that the discrepancies so found would be the same in each subsequent cycle. He carried out the comparison during his years as Astronomer Royal (1720–42), and the results were published posthumously in 1749. The attempt to employ them in correcting predictions drawn from the faulty Newtonian lunar theory (a cumbrous procedure!) appears to have been short-lived.

As for attempts to develop the lunar theory mathematically, in England these followed the Newtonian pattern, combining geometry with limiting ratios of infinitesimals (as in John Machin's propositions on the motions of the Moon's nodes, incorporated in the third edition of the *Principia*). How much might have been accomplished by such methods must remain a matter of speculation.

A limitation in the Newtonian procedures, as compared with the Leibnizian calculus which was being developed by mathematicians on the Continent (for instance Pierre Varignon, Jakob I and Johann I Bernoulli, Euler, Alexis Fontaine, Clairaut), was the absence of an explicit representation of the independent variable. Newton's "pricked letters", introduced in the 1690s to rival the Leibnizian fractional notation for the derivative, implied a single independent variable, usually time. The Leibnizian notation, by incorporating an explicit symbol for the independent variable, facilitated changing one such for another, the fractional form for the derivative suggesting the rules to be followed in this transformation. (Such shifts of variable would have been possible in the Newtonian notation, but required a little more doing.) Changes of variable were used from the beginning of the analytic development of perturbation theory, for instance in eliminating time as a variable from the differential equations so as to obtain an equation for the orbit in polar coordinates.

Yet, whatever the advantages of the Leibnizian calculus, it did not "lead one by the hand" to any obvious procedure for treating the three-body problem. As we have seen (Chapter 14), when G. W. Leibniz proposed the problem to Varignon, the latter succeeded only in proposing a way of coping with perturbations caused by a *fixed* centre of force.

The Continental mathematicians were largely committed to aethereal theories of gravity, and their attempts to reconcile such theories with Newton's inverse-square law no doubt absorbed time and effort that could have been expended on the perturbational problem. But the chief obstacle

20.1. Leonhard Euler.

to analytic development of perturbation theory appears to have been a technical one.

This theory as worked out by Euler, Clairaut, d'Alembert, Lagrange and Laplace depends on integration of trigonometrical functions. The development of the calculus of these functions was much slower than that of the other well-known transcendental functions, the logarithm and exponential. Trigonometrical tables were widely employed, but sines, cosines, and tangents tended to be regarded as lines in diagrams rather than as functions of an independent variable. Formulas for the derivatives of the sine, tangent, and secant were given by Roger Cotes in his *On the Estimation of Errors*, published posthumously in 1722, but this promising start was not followed up.

It was Euler (Figure 20.1) in 1739 who effectively introduced the calculus of trigonometrical functions to the community of Continental mathematicians, in papers in which he developed a general procedure for solving linear differential equations with constant coefficients. He later claimed this calculus to be his own invention, "in so far as the special signs and rules are [there] comprised". Already by the early 1740s, presumably from the reading of Euler's papers, the rules had become well-known to such practitioners of analysis as Daniel Bernoulli, Clairaut, and d'Alembert.

It was again Euler who, in the early 1740s, first attempted systematically to derive the lunar inequalities by means of the Leibnizian calculus. In his *Mechanica* of 1736, he had applied this calculus to many of the problems and theorems of Newton's *Principia*, but treatment of the three-body problem was conspicuous by its absence. By 1744 he had constructed lunar tables incorporating the inequalities caused in the Moon's motion by the Sun, as derived from the inverse-square law. What made the difference in the latter year, we hypothesize, was Euler's command of the calculus of trigonometrical functions.

Euler's lunar tables of 1744 were first published (probably in revised form) in his *Opuscula varii argumenti* of 1746, and then without an exposition of the calculations by which they were derived. In his preface Euler characterized these tables as follows.

I come now to my lunar tables. Their nature and basis of construction would require too much space to explain. I therefore only point out that they are derived from the theory of attraction which Newton with such happy success introduced into astronomy. Although it is claimed of several lunar tables that they are based on this theory, I dare to assert that the calculations to which this theory leads are so intricate, that such tables must be considered to differ greatly from the theory. Nor do I claim that I have included in these tables all the inequalities of motion which the theory implies. But I give all those equations which are detectable in observations and are above $\frac{1}{2}$ arc-minute. For I was able to carry the calculation to the point of identifying the arguments of the individual inequalities, and I determined the true quantity of many of the equations by the theory alone, but some I was forced to determine by observations.

Along with his lunar tables, Euler constructed solar tables, but these embodied only one perturbation deriving from gravitational attraction, namely the lunar inequality. Newton had mentioned this perturbation in Prop. 13 of Bk III of the *Principia* (second and third editions). It was the centre of gravity of the Earth and Moon, he stated, rather than the Earth itself, that moves in an ellipse about the Sun according to the area law. The rule thus enunciated is not quite exact on the assumption of the inverse-square law, but it is sufficiently so, and Euler followed it in his solar tables. He also employed Newton's erroneous value for the ratio of the masses of the Moon and Earth, namely 1:39.788, too large by a factor of more than 2, and so arrived at a lunar inequality in the Sun's apparent motion of about 15″ (the correct value is about 6″.5).

Besides the lunar perturbation, Euler incorporated in his solar tables the action of a non-gravitational force: the resistance of an aether. The existence of a subtle material filling the space between gross bodies was for Euler a matter of conviction. He regarded action-at-a-distance as contrary to reason, and in an essay of 1740, he asserted the only possible cause of gravitational attraction to be the centrifugal force developed in vortices. Daniel Bernoulli, writing to Euler in January 1742, objected that vortices in which the centrifugal force varied inversely as the square of the distance from the centre would be unstable. Euler bowed to this argument, and went on to accept an explanation of gravity that Bernoulli had proposed in his *Hydrodynamica* of 1738, according to which gravitational force is the result of variation in aethereal pressure, arising from differences in velocity. Euler's commitment to the aether was reinforced by his theory of light, for he believed that the transmission of light could be plausibly explained only by means of waves in an aether. The aether thus filled interplanetary space. As it did not move with the planets it must necessarily resist their motions.

In an essay of 1746, Euler derived the effects of such aethereal resistance. The planet's velocity, his differential equations showed, would decay exponentially; hence the orbit would gradually shrink and the period diminish. The line of apsides would remain stationary, but the orbital eccentricity would decrease. That astronomers had reported no evidence for such effects, Euler suggested, was due to the low density of the aether. In comparing ancient and modern solar observations, he believed he had found evidence that the Earth's motion was subject to the secular changes that aethereal resistance would cause: the orbital eccentricity was decreasing, and the tropical year diminishing, the latter at a rate of five seconds per century. In 1753 Tobias Mayer would persuade him that he was wrong, and that within the limits of observational error, the tropical year had remained constant.

In his solar tables of 1746, Euler took the solar apogee to be fixed with respect to the stars; the evidence that astronomers had previously adduced for assigning a secular motion to the apogee, he urged, was insufficiently firm. This meant that he was assuming perturbations of the Earth's motion by other planets to be negligible. If there were any motion of the apogee, he remarked, he would prefer to attribute it to the action of some comet passing close to the Earth. Thus at the time of drawing up his solar tables of 1746 Euler believed the Earth's motion to be determined almost exactly by the inverse-square force due to the Sun, the sole modifications being due to the lunar inequality and the resistance of the aether.

He was soon to encounter facts that would cause him to change his mind, and to conclude that perturbations of the planets by one another could not be assumed to be negligible. At the same time, he was entering a phase in which the exactitude of the inverse-square law of gravitation no longer appeared to him a reliable assumption.

The Paris Academy's contest of 1748

The topic for the Paris Academy's contest of 1748 was chosen at a meeting of the prize commission in March 1746. The astronomer Pierre Charles Le Monnier had presented evidence that Jupiter and Saturn were subject to observationally detectable inequalities, attributable, he believed, to the mutual attractions of the two planets. The prize, it was decided, should be offered for a "theory of Jupiter and Saturn explicating the inequalities that these planets appear to cause in each other's motions, especially about the time of their conjunction".

That the mutual action would be especially detectable around the time of conjunction was suggested by Newton's theory of gravitation. As Newton put it in Prop. 13 of Bk III of the *Principia* (second and third editions), from the action of Jupiter upon Saturn "... arises a perturbation of the orbit of Saturn in every conjunction of this planet with Jupiter, so sensible, that astronomers are puzzled with it". The error, Newton stated, could be largely avoided by placing the focus of Saturn's ellipse in the common centre of gravity of Jupiter and the Sun; the anomaly in the mean motion, he claimed, would not exceed 2' per year.

Astronomers had long been puzzled by the anomalies in the motions of Jupiter and Saturn, but these did not appear to be confined to the times of conjunction. In a letter to Michael Mästlin in 1625 Johannes Kepler had reported finding, by a comparison of ancient observations reported by Ptolemy and the sixteenth-century Tychonic observations with those of Bernhard Walther in the late fifteenth century, that "the mean motions are no longer mean", but subject to sizeable long-term variations. He assumed these inequalities to be periodic, but concluded that their periods and sizes could be determined only after good observations had been accumulated over a period of several centuries. In his *Rudolphine Tables*, he calculated the mean motions of the two planets on the basis of the Ptolemaic and Tychonic observations alone.

That the values so obtained were unsatisfactory was recognized by such seventeenth-century astronomers as Jeremiah Horrocks, Noël Durret, Maria Cunitia, and Ismael Boulliau. Jupiter was moving more rapidly and Saturn more slowly than Kepler's numbers implied. No mere correction of the mean motions, Horrocks found, would accommodate the ancient observations.

John Flamsteed struggled with the problem for decades. During the 1670s and 1680s he tried to solve it by correcting Kepler's numbers. For Jupiter he obtained a theory fitting the observations for a number of years, but then it ceased to be accurate. In the early 1690s Newton was proposing "to conciliate the orbits of Jupiter and Saturn to the heavens by a consideration of physical causes", and Flamsteed furnished him with observations for this purpose. The apsides of the two planets, Newton told him, were "agitated by a libratory motion". This suggests that Newton at this time was attempting to cope with the anomalies by following the analogy of Jeremiah Horrocks's lunar theory, with its oscillating apse and eccentricity.

By the late 1690s Newton had apparently abandoned his efforts, and Flamsteed now set out, with the aid of calculators, to solve it on his own. The anomalies, he believed, should be cyclical, with the period of 59 years in which Jupiter and Saturn return very nearly to the same configuration with respect to each other and the Sun. Repeated attempts to apply this or other periods, however, led nowhere. By 1717 Flamsteed had given up in despair. As he wrote to Abraham Sharp, one of his calculators:

... my hopes that there might be a restitution of the inequality in Saturn and Jupiter, after 2 revolutions of Saturn and 5 of Jupiter, are vanished; and the doctrine of gravitation and its effects are not as yet so perfectly understood as we imagined.

Astronomers in Paris – G. F. Maraldi (Maraldi I), Jacques Cassini, Le Monnier – encountered similar difficulties: no values for the mean motions would satisfy the observations of all ages, nor did the variations follow a discernible pattern.

In April 1746 Jacques Cassini (Cassini II, 1677–1756), who had previously been a partisan of planetary vortices, presented to the Academy a memoir aiming to show that a long-term acceleration of Jupiter and deceleration of Saturn could arise from the gravitational attractions of the two planets. The argument is qualitative, depending on the non-concentric configuration of the orbits which suggests an incomplete compensation of the accelerations and decelerations before and after conjunction. Cassini recognized that a full demonstration would require deriving the continuous action of the perturbative forces in all parts of the orbits. He assumed the anomaly to have a period of 84 000 years, the time for the two lines of apsides, advancing at their mean rates, to move out of and back into coincidence.

The possibility of secular variations in the mean motions of Jupiter and Saturn was not mentioned in the setting of the prize problem for the Paris contest of 1748. It was to assume more importance afterwards, as will become apparent in later chapters.

The contest of 1748 was significant because it engaged mathematicians trained in the Leibnizian calculus in the search for a way to deal with the problem of perturbation. In the essay that Daniel Bernoulli (1700–82) submitted (we possess only the draft), he remarks that, with the advent of Newton's system, astronomy has become inseparably connected with mechanics. Consequently

it is natural that what remains for us to discover cannot fail to be extremely difficult. Such above all is the question proposed by the illustrious Academy of Sciences for the prize of the year 1748. It is so to such an extent that the great Newton himself, foreseeing no doubt the terrible calculations that the solution demands, did not wish to undertake it except by way of some general reflections...,

The prize in 1748 went to Euler; Daniel Bernoulli's essay received the *proxime accessit* (runner-up). But the commission was not satisfied. Euler had failed to give the derivation of his differential equations (this he had done earlier in a memoir submitted to the Berlin Academy in 1747); his application of his solution to observations spanning the years from 1582 to 1745 left errors as high as 8' and 9'. Daniel Bernoulli had not even attempted observational comparisons. The commission therefore proposed the same topic for the contest of 1750; Euler was again one of the contestants, but no prize was awarded. Then once more, for the contest of 1752, the commission proposed the same problem. This time Euler was the winner, Rudjer J. Bošković (1711–87) received the *proxime accessit*, while Daniel Bernoulli's submission in this year went unmentioned. In none of the essays was there an attempt at an empirical test. Euler excused his failure to carry out such a test on the ground that the available observations were insufficiently accurate.

What influence did these various prize essays have on later developments? Bernoulli's were never published. Bošković caused his essay to be published in Rome in 1756, but neither he nor anyone else appears to have undertaken to apply its theory to planetary observations. (Certain features of both Bernoulli's and Bošković's essays may have influenced Clairaut in his later calculation of the return of Halley's Comet; see the appendix to Chapter 19.) The theories of Euler's two essays contained calculative errors and lacked important terms. On the other hand, they contained seminal ideas on which much of the further development of celestial mechanics would turn. The essay of 1748 was published in Paris in 1749, and on its basis Lagrange would make important discoveries. The essay of 1752 was not published until 1769, but then the disagreement of its results with those of Lagrange triggered Laplace's entry into the fray.

Advances in celestial mechanics during the late 1740s were not the work of Euler alone; beginning in 1746 Clairaut and d'Alembert had taken up the struggle with the three-body problem. Both were members of the prize commission. From the moment the topic for the prize of 1748 was agreed upon, each of them, unbeknownst to the other, set out to derive and solve differential equations representing the motion of a planet or satellite subject to perturbation. As members of the Paris Academy they were ineligible to enter the contest. Their motivation was explained a year later in a letter of the Marquise du Châtelet to François Jacquier: "M. Clairaut and M. d'Alembert are after the system of the world, understandably they do not wish to be forestalled by the essays for the prize".

By late 1746 both Clairaut and d'Alembert had formulated differential equations and a general procedure for their resolution. D'Alembert sent his formulation to the Berlin Academy; Clairaut deposited his in a sealed envelope with the Secretary of the Paris Academy. During 1747 there were more sealed envelopes, from d'Alembert as well as Clairaut. The race came out into the open in the summer, when both of them read memoirs on the subject to the Academy. These appeared in the Academy's *Mémoires* for 1745, published in 1749. More mature accounts of their respective procedures were given in Clairaut's *Théorie de la lune*, which won the prize of the St Petersburg Academy for 1750 and was published in 1752, and in Vol. 1 of d'Alembert's *Recherches sur différens points importans du système du monde*, which also deals with the lunar theory and was published in 1754.

In our account of these several works, it will become apparent that the lunar and planetary problems differ in an important respect. In the case of the Moon the distance of the perturbing body (the Sun) varies only slightly as the Moon circles the Earth, and the variations can be approximated accurately in an expression of a few terms. In the case of Saturn perturbed by Jupiter the distance between the two planets varies by a factor of more than 3 from opposition to conjunction, and because the orbits are not concentric, the force

exerted by Jupiter on Saturn at conjunction can vary from one part of Saturn's orbit to another by a factor of more than 1.6. As we shall see, it was Euler who proposed the artifice whereby these wild variations were finally to be coped with successfully.

Abandoned initiatives: Daniel Bernoulli and Bošković on the perturbations of Saturn and Jupiter

The difficulty of the problem of planetary perturbation can be seen in the laborious struggles of Daniel Bernoulli and Bošković to cope with it. The starting-point is necessarily a false supposition, since the perturbations of, say, Saturn by Jupiter depend on the changing distance between the two planets, hence on the perturbations. Bernoulli set out from two false suppositions: that the Sun remains at rest, and that the orbits the planets would follow if unperturbed would be concentric with the Sun. The first of these suppositions he defended as Keplerian, and compatible with Newton's system if the Sun's inertial mass were effectively infinite, and its gravitational mass of the magnitude Newton assigns to it. In his essay of 1752 he withdrew this supposition as unacceptable, excused it in the earlier essay as imposed by lack of time, and proceeded to compute the motions of the Sun about its common centre of gravity with Jupiter and Saturn, thus rectifying the earlier omission. The second supposition he admitted, already in his first essay, to be a source of inaccuracy, since the orbits, although departing only very slightly from circularity, are decidedly not concentric.

But now the mathematical difficulties began. Bernoulli derived a second-order differential equation for the departure (a), in the radial direction, of Saturn from its concentric circle. The equation was only approximately true under the initial assumptions. The independent variable was the heliocentric angle (σ) of elongation between the two planets, but the equation also contained the expressions $1/z$ and $1/z^3$, z being the linear distance between Jupiter and Saturn. To express these factors in terms of σ, Bernoulli first calculated the numerical values of $1/z$ and $1/z^3$ at $10°$ intervals starting with conjunction and going to opposition, then derived interpolation formulas for the intermediate values in terms of powers of σ for the six 30° intervals from conjunction to opposition. To extend the interpolation formulas over wider intervals, he found, would lead to unacceptable inaccuracies. He was thus led to six differential equations for α, which he solved by the method of undetermined coefficients, obtaining polynomial expressions for α in terms of powers of σ multiplied by numerical coefficients for each of the six intervals. Any error was cumulative, since the accuracy of the expressions for the later 30° intervals depended on that for the earlier intervals.

At 180° of elongation, it emerged, $d\alpha$ was not zero, for Saturn at this point was moving away from the Sun. The laborious calculations needed to be continued, since no sign of periodicity had emerged. "After all the fatigues I had surmounted and all the success I had had in my investigation," Bernoulli remarked, "I would rather have abandoned my enterprise than undertake these new calculations." Fortunately he thought of an artifice that enabled him to avoid the drudgery.

The values of α after 180° of elongation would be symmetrical with those before if the tangent were slightly turned at this point so as to make an equal angle on the other side of the radius vector. Such a change would in effect shift Saturn's circle into an eccentric position with respect to the Sun, and by subtracting out the changes in radial distance that such a shift entails, Bernoulli was able to determine the values of α for succeeding intervals. The accompanying diagram (Figure 20.2) shows his graph of α from first conjunction (C′) to fourth conjunction (C‴). The numbers in the graph, when divided by 211, represent fractions of the radius of the circle initially assumed for Saturn. The dotted lines have to do with the artifice just described. At the fourth conjunction, about 59 or 60 years after the first, both α and $d\alpha$ are nearly zero, and Bernoulli believed he had found a fundamental periodicity. "Nature", he averred, "tends in all its operations to harmonious periods." With α known, he could compute Saturn's orbital velocity.

That the initial assumption of concentric orbits was wrong, Bernoulli was quite aware. (He did not realize, however, that if he chose non-concentric orbits to start with, the periodicity he believed he had found would disappear.) He thought of redoing his investigation with more realistic initial

conditions, but the labour appeared daunting. As he wrote to Euler in 1749,

I have much less hope than ever of deriving the irregularities of Saturn from mechanical principles; but I can assure you that I see sufficiently into this subject to hope that with equal effort I could bring forth as much as anyone else. An exact solution is impossible and all approximations so dangerous that it would require an unsurpassable effort to determine the irregularities with sufficient exactitude and certainty.

And he went on to express wonder at Euler's complete confidence in having solved the problem with the utmost precision.

Bošković's procedure, unlike Bernoulili's, involved no differential equations or series approximations. Newtonian in style, it relied on geometry and ratios of infinitesimals. Even astronomers untutored in the integral calculus, Bošković boasted, could follow it.

At each moment the perturbed planet was to be conceived as moving in the elliptical orbit in which it would continue if the perturbing force were just then removed. The changes in this ellipse due to the perturbing force, during the time the planet traversed a tiny arc of its path, were then to be computed by means of formulas which Bošković derived geometrically using infinitesimals. These formulas gave the variations in transverse axis, eccentricity, aphelion, and rate of description of areas caused by infinitesimal changes in the planet's speed and direction. Each formula involved the ratio of the perturbing force to the solar force for the momentary locations of the two planets, the sine or cosine of the angle between the directions of the tangent and perturbing force, and in addition sines or cosines of certain angles, and certain ratios of lengths, in the unperturbed ellipse.

To obtain the changes in the elliptical elements over a finite time, Bošković proposed treating the daily variations as infinitesimal (Saturn moves about 2′, Jupiter about 5′ per day), and adding them up graphically. The abscissas would be lengths of the planet's path; the ordinates would be the diurnal variations of the element whose change was sought. The net area enclosed by the graph was to be approximated using formulas from Cotes's *Harmonia mensurarum* (1722), which Bošković had come to know from a French commen-

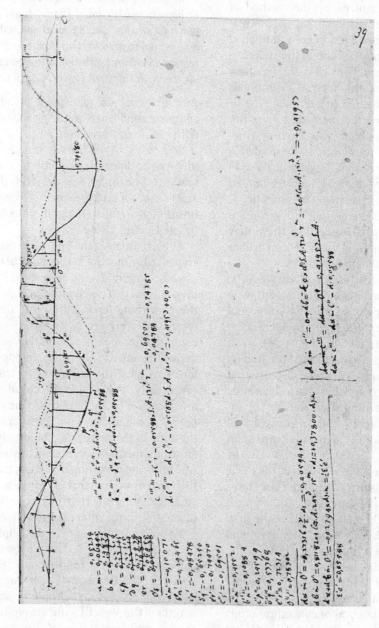

20.2. Daniel Bernoulli's diagram for the departure in the radial direction of Saturn from its concentric circle.

tary by Charles Walmesley. Boškovič proposed computing the variations over a period of 60 years, after which time he expected the pattern of variations to repeat very nearly. He suggested various calculational shortcuts; but the labour, we may be sure, would have been immense.

Boškovič wrote his *De inaequalitatibus quas Saturnus et Jupiter sibi mutuo videnter inducere* during the years 1750–52, while on expedition with Christopher Maire to measure a degree of the meridian in the papal states. The essay contains neither illustration nor sample of the vast programme of calculations it envisages. Boškovič excused the omission as due to the pressure of his other work, and to uncertainty as to the mass of Jupiter, which from data given by Cassini II would come out one-fifth greater than Newton's estimate.

Both Bernoulli and Boškovič recognized that there was a problem in determining the initial orbit to which the calculated perturbations were to be applied. Boškovič proposed first determining orbital elements from three observations by a procedure then standard. The elements thus obtained were initially to be taken as applying at the time of one of the observations, say the first. The times to the other positions observed were then to be corrected by subtracting out the effect of perturbation. Finally, with the original data and the corrected times, new orbital elements were to be computed, and these, Boškovič held, could be assumed to be almost exactly applicable at the time of the first observation. Bernoulli seems to have had in mind some similar procedure of "correcting" the observations by subtracting out the effect of perturbation. Neither author had in mind a way of coping with observational error. It would be Euler who first proposed a technique whereby the orbital elements could be corrected for the effects of both perturbation and observational error.

Euler on the inequalities of Saturn: the first try

It was the Paris contest of 1748 that led to Euler's first published memoirs on the perturbational problem: his "Recherches sur le mouvement des corps célestes en générale", presented to the Berlin Academy in June 1747 and published in the *Mémoires de Berlin* in 1749, and his "Recherches sur la question des inégalités du mouvement de Saturne et de Jupiter", the prize-winning essay of 1748, which was published (as a book) in Paris in

1749. In the first of these Euler derived the differential equations used in the second, which despite its title concerned Saturn's inequalities alone. In the present section we focus on the problem of the inequalities in longitude as dealt with in these two essays.

In both essays, Euler calls the exactitude of the inverse-square law into doubt, while nevertheless basing the mathematical development on this law. The reader is referred to Chapter 16 for an account of how Euler's doubt was eventually put to rest.

Euler began his development by applying Newton's second law in rectangular coordinates:

$$2\frac{d^2x}{dt^2} = \frac{X}{m}, \quad 2\frac{d^2y}{dt^2} = \frac{Y}{m}, \quad 2\frac{d^2z}{dt^2} = \frac{Z}{m}, \quad (1)$$

where m is the mass of the perturbed planet, and X, Y, Z are the components of force acting on it in the coordinate directions. (The factor 2 is due to Euler's choice of units, and eventually drops out of the calculation.) Taking the equations of motion in rectangular coordinates as starting-point is nowadays a standard procedure, but in Euler's use of it here we have its first appearance. He identified the x–y plane with the orbital plane of the perturbing planet (Jupiter); z thus measured the small latitudinal departures that Saturn makes from this plane.

Next, Euler transformed the first two equations of (1) into the polar coordinates used by astronomers by setting $x = r \cos \phi$, $y = r \sin \phi$; also, he replaced X/m by $(-P \cos \phi + Q \sin \phi)$, and Y/m by $(-P \sin \phi - Q \cos \phi)$, where P and Q are accelerations directed so as to diminish, respectively, r and ϕ. The differential equations of motion in the x–y plane thus became:

$$d^2r - rd\phi^2 = -P\frac{dt^2}{2}, \quad (2)$$

$$2drd\phi - rd^2\phi = -Q\frac{dt^2}{2}. \quad (3)$$

The third equation of (1) also got transformed, but we postpone our discussion of this step to the next section.

If the perturbed planet was to be Saturn, then the radial acceleration P and the transradial acceleration Q had to be expressed in terms of the attractions due to the Sun and Jupiter. Here account had also be taken of the forces exerted by Jupiter and Saturn on the Sun, in order that the Sun could be regarded as at rest; hence the forces on it had to be transferred to Saturn with their directions reversed. Figure 20.3 shows the geometry involved. Saturn was thus considered to be acted upon by four forces:

two forces in the direction $S\odot$, yielding

$$\frac{(\odot + S)\cos^2\psi}{r^2};$$

a force in the direction SJ, $= J/v^2$;
a force in the direction HR, $= J/r'^2$.

By taking components of these forces in the direction of the radius $H\odot$ and in the transradial direction HN, Euler obtained the following expressions for P and Q:

$$P = \frac{(\odot + S)\cos^3\psi}{r^2} + \frac{Jr}{v^3}$$
$$+ \frac{J\cos(\phi' - \phi)}{r'^2} - \frac{Jr'\cos(\phi' - \phi)}{v^3},$$

$$Q = \frac{J\sin(\phi' - \phi)}{r'^2} - \frac{Jr'\sin(\phi' - \phi)}{v^3}.$$

To replace $dt^2/2$ in equations (1) and (2), Euler set the inverse-square force of the Sun on Jupiter at its mean distance a' equal to Jupiter's mean centripetal acceleration: $\odot/a'^2 = 2a'dM'^2/dt^2$, where M' is Jupiter's mean anomaly; hence $dt^2/2 = a'^3dM'^2/\odot$. Next, he replaced the ratios of the planetary to the solar masses by single letters: $S/\odot = n$, $J/\odot = n'$. (In the foregoing we have for clarity's sake replaced Euler's symbols by indicial notation, which was first introduced by Lagrange.) Equations (2) and (3) thus took the form

$$d^2r - rd\phi^2$$

$$= -a^3dM^2 \begin{bmatrix} \dfrac{(1+n)\cos^3\phi}{r^2} + \dfrac{n'r}{v^3} \\[2ex] + \dfrac{n'\cos(\phi' - \phi)}{r'^2} \\[2ex] - \dfrac{n'r'\cos(\phi' - \phi)}{v^3} \end{bmatrix}, \qquad (2')$$

$$2drd\phi + rd^2\phi$$

$$= -n'a'^3dM'^2\sin(\phi' - \phi)\left[\frac{1}{r'^2} - \frac{r'}{v^3}\right]. \qquad (3')$$

These differential equations provided an exact expression of the inverse-square forces, but in solving them Euler found it necessary to resort to approximation. Specifically, he replaced the variables r, r', ϕ, ϕ', and v in the right-hand members by

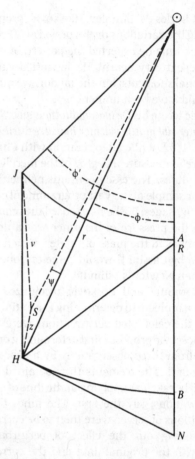

20.3. Euler's figure (with the lettering modified for clarity) to illustrate his theory of the perturbations of Saturn by Jupiter:

$\odot$ = Sun's position (and also its mass)
J = Jupiter's position (and also its mass)
S = Saturn's position (and also its mass)
$r = \odot H$ = the curtate solar distance of Saturn (the projection of Saturn's radius vector on the plane of Jupiter's orbit)
v = distance between Jupiter and Saturn
$r' = \odot J$ = radius vector of Jupiter
ϕ = longitude of Saturn
ϕ' = longitude of Jupiter
ψ = latitude of Saturn (from plane of Jupiter's orbit)
$z = SH$ = distance from Saturn to plane of Jupiter's orbit
HR = line drawn through H parallel to $\odot J$
HN = line drawn through H at right angles to $\odot H$

approximate expressions in terms of M', the mean anomaly of Jupiter. Having exact formulas to start from, he was in a better position than a Bernoulli or a Bošković to keep track of different orders of approximation.

In a first approximation, r and r' could be identified with a and a', the mean solar distances of Saturn and Jupiter. In effect the planets were here being conceived to move in circles concentric to the Sun, hence uniformly except insofar as they perturbed each other's motions; ϕ and ϕ' became identical with the mean motions of the two planets, so that $\phi' = M' + k'$, and $\phi = k''\phi'$, where k', k'' are constants.

What about v, the distance between the two planets? Its only occurrence was as the inverse cube, v^{-3}, which could be expressed as

$$[a^2 + a'^2 - 2aa' \cos(\phi' - \phi)]^{-\frac{3}{2}},$$

or

$$[a^2 + a'^2]^{-2/3}[1 - g \cos(\phi' - \phi)]^{-\frac{3}{2}},$$

where

$$g = \frac{2aa'}{a^2 + a'^2}.$$

If a were much greater or less than a', as in the case of the Moon perturbed by the Sun where the ratio of the two distances is 389:1, v^{-3} could be accurately approximated by two or three terms of a Taylor expansion:

$$v^{-3} = [a^2 + a'^2]^{-3/2}\left[1 + \frac{3}{2}g \cos(\phi' - \phi) + \dots\right].$$

In the case of planets perturbing planets, on the contrary, a and a' were much closer to each other in size, and the Taylor series converged much less rapidly. Thus for Jupiter and Saturn, $a'/a = 0.545$, and the series would have to be continued to some sixty terms to yield the same precision as that given by three terms in the case of the Moon. An additional difficulty was that when the expansion was carried past the second term, it contained powers of $\cos(\phi' - \phi)$, which would have to be transformed one-by-one into integrable form.

The difficulty presented by the variation in distance between the two planets, we recall, had led Daniel Bernoulli to resort to elaborate interpolations and successive integrations in numbers, each dependent on the preceding. For a time Euler thought it might be necessary to turn (like Boškovié later) to numerical quadratures. But at length he thought of a remedy.

For $\phi' - \phi$, let us write θ. The powers of $\cos\theta$ in the Taylor expansion, Euler discovered, could be replaced by expressions in terms of cosines of

multiples of θ (he was the first to develop these expressions). Thus he transformed the Taylor series into something quite new, a trigonometric series: $A + B \cos\theta + C \cos 2\theta + \dots$. Here the terms were immediately integrable.

But now a new difficulty emerged: each of the coefficients A, B, C, $\dots$ was itself an infinite series. To meet this difficulty, Euler first showed – by logarithmic differentiation with respect to θ of his general formula for the series,

$$s = (1 - g \cos\theta)^{-\mu}$$

– that each coefficient after the first two was expressible in terms of the two preceding coefficients. Then he provided two different procedures for approximating A and B, one of them an approximation to Fourier integrals by numerical integration. We have here the opening chapter in the pre-history of Fourier analysis.

With the introduction of a trigonometric series for v^{-3}, the right-hand members of (2') and (3') assumed integrable form, and Euler proceeded to integrate the equations by the method of undetermined coefficients. The calculation was straightforward, and Euler's result may be written:

$$r = a[1 + 0.000\,854\,2 \cos\theta + 0.000\,144\,9 \cos 2\theta \\ + 0.000\,033\,5 \cos 3\theta \\ + 0.000\,010\,5 \cos 4\theta + \dots]; \tag{4}$$

$$\phi = \text{const} + k''M' + 4'' \sin\theta - 32'' \sin 2\theta \\ - 7'' \sin 3\theta - 2'' \sin 4\theta - \dots. \tag{5}$$

Daniel Bernoulli, in a letter to Euler of January 1750 and in his essay for the contest of 1752, objected that this result could not be correct, because it makes r and ϕ functions of θ alone, whereas his own calculations had shown that r is significantly different at successive conjunctions, going for instance from a at the first conjunction to $a[1 - 0.003\,29]$ at the second. He accused Euler of paralogism. The relations between Bernoulli and Euler had by 1750 become strained; whether Bernoulli ever received an answer is unclear. Attempting to reply for Euler we would say: any assault on the perturbational problem involves assuming approximate values to obtain other, more closely approximating values of the same variables. Euler's procedure thus far was such as necessarily to yield results periodic with θ, but a further stage of approximation was to follow. Would these successive steps converge to the result implied by Newton's law? Euler in his essay of 1748 appeared confident that the series he was using were convergent, although he had no proof of this (tests and proofs of convergence were not to be introduced till the next century). Later on, as we shall see, he would become doubtful of the reliability of trigonometric series for computing perturbations when the apparent convergence was slow.

In two further stages of approximation Euler took into account the orbital eccentricities, first Saturn's, then Jupiter's; in each case he treated the orbit of the second planet as circular and non-eccentric. To express the motion of a planet on an eccentric orbit, he used Kepler's eccentric anomaly. Thus for Saturn,

$$r = a(1 + e \cos E),$$

where e is the eccentricity and E the eccentric anomaly. E is related to the mean anomaly M by Kepler's equation

$$M = E + e \sin E,$$

and to the true anomaly ϕ by

$$d\phi = \frac{a}{r} \sqrt{(1 - e^2)} \, dE.$$

There were corresponding equations for Jupiter (in our notation, the letters are the same but distinguished by prime marks). The inverse cube of v had to be developed further than before to take account of the variability of r and r'; this entailed developing new terms in the series for v^{-3}, involving the determination of the coefficients in the trigonometric series

$$(1 - g \cos \theta)^{-5/2} = P + Q \cos \theta + R \cos 2\theta \\ + S \cos 3\theta + T \cos 4\theta + \dots.$$

The result of integrating the equations that had been elaborated in this way was that the terms obtained in the previous calculation turned up again, but in addition there were new terms proportional to the first powers of the eccentricities. In particular, Euler found two especially large terms:

$$-257'' \sin(\theta - E) - 243'' \sin(2\theta - E').$$

Unfortunately, the two coefficients are wrong: there were errors of sign in Euler's derivation. (In the text as given in *Leonhardi Euleri opera omnia series II* Vol. 25, the mistakes occur on p. 80, where they are indicated by the editor, and on p. 100, where R'' should be, not $+12.63013$, but -12.63013.) When these errors are corrected, the

two coefficients become $-182''$ and $+413''$, respectively, which are in close agreement with the values later obtained by Laplace and his successors.

A different difficulty emerged in the term in the radius vector proportional to $\cos(\theta - E')$: the coefficient proved to be a fraction with zero for denominator. Euler took this to be an indication that, in addition to the terms he had already assumed for the integral, there was another term proportional to the time, or to E' understood now as an angle increasing indefinitely. Thus in the longitude ϕ he concluded that there were two terms with argument $(\theta - E')$, namely

$$-Q E' \cos(\theta - E') - T \sin(\theta - E');$$

the constants Q and T, indeterminate in his derivation, would have to be obtained by a comparison with observations. The presence of a term with coefficient proportional to E' would explain why astronomers had obtained results for the period of Saturn differing by as much as three or four days. It meant that all the empirical constants would in time need to be re-determined: no eternally valid formula was attainable.

There was another possible interpretation which Euler here mentioned but did not pursue. In obtaining the terms with argument $(\theta - E')$ he had assumed the line of apsides of Jupiter's orbit to be fixed. But he had already found that, when Saturn's orbit is taken to be eccentric, its orbital apse moves owing to the perturbations caused by Jupiter. By symmetry the apse of Jupiter's orbit as perturbed by Saturn must also be in motion. But if the Jovian apse moves, the coefficient of $\cos(\theta - E')$ no longer has zero for denominator. D'Alembert, writing to Euler after the contest was finished, asked why he had not adopted this alternative interpretation. Eventually Euler did so: in his prize essay of 1752 the changes in orbital apse and eccentricity of the two planets were taken as simultaneous and mutually implicated.

In a supplement to his memoir of 1748, Euler observed that the angle $(\theta - E')$ is given (if terms of the order of the eccentricities are ignored) by $-[E + (A - A')]$, where A' and A are the longitudes of the two aphelia, and $(A - A')$ is an angle of about $78°$ which increases slowly over the centuries. It follows that $\cos(\theta - E')$ is approximately replaceable by $\sin E$, and the term containing $\cos(\theta - E')$

may thus be viewed as contributing to Saturn's equation of centre. The gradual change in the angle $(A - A')$ implies (when proper account is taken of signs) that Saturn's equation of centre is slowly decreasing. To the question of the secular variations in the orbital elements, Euler was to return in his memoir of 1752.

In the main body of his memoir of 1748, however, Euler set himself the task of determining the constants Q and T by means of observations. There were other constants, as well, to be empirically determined. In 1740 Euler had presented to the St Petersburg Academy a memoir on the differential correction of orbital elements, and he now invoked this method to correct the orbital elements of Saturn as given in Cassini's tables, since (in his words) "the inequalities caused by Jupiter have been there enveloped in the eccentricity and position of the orbit of Saturn". All told, there were eight unknowns to be determined by the observations; eight observations and eight simultaneous equations would have sufficed for the determination. Recognizing, however, that the observations inevitably contained error, Euler introduced the use of multiple 'equations of condition' to bring redundant data to bear on the determination of the constants.

This is an important step – a first major step towards bringing planetary astronomy under the aegis of statistics. Euler was here proposing to resolve a problem the difficulty of which Bernoulli and Bošković had failed to recognize: the problem of identifying the unperturbed orbital motion to which the perturbations were to be applied. Euler's solution was to fit both the theoretically derived perturbations and the empirically determined orbital elements to multiple observations, by adjusting the constants of doubtful accuracy so as to minimize the errors.

The method of least squares had not yet been invented, and the procedure that Euler used in its stead was crude: he sought to add equations together in such a way as to render the coefficient of one unknown large, and the coefficients of the remaining unknowns small. And the success of his attempt was limited: for 36 observed oppositions from Tycho's time to 1745 he reduced the error in the theory to about $3'.3$ on the average. A few of the earlier observations contained errors larger than this, but the chief reason he could not do

better was that he had not derived, and indeed not suspected the existence of, the sizeable inequalities in the motion of Saturn that are proportional to the squares and cubes of the eccentricities. Laplace would be the first to derive these inequalities, and in a test of his theory against 43 observations of Saturn from the period 1582–1786, using basically the same statistical procedure as Euler (but see Chapter 27), would succeed in lowering the errors to about $50''$ on the average. The agreement became still better after J.-B. J. Delambre had corrected the observations for aberration and nutation.

During the 1750s Tobias Mayer deployed the method of equations of condition more successfully than Euler (see Chapter 18). According to Delambre, Mayer derived the method from Euler, and Laplace had it from Mayer's writings.

Euler on the secular inequalities

We return now to the third equation of (1), namely

$$2\frac{d^2z}{dt^2} = \frac{Z}{m}.$$

This equation, we recall, determines the perturbations in Saturn's small latitudinal departures from the plane of Jupiter's orbit. In his memoir of 1748, Euler undertook to transform this equation into a result more in accord with astronomical usage. Astronomers since Kepler had viewed each planet as moving in an orbital plane inclined to the ecliptic, its position being specified in terms of the inclination and longitude of the boreal node (the point at which the planet moves from south to north of the ecliptic). The variable z is expressible in terms of the inclination ρ and the longitude of the node π by

$$z = r\sin(\phi - \pi)\tan\rho. \tag{6}$$

If all perturbations were absent, both ρ and π would be constant, and in forming the differential dz we would consider z to depend solely on the variables r and ϕ. But if perturbations are present, ρ and π become variable, and the differential dz comes to depend on $d\rho$ and $d\pi$ as well as dr and $d\phi$. Yet even in this case, Euler stated, it is permissible to treat dz as depending only on dr and $d\phi$, for ρ and π are changing much more slowly than r and ϕ, and can be regarded as constant during the time dt that the variation dz occurs. Forming dz in both ways, Euler set the two results equal and so obtained

$$d(\ln \tan \rho) = \frac{d\pi}{\tan(\phi - \pi)}. \tag{7}$$

A second differentiation of (6), followed by substitution of the result in the third equation of (1), then yielded an expression for $d\pi$ in terms of the force components P, Q, Z.

We have here the basic step in the method of variation of orbital elements: the formula valid in the two-body case is assumed to be valid in the three-body case, provided that the constants of integration – the orbital elements used to specify the size, shape, and position of the orbit in the two-body case – are taken as variable. At each moment the planet or satellite is imagined as moving on an ellipse; but the ellipse itself is changing position, size, and shape, owing to the perturbations.

An outcome of Euler's investigation of $d\rho$ and $d\pi$ in the case of Saturn (with the plane of reference being taken as Jupiter's orbital plane) was that the node of Saturn's orbit steadily regressed. A similar result must hold for every planet: the nodes of its orbit on the orbital plane of any perturbing planet must undergo a steady motion. In particular this applied to the nodes of the ecliptic on the orbital planes of the planets other than the Earth. Euler undertook a systematic investigation of the resulting motion of the ecliptic in a memoir completed in 1754 and published in 1756, "De la variation de la latitude des étoiles fixes et de l'obliquité de l'écliptique". The question whether the obliquity of the ecliptic was decreasing had long been disputed. Euler demonstrated that the perturbationally induced motion of the nodes implied a secular diminution. His value for it, which depended on hypothesized masses for Mercury, Venus, and Mars, was 47".5 per century, very close to the present-day value.

Euler's prize essay of 1752 applied the idea of mutually induced changes in orbital elements to the perturbations in longitude of Jupiter and Saturn. This time, no *arcs de cercle* emerged. Instead, the apse of each planet was progressing, and each planet was inducing eccentricity in the orbit of the other, so that the observed eccentricity in each case was a conflation of a proper eccentricity and an induced eccentricity. Thus for the *radii vectores* of the two planets Euler obtained expressions of the form

$$r' = a'[1 + e'\cos(\phi' - A') + ke\cos(\phi' - A)],$$
$$r = a[1 + e\cos(\phi - A) + k'e'\cos(\phi - A')],$$

where A, A' were now the longitudes of the two *proper* aphelia; e, e' were the two *proper* eccentricities; and k, k' were two constants for each of which Euler obtained (after neglecting many smaller terms and simplifying) a quadratic equation. Unfortunately, Euler's computation was riddled with error; the two quadratic equations were really one, and the correct values for k, k' were of opposite sign, whereas Euler's mistaken values were both positive. In what amounts to an addition of vectors, Euler showed that the squares of the *apparent* (that is, observationally determined) eccentricities were

$$f^2 = e^2 + k'^2 e'^2 + 2k'ee'\cos(A - A'),$$
$$f^2 = e'^2 + k^2 e^2 + 2kee'\cos(A - A').$$

The angle $(A - A')$ was approximately 78° and increasing, and Euler therefore concluded that the apparent eccentricities of both orbits were decreasing. But in fact Saturn's orbital eccentricity is decreasing while that of Jupiter is increasing (for k is really negative).

Another of Euler's inferences, based on the change in the angle $(A - A')$, was that the mean motions of both planets were slowly accelerating. Euler's memoir of 1752 was not published until 1769, and in the meantime Lagrange had published a deduction purporting to show that the mean motion of Jupiter was accelerating while that of Saturn was decelerating, in agreement with observations. The contradiction between the two results was to lead Laplace to re-do the derivation, with an outcome to be reported in Chapter 22.

Euler employed the variation of orbital elements in several later works, including the *Additamentum* to his *Theoria motus lunae* of 1753 and his memoir on the planetary perturbations of the Earth's motion, *Investigatio perturbationum quibus planetarum motus ob actionem eorum mutuam afficiuntur*, which won the prize of the Paris Academy for 1756. But there was one orbital element whose variation he did not examine, namely the epoch. It was left to Lagrange to complete the method by including the variation of this element in the analysis.

Clairaut's procedure for determining perturbations

Clairaut's method, as described in his *Théorie de la lune* of 1752 as well as in earlier papers, is different from Euler's, and worthy of a separate account. It lent itself with a certain clarity to Clairaut's discovery of the full motion of the lunar apse. (For an account of this discovery the reader is referred to Chapter 16.) It was the method that Lalande chose to incorporate in his *Astronomie* (editions in 1764, 1771, and 1792), the chief text from which European astronomers learned astronomy in the last third of the eighteenth century. In carrying out his integration of the equations of motion, Clairaut introduced a number of innovations that were put to use by Euler, Lagrange, and Laplace.

Clairaut began by supposing the orbit of the perturbed body to lie in a single plane; thus he left out of account the perturbations in latitude. His attempt to reintroduce the latitudinal perturbations at a later stage entailed an illogicality, for which d'Alembert took him to task. The initial simplification allowed him to reduce the problem to the following two equations, which differ little from the Eulerian equations we labelled (2) and (3) above:

$$r \, d^2\phi + 2 \, dr \, d\phi = Q \, dt^2; \tag{8}$$

$$r \, d\phi^2 - d^2 r = \left(\frac{M}{r^2} + P\right) dt^2, \tag{9}$$

where M represents the sum of the masses of the central and the perturbed body.

Next, in a series of moves that won Euler's admiration, Clairaut obtained an integrated form of (8). After multiplying by r, he found in a first integration

$$r^2 \frac{d\phi}{dt} = f + \int Qr \, dt, \tag{10}$$

where f is a constant of integration. To obtain an expression for the right-hand member of (10) from which dt is absent, he multiplied the equation by $Qr \, dt$, integrated, completed the square, and extracted the square root:

$$f + \int Qr \, dt = [f^2 + 2\int Qr^3 \, d\phi]^{\frac{1}{2}}. \tag{11}$$

Substitution of (11) into (10) then gave

$$dt = \frac{r^2 d\phi}{[f^2 + 2\int Qr^3 \, d\phi]^{\frac{1}{2}}} \tag{12}$$

$$= \frac{r^2 d\phi}{f\sqrt{1 + 2\rho}},$$

where

$$\rho = \frac{1}{f^2} \int Qr^3 \, d\phi.$$

Equation (12) permits the elimination of dt from equation (9), yielding an equation for the orbit in which r is a function of ϕ alone. By twice integrating the latter equation, Clairaut then obtained his solution in the form

$$\frac{f^2}{Mr} = 1 - g \sin \phi - c \cos \phi \tag{13}$$

$$+ \sin \phi \int \Omega \cos \phi \cdot d\phi - \cos \phi \int \Omega \sin \phi \cdot d\phi,$$

where g and c are constants and

$$\Omega = \frac{\dfrac{Pr^2}{M} + \left(\dfrac{Qr}{M}\right)\left(\dfrac{dr}{d\phi}\right) - \left(\dfrac{2}{f^2}\right)\int Qr^3 d\phi}{1 + \left(\dfrac{2}{f^2}\right)\int Qr^3 d\phi}. \tag{14}$$

Equation (13), when the terms involving Ω are deleted, reduces to the equation of a conic section:

$$\frac{p}{r} = 1 - c \cos(\phi - a), \tag{15}$$

where the parameter $p = f^2/M$, the eccentricity $c' = [g^2 + c^2]^{\frac{1}{2}}$, and $\tan a = g/c$.

Suppose now that Ω can be expressed in the form

$$A \cos n\phi + B \cos p\phi + \dots,$$

where n, p, etc. are numerical factors multiplying the angle ϕ. Clairaut showed that in this case the final two terms of (13) will be expressible in the form

$$+ \frac{A}{n^2 - 1} [\cos \phi - \cos n\phi]$$

$$+ \frac{B}{p^2 - 1} [\cos \phi - \cos p\phi]$$

$$+ \dots.$$

To reduce Ω to the required form, it was necessary to replace r where it occurred in (14) by an approximate expression in terms of ϕ. It would be a mistake, Clairaut urged, to use formula (15), which represents the fixed ellipse resulting when the perturbing forces are absent, since the observations teach us that the line of apsides of the orbit rotates about the central body: in a few revolutions the formula would be inaccurate. Instead, Clairaut proposed using

$$\frac{k}{r} = 1 - e \cos m\phi, \tag{16}$$

which represented a rotating ellipse. Here k, e, and m were constants yet to be determined, and $m\phi$ was the real anomaly, completing 360° when the planet or satellite returned to its apse. Equation (16) would be found to be justified, Clairaut said, if the larger terms of equation (13), once the substitutions were carried out, proved to have the form of (16), permitting k, e, and m to be identified in terms of the other constants of the theory, while the remaining terms were relatively small by comparison. This turned out to be the case.

Besides r, the formula (14) for Ω contained the perturbing forces P and Q, and these, too, needed to be expressed as functions of ϕ. The resulting expression for Ω contained products of sines and cosines of several multiples of ϕ; using trigonometric identities Clairaut converted the products into sums and differences. With Ω thus transformed, the integrations required in the last two terms of (13) were easily performed.

After r had been determined in terms of ϕ, the time corresponding to a given value of r could be determined by integrating equation (12).

Such is the outline of Clairaut's procedure. It involved approximations at every stage. In order to abbreviate the calculation, Clairaut introduced numerical substitutions appropriate to the lunar case before completing his derivation, and so failed to attain a fully algebraic formulation of his theory. His reliance on empirical results to help resolve the differential equations – particularly his use of equation (16) as qualitatively justified by observations – was arguably objectionable from a mathematician's standpoint, and d'Alembert did not fail to urge the objection. Clairaut's rebuttal was that his method was easier to apply than the methods of d'Alembert and Euler; and such followers as Lalande agreed.

D'Alembert's procedure for determining perturbations

D'Alembert's analysis, as presented in his mature lunar theory of 1754, was no doubt superior to Clairaut's from a logical point of view; it avoided the faults that d'Alembert had found in Clairaut's theory. D'Alembert took the ecliptic as his basic plane of reference. He described the Moon's motion by means of the projection of the orbital motion onto the plane of the ecliptic, together with the momentary inclination of the Moon's orbital plane to the ecliptic, and position of the line of nodes. He warned that the ecliptic could not be identified as a plane through the Earth's orbit since, owing to the perturbations in the Earth's motion caused by the

Moon, this orbit was not in fact confined to a single plane.

The warning was apropos, but in several writings d'Alembert mistakenly identified the ecliptic as the plane in which the Earth is moving when the Moon, the Sun, and the Earth's line of motion are all in the same plane; and he deduced from this identification that the Sun would be seen to depart from the ecliptic by as much as $15''$. An anonymous and largely laudatory review of the second part of his *Recherches sur différens points importans du système du monde* (1754) in the *Journal des sçavans* for December 1754 (by Clairaut?) urged that the ecliptic be identified rather as the plane traced out by the centre of gravity of the Earth and the Moon; the apparent departure of the Sun from this plane is never more than $1''$. This definition is in fact more in accord with the ecliptic that astronomers determine observationally, but d'Alembert refused to accept the criticism.

For the equation of the projection of the orbit onto the ecliptic, d'Alembert obtained

$$\frac{\mathrm{d}^2 u}{\mathrm{d}\phi^2} + u - \frac{1}{u^2 g^2}\left[\frac{S - \left(\frac{Q}{u}\right)\left(\frac{\mathrm{d}u}{\mathrm{d}\phi}\right)}{1 + \frac{2}{g^2 h^2}\int\left(\frac{Q}{u^3}\right)\mathrm{d}\phi}\right] = 0, \quad (17)$$

where $u = 1/r$, the reciprocal of the projection of the radius vector onto the ecliptic; S is the sum of all the radial forces, that is, $M/r^2 + P$; g is the initial velocity; and h is the sine of the angle that the Moon's initial direction of motion forms with r.

In preparing to solve (17) to a first order of approximation, d'Alembert replaced u by $K + u'$, where K was the mean reciprocal radius vector, and u', because of the near-circularity of the orbit, was a variable that would always remain small. The substitution made it possible to put (17) in the form

$$\frac{\mathrm{d}^2 u'}{\mathrm{d}\phi^2} + N^2 u' + R = 0, \quad (18)$$

where R is a function of u', $\mathrm{d}u'/\mathrm{d}\phi$, and sines and cosines of multiples of ϕ, and N^2 is a constant differing from 1 by a small quantity of the order of magnitude of the perturbing force.

The reason that N^2 is not identically 1, as substitution of $K + u'$ for u in the second term of (17) might lead one to expect, is that terms linear in u' emerge from the third term of (17). Thus amidst the terms giving his first approximation to the third term of (17), d'Alembert found $n'^2/2g^2u^3$, where n' is the mean rate of motion of the Sun, and on substitution of $K + u'$ for u, this became

$$\frac{n'^2}{2g^2(K+u')^3} = \frac{n'^2}{2g^2K^3}\frac{1-3u'}{K},$$

which contains a term linear in u'.

In his integration of equation (18), d'Alembert employed an exponential integrating factor with complex exponent; he then converted the complex exponential terms into sines and cosines. His first-order solution for u' thus took the form

$$u' = D \cos N\phi - \frac{H}{N^2} \qquad (19)$$

$$+ \frac{H}{N^2}\cos N\phi + Km'^2\cos 2(1-m')\phi \ldots.$$

Here H is a constant of integration, m' is the ratio of the Sun's mean motion to the Moon's mean motion, and N is given by $(1 - 3m'^2/4 + \ldots)$. In contrast to Clairaut, d'Alembert has here derived the fact of the apsidal motion from the differential equation, rather than assuming it as an observational result.

In his memoir of 1747, d'Alembert had recognized that N as just calculated gives but half of the observed apsidal motion, but in opposition to Clairaut had supposed the discrepancy to be caused by some non-gravitational force, say magnetism, rather than indicating a need to revise the inverse-square law. In his mature lunar theory of 1754, d'Alembert carefully distinguished the orders of magnitude involved in successive approximations. Equation (19) gives a first-order approximation to u', from which a first-order value of $du'/d\phi$ can be calculated. When these values of u' and $du'/d\phi$ are substituted into the term R of (18) and the integration is carried out, there results a new and more exact expression for u'. The process can be repeated as many times as may be necessary to attain the same precision in the calculations as is found in the observations. D'Alembert carried the iterations far enough to obtain the first four terms in the motion of the apogee, namely 1° 30′ 37″, 1° 3′ 21″, 23′ 30″, 5′ 5″, which add to 3° 2′ 33″, just 1′ 4″ less than the observational value. He gave the algebraic expression for the second term in N, namely $-\frac{225}{32}m'^3$, and formulas from which the algebraic expressions for the next two terms could be deter-

mined. Here as elsewhere he carried the algebraic articulation of the theory much farther than Clairaut.

Although the result for the apsidal motion appears to confirm the Newtonian law, d'Alembert warned that it is impossible to demonstrate rigorously the exactitude of the formulas he deduced. Further steps in the iterative process might fail to be convergent. Whether the resulting tables were empirically reliable could be asserted only after a long series of observational tests. D'Alembert himself provided a detailed comparison between his own and the Newtonian tables, as given in P.-C. Le Monnier's *Institutions astronomiques*. He objected to the tables of Tobias Mayer because Mayer provided no explanation of their theoretical and observational basis; the empirical success of such tables, he complained, told nothing about the adequacy of the inverse-square law. In giving a detailed algebraic development of Newton's theory, with clearly defined stages of approximation, d'Alembert prepared the way for an exact test as well as for more precise approximations. But the research programme thus laid out does not appear to have been pursued. As was pointed out in Chapter 18, the lunar tables that came to be generally accepted during the late eighteenth century were those of Tobias Mayer, the obscurity of their theoretical basis notwithstanding.

Later contributions to perturbation theory by Clairaut, d'Alembert, and Euler

As we saw in Chapter 18, Clairaut applied his analytic procedure to the determination of the planetary and lunar perturbations of the Earth's motion, and his results were incorporated in Lacaille's solar tables. In 1757–58, aided by J.-J. L. de Lalande and Mme Nicole-Reine Lepaute, Clairaut used numerical integration to compute the return of Halley's Comet (see Chapter 19). Between 1758 and 1762 he defended himself in an acrimonious controversy initiated by d'Alembert over both this computation and his method of determining lunar and planetary perturbations. He died in 1765 at age 52.

D'Alembert's later memoirs on the lunar theory, the precession of the equinoxes, the shape of the Earth, and other topics relevant to celestial mechanics, were published in his *Opuscules mathématiques* (eight volumes, 1761–80). These memoirs generally presuppose d'Alembert's choice of symbols in earlier writings, and so are not easy reading. They do not appear to contain major discoveries, as d'Alembert himself was admitting in the last volumes. But Lagrange read them all, and claimed to profit from them.

Euler, in contrast, continued to make contributions of considerable importance. In two memoirs completed in 1762 and 1763, and printed in the volumes of the St Petersburg memoirs for 1764 and 1765, he gave the first treatment of what has since come to be known as "the restricted problem of three bodies". Two of the bodies are assumed to be much more massive than the third, which is thus taken to have no perturbing effect; consequently the first two bodies move in exact accord with the solution of the two-body problem. Euler established that there are collinear solutions, in which the massless 'Moon' stays in constant opposition or conjunction with the 'Sun', or moves in small oscillations about the point of opposition or conjunction. (This discovery is sometimes attributed to Lagrange, who published the result only in 1772.)

Euler further showed that, if the 'Sun' and 'Earth' of the restricted three-body problem be identified with our Sun and Earth, then the 'Moon' could be no more than four times the distance of our actual Moon from the Earth, and remain in constant conjunction or opposition. Were its relative distance (compared with the Earth–Sun distance) greater, it would cease to follow the laws of satellites. But – unless its distance were to become much greater – it would not yet orbit the Sun in the manner of a planet, but rather move in so complicated a manner as to defy mathematical analysis. It was Euler's conviction, asserted more than once, that the Creator in designing the universe had taken into account the weakness of the human intellect, and made the celestial bodies either planets or satellites, so that none of them fell into the incomprehensible in-between class.

During the early 1760s Euler became distrustful of the trigonometrical series he had introduced a decade and a half before. In the case of several planets these series did not appear to be very convergent, and Euler feared that the neglected terms might add up to a significant quantity. He therefore proposed (in the Berlin *Mémoires* for 1763, published in 1770) to follow Clairaut's

procedure in dealing with Halley's Comet, and use numerical integration; in contrast to Clairaut, however, he would apply the numerical integration directly to the differential equations. The position and velocity of a planet having been determined observationally for some moment of time, the increments to the x, y, and z position coordinates and velocity components were to be calculated from the differentials during successive small intervals of time, taking account of all the forces to which the planet was subject. The error could be expected to increase as the calculation was continued, so that at some point it would be necessary to determine a new starting-point from observation. The initial position and velocity were to be deduced from a sequence of observations by means of the calculus of finite differences – a method later applied by Laplace to determine positions and velocities of comets.

Euler believed the method to be capable of high accuracy. As he pointed out, it was applicable whether the perturbing bodies were one or many.

In the early 1770s, in studies of both lunar and planetary perturbations, Euler introduced a coordinate frame rotating with the mean speed of the perturbed body. Thus in the new lunar theory which he constructed during these years, and for which he received prizes from the Paris Academy in 1770 and 1772, a line rotated about the Earth in the plane of the ecliptic with the Moon's mean speed. The vertical distance from the Moon to the plane of the ecliptic gave the z-coordinate; a line dropped from the foot of this vertical at right angles to the rotating line gave the y-coordinate; and the distance so determined along the rotating line was set equal to $1 + x$, the Moon's mean radius vector being normalized to equal 1. The variations in x, y, z to be determined from the differential equations in these variables were in this way made very small. During the next century, rotating coordinates became standard in treatments of the three-body problem; but they were generally made to rotate with the mean speed of the perturbing rather than that of the perturbed body.

For the prize contests of 1770 and 1772, the Paris Academy has asked the contestants

to perfect the methods on which the lunar theory was founded, and in this way to determine the equations of this satellite that were still uncertain, and in particular to examine whether by this theory one could account for the secular equation in the movement of the Moon.

The gradual acceleration of the Moon's mean motion since ancient times, first detected by Halley, was by this time a well-established fact. But Euler in his deductions found no secular change in the mean movement of the Moon, and – believing his method to be a net from which no detectable inequality could escape – concluded that no such change could be produced by the forces of gravitational attraction. In his prize memoir of 1772 he concluded: "there no longer remains any doubt that this secular equation ... is the effect of the resistance of the milieu in which the planets move". He added, however, that lunar tables could not be made accurate to within less than 30 arcseconds without taking the planetary perturbations of the Moon into account.

In a memoir of 1771 on the perturbations of the Earth due to Venus, besides employing rotating coordinates, Euler espoused the idea of plotting the functions to be integrated at 5° intervals of the angle between the *radii vectores* of Venus and the Earth, and performing the integrations graphically, so as to avoid reliance on trigonometric series. By this time Euler was totally blind, and the calculations were performed by Anders Johan Lexell, a colleague at the St Petersburg Academy. The table of perturbations that emerged proved to be in sharp disagreement with the tables then in use, those of N.-L. de Lacaille and Mayer. Laplace, at the urging of Lalande, wrote to Euler and Lexell to ask their opinion as to the cause of the discrepancy. Lexell, as he reports in a memoir of 1779, discovered that his calculation contained a significant error, and on performing the whole series of quadratures over again, this time at 1° intervals, obtained results in agreement with those in the earlier tables.

Euler, like d'Alembert, died in 1783, two years before Laplace's announcement that he had at last derived from Newton's law the anomalies in the motions of Saturn and Jupiter that had puzzled the members of the prize commission of the Paris Academy thirty-nine years before. Two years later he would claim to have similarly derived the secular acceleration of the Moon – a claim that would turn out to be only partially justified, as we shall see in Chapter 22.

Further reading

Alfred Gautier, *Essai historique sur le problème des trois corps* (Paris, 1817)

Victor J. Katz, The calculus of trigonometric functions, *Historia Mathematica*, vol. 14 (1987), 311–24

F. Tisserand, *Traité de mécanique céleste*, vol. 3: *Exposé de l'ensemble des théories relatives au mouvement de la Lune* (Paris, 1894)

Curtis Wilson, Perturbations and solar tables from Lacaille to Delambre: the rapprochement of observation and theory, *Archive for History of Exact Sciences*, vol. 22 (1980), 53–304

Curtis Wilson, The great inequality of Jupiter and Saturn: from Kepler to Laplace, *Archive for History of Exact Sciences*, vol. 33 (1985), 15–290

21

The work of Lagrange in celestial mechanics

CURTIS WILSON

Joseph Louis Lagrange was born on 25 January 1736 into a well-to-do French–Italian family of Turin, Italy. His father, who was treasurer to the King of Sardinia, lost the family fortune in speculation, so that the young Lagrange was forced to think of embarking upon a career. In later life Lagrange remarked that, if he had been rich, he probably would not have devoted himself to mathematics. In his teens, the chance reading of an essay by Edmond Halley aroused his interest; he was soon awakened to his own extraordinary talent for mathematical analysis. At 18 he sent Leonhard Euler an account of his first important invention, a new mode of analysis which Euler later named the calculus of variations. By age 19 he was teaching mathematics in the artillery school of Turin. By his thirtieth year he had produced a prodigious body of work, and earned the profound respect of Jean le Rond d'Alembert in Paris – an arbiter of fate for up-and-coming mathematicians – as well as of Euler in Berlin. Already he had won two of the contests of the Paris Academy of Science (later, he would garner three more of these lucrative prizes).

In his native city Lagrange helped found a Royal Academy of Sciences; however, it received only meagre financial support, and Turin remained a scientific backwater. In 1766 Euler was leaving Berlin to return to St Petersburg, and through his recommendation, and more especially that of d'Alembert, Frederick II of Prussia was led to designate Lagrange as Euler's successor in the Berlin Academy, heading the mathematical–physical section. The greatest king in Europe, Frederick announced, wished to have the greatest mathematician in Europe at his court. Lagrange remained in Berlin for twenty-one years, but in 1787, on the death of Frederick II, chose to accept Louis XVI's invitation to come to Paris. He spent the last twenty-six years of his life there as a member of the Academy of Sciences and its successor, the Institut National. His death came on 10 April 1813, before he had completed revisions for the second edition of Vol. 2 of his masterwork, the *Mécanique analytique*.

Displacements from one city to another altered little the even tenor of Lagrange's daily life. It regularly included a solitary afternoon walk and a morning and evening of study and writing. Unlike d'Alembert who loved controversy and Euler who could be irascible, Lagrange was a peace-loving man; Frederick II described his character as "full of sweetness and modesty". However, the Parisian mathematician Jean-Charles Borda, himself a stubborn man, remarked after an argument with Lagrange that he had never known anyone more obstinate.

Lagrange's mathematical memoirs had an elegance often commented on. According to the mathematician J. B. J. Fourier, writing in 1829,

All [Lagrange's] compositions are remarkable for a singular elegance, for the symmetry of the forms and the generality of the methods, and, if one may so speak, the perfection of analytic style.

Lagrange's contributions to celestial mechanics were of the first importance. He studied intensively the earlier attempts to deduce the consequences of Newton's gravitational law analytically, penetrated the difficulties to which they led, and showed how these difficulties could be overcome. He perfected the Eulerian methods, giving them an elegant form. Many of the algorithms he introduced came to be widely adopted. His solutions supported and inspired a vision of a stable solar system in which all perturbations could be expected to be oscillatory and confined within narrow bounds. His achievement prepared the way for much later

21.1. Joseph Louis Lagrange

work, above all that of Pierre-Simon Laplace (1749–1827). In Laplace's mathematical development, Lagrange's influence was pervasive. In a letter to Lagrange of November 1778, Laplace wrote:

No one reads you with more pleasure than I, because no geometer [i.e., mathematician] appears to me to have carried to as high a point as you all the parts that go to make a great analyst. Permit me this avowal of my gratitude and respect, since it is principally by an assiduous reading of your excellent works that I have formed myself.

That Lagrange and Laplace felt themselves to be rivals we cannot doubt, but the rivalry never became open or ugly. Lagrange once wrote to Laplace: "If you are not jealous because of your success, I am not less so because of my character."

In reviewing Lagrange's contributions to celestial mechanics, we shall mainly follow a chronological order. However, in some cases he returned to a topic after a lapse of years to modify his original

approach or conclusions, and in these cases we shall take up consecutively his several assaults on the same problem.

The libration of the Moon

Lagrange's first memoir on celestial mechanics was the prize-winning essay of the Paris Academy's contest of 1764 concerning the libration of the Moon. The task imposed by the Academy was to explain (a) "the physical reason why the Moon always presents to us approximately the same face", and (b) "how to determine by observation and theory whether the Moon's axis is subject to a movement like that which affects the Earth's axis and causes the precession of the equinoxes and the nutation". Lagrange resolved the first of these problems convincingly, but his result for the second failed to account for the appearances. Sixteen years later, however, he returned to the question and achieved a satisfactory resolution.

Isaac Newton had dealt with the Moon's libration in Props. 17 and 38 of Bk III of the *Principia*. Proposition 17 accounted for the Moon's *optical* libration in longitude. Suppose the Moon's rotation about its axis to be uniform, with a period equal to its orbital period about the Earth. The orbital motion, being eccentric and in accord with the law of areas, is not quite angularly uniform about the Earth's centre. The consequence is that between perigee and apogee the Moon will appear to librate about 6° to one side or the other from the face it shows at perigee and apogee.

Besides the optical libration, Newton hypothesized that there would be a *physical* libration, as he explained in Prop. 38 and its corollary. If the Moon had been originally fluid, the Earth's attraction would raise a lunar tide with a height which he computed to be 93 feet. (The result is too small, principally because Newton's value for the mass ratio of Moon to Earth was too large.) The figure of the Moon would thus be a spheroid, whose greatest diameter when extended would pass through the centre of the Earth, and exceed the diameter in the direction of its orbital motion by 186 feet. (In fact, the excess is about 0.25 miles or 0.4 km.) "Such a figure," Newton claimed, "the Moon possesses, and must have had from the beginning." In consequence of this shape, the Moon would be in equilibrium only if its long axis passed through the Earth; if displaced from this orientation it would librate to either side, the period of the oscillations being very long because of the weakness of the forces exciting them. These qualitative conclusions were to be corroborated in Lagrange's analysis.

Gian Domenico Cassini (Cassini I, 1625–1712) was the first to give a precise description of the lunar libration, basing it on measurements of the positions of the lunar maria within the visible face of the Moon; his conclusions – but not his observations – were published posthumously in the *Mémoires* of the Paris Academy for 1721. The Moon's equator, he found, was constantly inclined to the ecliptic at an angle of 2°.5, and to the plane of its orbit at an angle of 7°.5; the axis of rotation was always in a plane parallel to the great circle of the celestial sphere passing through the poles of the orbit and those of the ecliptic. Further, he found the Moon's period of rotation to be just equal to its Draconitic or nodical month, namely 27 days, 5 hours and a fraction, in which the Moon returns to the node of its orbit with the ecliptic.

These conclusions, with some modification, were confirmed by Johann Tobias Mayer (1723–62) in an treatise published in the *Kosmographische Nachrichten* of Nuremberg for 1750. In deriving his results, Mayer used numerous observations of the positions of the lunar markings, and analyzed them by means of Euler's method of equations of condition (see p. 101 in Chapter 20). He found the inclination of the lunar equator to the ecliptic to be 1° 29′ rather than 2° 30′, and the intersection of these two planes to be always parallel, or nearly so, to the line of nodes of the lunar orbit. Like Cassini, he asserted a precise equality between the Moon's period of axial rotation and the Draconitic month.

In the formulation of the topic for the contest of 1764, d'Alembert no doubt had a hand. In 1756 he had presented to the Paris Academy a memoir on the precession and nutation of a spheroid with dissimilar meridians (that is, an ellipsoid with three different axes); this was published in the Paris *Mémoires* for 1754. He applied his formulas to the Moon in the second volume of his *Opuscules mathématiques*, published in 1761; here he derived values for the precession and nutation of the lunar axis, but they did not appear to be in agreement with observations. Lagrange was familiar with both of d'Alembert's memoirs, as well as the memoirs giving the observational results of Cassini

and Mayer, when he wrote his prize essay of 1764.

In this essay Lagrange started from the principle that he would later make fundamental in his *Mécanique analytique* (1788): d'Alembert's principle of dynamics, as formulated by means of the principle of virtual work. According to d'Alembert's principle, a dynamical system may be viewed as a system of forces in static equilibrium, if the products 'mass × acceleration' for each particle, taken negatively, are added to the other forces. According to the principle of virtual work (Lagrange called it the principle of virtual velocities), if to a system of bodies in static equilibrium we give a small movement, and then form the product of each force by the component of displacement in the direction of the force, the sum of these products will be zero.

Lagrange was here formulating his dynamical principle for the first time, and claiming that it comprehended the solution of all problems concerning the motion of bodies. In symbolizing the virtual displacements he used the special differential operator 'δ', which he had introduced in his first letter to Euler on the calculus of variations. An important feature of the new calculus was the commutativity of the operator 'δ' with the ordinary differential operator 'd', employed in symbolizing real velocities and accelerations.

Applying his dynamical principle to the case of the Moon attracted gravitationally by the Earth and the Sun, Lagrange obtained the formula

$$\frac{1}{dt^2}\int \alpha[d^2X\delta X + d^2Y\delta Y + d^2Z\delta Z] \qquad \text{(A)}$$

$$+ T\int\left(\frac{\alpha}{R^2}\right)\delta R + S\int\left(\frac{\alpha}{R'^2}\right)\delta R' = 0.$$

Here X, Y, Z are rectangular coordinates of a particle α of the Moon's mass; R and R' are the distances of the particle from the Earth and the Sun, and T and S are the masses of the Earth and the Sun; δX, δY, δZ, δR, $\delta R'$ are the variations in the coordinates and distances brought about by a small 'virtual' motion in the system; and the integrals are taken with respect to all particles α in the body of the Moon.

Lagrange's next step was to transform (A) by a change of variables. Replacing X, Y, and Z, he introduced six new variables, three of them specifying the Moon's orientation with respect to the ecliptic, and three specifying the position of an arbitrary mass-particle of the Moon with respect to axes fixed in the Moon's body. The variations of the first three accounted between them for all possible shifts in the position of the Moon's body about its centre. The expressions by which they were multiplied in the final form of (A) could be set separately equal to zero, yielding three differential equations by which the Moon's motion about its centre was determined.

In the transformation of (A) just described, Lagrange assumed that if the external forces due to the Earth and Sun were absent, the Moon would rotate uniformly about its polar axis. To show that this assumption was not very restrictive he proved that in every rigid body there are three mutually perpendicular axes about which the centrifugal forces cancel, so that the body can rotate without wobbling. Euler had informed Lagrange of his discovery of this result in a letter of October 1759, but Lagrange evidently developed his proof of it independently. Euler's proof, first presented to the Berlin Academy in July 1758, was not published till 1765, when it appeared in both the Berlin *Mémoires* for 1758 and his *Theoria motus corporum solidorum*.

For one of the second-order equations he had derived from (A), Lagrange was able to obtain an approximate solution giving the libration of the Moon in longitude:

$$\theta = -a \sin mV \qquad (1)$$
$$+ C \sin V \sqrt{\frac{3M}{H}}$$
$$+ \left(\frac{N}{2M}\right) \left\{1 - \cos V \sqrt{\frac{3M}{H}}\right\}.$$

Here θ is the angular distance between the apparent centre of the Moon, and a 'prime meridian' of the Moon, which Lagrange chose so that its plane contained the radius vector from Earth to Moon when the Moon's mean place coincided with its true place. V is the Moon's mean motion in longitude, and m is a number slightly less than 1, such that mV is the Moon's mean anomaly. The term $(-a \sin mV)$ is the main term in the Moon's equation of centre, and thus represents the main term in the optical libration. C is an arbitrary constant. H, M, and N are constants determining the Moon's shape and distribution of mass, M and N being very small relative to H, with N (the product of inertia about the polar axis) possibly

equal to zero, and M/H giving the ellipticity of the Moon's equator.

The last two terms of the equation for θ represent the physical libration, which will have as period the fraction $(3M/H)^{\frac{1}{2}}$ of the Moon's sidereal period. Because C is arbitrary, Lagrange was able to conclude that the Moon's initial velocity of rotation need not have been exactly equal to its mean motion in longitude, provided only that the difference was relatively small, and that M was not zero or negative. On the assumption of the Moon's homogeneity, Lagrange found for the physical libration a period of about 119 months.

The second and third equations derived from (A) were also second-order differential equations, but the coefficients of the terms containing the second-order differentials were relatively small, and Lagrange decided that these terms could be dropped. Integration of the resulting first-order equations yielded formulas for the inclination of the lunar equator to the ecliptic, and for the longitude of the descending node of the lunar equator where it intersects a plane through the Moon's centre parallel to the ecliptic. Cassini and Mayer, as we have seen, claimed that the longitude of the descending node of the lunar equator was coincident, or nearly so, with the longitude of the ascending node of the lunar orbit; but Lagrange could find in his analysis no reason for this coincidence.

He returned to the problem of the lunar libration in 1780, in a memoir presented to the Berlin Academy. In this second assault he made use of several new techniques.

Once more he took as starting-point his fundamental dynamical principle. But by 1780 he had discovered an algorithm which expedites the derivation of the differential equations of motion. Let the generalized coordinates adopted in formulating the problem (they should be as few as possible, and independent of one another) be $\phi, \phi', \ldots$, and let T and V be the functions of these coordinates that we now know as the kinetic and potential energy. Then the differential equation for any one of the variables ϕ will be given by

$$d \frac{\delta T}{\delta d\phi} - \frac{\delta T}{\delta \phi} + \frac{\delta V}{\delta \phi} = 0. \qquad (B)$$

The operator 'δ' here signifies partial differentiation.

For the coordinates of the centre of gravity of the Moon, Lagrange adopted the rotating coordinate system used by Euler in his *Nouvelle théorie de la lune* of 1772: the curtate radius vector, rotating with the Moon's mean speed in the plane of the ecliptic, is $1 + x$, where the mean radius vector has been normalized to equal 1; y is at right angles to the radius vector in the plane of the ecliptic; and z is at right angles to the plane of the ecliptic. The variables x, y, z are advantageous because they remain very small, so that their powers and products of more than two dimensions can be neglected. For Lagrange there was the additional advantage that Euler had already calculated their values "with great precision", insofar as they depended on the action of the Earth and the Sun on the Moon, regarded as a point. (In his calculation, Euler had not taken into account the non-sphericity of the Moon, an omission that Lagrange corrected in the present memoir, but without finding any notable effect.)

The most important innovation in the memoir of 1780 was a new choice of variables for the determination of the motions of the Moon's body about its center of gravity. Let ω be the inclination of the Moon's equator to the ecliptic; ψ the longitude of the ascending node of the lunar equator with a plane through the Moon's centre parallel to the ecliptic; ϕ the angular distance of a prime meridian on the Moon from this node. For two of his new variables, Lagrange introduced

$$s = \tan\left(\frac{\omega}{2}\right) \cdot \sin \phi,$$

$$u = \tan\left(\frac{\omega}{2}\right) \cdot \cos \phi.$$

Since according to the observations ω remains less than about 2°, s and u will remain small. An effect of the new definitions is to avoid explicit occurrence of trigonometric functions in the differential equations. Lagrange had used a similar trick in his memoir of 1774 on the secular equations of the nodes and inclinations, to be dealt with in a later section.

To obtain his final new variable, Lagrange set $\theta = \phi + \psi$, and then noted that θ would be equal to $180° + t$, where t is the mean motion of the Moon, if there were no physical libration. Then if r is the physical libration, $\theta = \phi + \psi = 180° + t + r$, and r will always remain very small. The motions of the Moon about its centre will be completely determined in terms of r, s, and u.

The three second-order differential equations that resulted from Lagrange's algorithm were linear in r, s, and u, and involved as constants the moments and products of inertia of the Moon about its polar axis and two axes in the plane of its equator, one of them being directed toward the Earth, and the other at right angles thereto. The equations showed that the products of inertia must be very small or zero, and Lagrange set them equal to zero, thereby adopting the hypothesis that the three axes mentioned were natural axes of rotation.

Lagrange's solution for r showed that it must be oscillatory, for the only alternative was that it would contain a term proportional to t, and so increase without bound, contrary to observation. Here as in the memoir of 1764, Lagrange saw the physical libration, however small, as explaining how the Moon's mean rotational and orbital velocities would have become equal, if not created exactly so at the beginning.

The main achievement of the memoir of 1780 was to obtain solutions for s and u while taking account of the second-order differentials. In a discussion of these solutions, Lagrange established that the angular distance of the Moon from the mean ascending node of its orbit must be equal to the mean angular distance of the Moon's prime meridian from the node of the lunar equator. Thus the equality that had appeared as an arbitrary fact in his earlier memoir was at last shown to be a consequence of the theory.

Perturbations in orbital motion: two memoirs of 1766

Lagrange's second memoir on celestial mechanics was his *Recherches sur les inégalités des satellites de Jupiter*, which won the Paris Academy's prize of 1766. This extensive memoir shows Lagrange in full command of the earlier work on celestial mechanics of Alexis-Claude Clairaut (1713–65), Euler, and d'Alembert, and arriving independently at the conclusion that Euler had set forth in his still unpublished memoir of 1752 on the inequalities of Jupiter and Saturn: the discovery that the orbital elements of a perturbed planet, as determined by observation, include major components deriving from perturbation.

We first take notice of the symmetry that Lagrange gave to the equations of motion. Let r be the curtate radius vector in the plane of reference (the orbital plane of Jupiter); u its reciprocal; ϕ, the longitude measured in this plane; p, the tangent of the latitude; F, the force that Jupiter exerts on a satellite at unit distance; R, Q, and P the components of the perturbing force in the three coordinate directions; and c an arbitrary constant. Lagrange's equations for u and p are:

$$\frac{d^2u}{d\phi^2} + u - \frac{F(1+p^2)^{3/2} + Rr^2 + Qr\left(\frac{dr}{d\phi}\right)}{c^2 + 2\int Qr^3 d\phi} = 0, \quad (2)$$

$$\frac{d^2p}{d\phi^2} + p - \frac{r^3\left[P - pR + Q\left(\frac{dp}{d\phi}\right)\right]}{c^2 + 2\int Qr^3 d\phi} = 0. \quad (3)$$

If R, Q, and P are set equal to zero, the solutions of these equations become

$$u - \frac{F}{c^2} = \rho\cos(\phi - a), \quad (4)$$

the equation of an ellipse of which c^2/F is the semi-parameter, $\rho c^2/F$ the eccentricity, and a the longitude of the inferior apse; and

$$p = \lambda\sin(\phi - \epsilon), \quad (5)$$

where λ is the tangent of the inclination and ϵ the longitude of the node. But even if the perturbing forces are not zero, Lagrange observed, these solutions will still be valid if ρ, a, λ, and ϵ are allowed to vary. Euler had given differential equations for the variations of the inclination and node; Lagrange set them forth alongside the entirely parallel differential equations for the variations of the eccentricity and aphelion. But in determining the perturbations in the present memoir, he did not make use of this idea (it would be the major theme of his later work in perturbation theory). Instead, he employed the method of 'absolute perturbations', in which the perturbations are regarded as independent of the orbital elements.

For this purpose, Lagrange set

$$r = a(1+nx), \quad \phi = \mu t + ny, \quad p = nz, \quad (6)$$

where a and μ are the mean values of r and $d\phi/dt$, and n is a very small coefficient. These expressions were substituted into the equations of motion, yielding differential equations in x, y, and z. Terms

involving n to a higher power than the second were discarded.

Next, Lagrange developed expressions for the perturbing forces R, Q, and P, due to the action of the Sun and the mutual actions of the satellites. These force components involved the inverse third and fifth powers of the distances between the satellites, which Euler had first shown how to approximate as trigonometric series. Lagrange gave an elegant method of deriving the coefficients of these series by way of De Moivre's theorem,

$$(\cos\theta + i\sin\theta)^n = \cos n\theta + i\sin n\theta.$$

The resulting expressions for R, Q, and P were series, to be substituted into the differential equations for x, y, and z.

The first step in obtaining solutions to the differential equations was to solve the simpler equations obtained when all the terms involving the factor n were deleted. Lagrange carried out this step for the first satellite; since he was using what amounts to indicial notation (he used 'traits' or prime marks, which we here replace by subscripts), the solutions for the other satellites could at once be obtained by an appropriate change in indices. The reduced equations in x and z had the general form

$$\frac{d^2u}{dt^2} + M^2u + T = 0, \quad (7)$$

where M is a constant and T is a function composed of sines and cosines of multiples of the time t. Lagrange gave the solution in exactly the form used by Clairaut; thus the terms

$$B\cos pt + b\sin pt$$

in T led to the following terms in the solution:

$$\frac{B}{p^2 - M^2}[\cos pt - \cos Mt] \quad (8)$$

$$+ \frac{b}{p^2 - M^2}[\sin pt - b\sin Mt].$$

The solution would be invalid, of course, if $p = M$.

In the first approximation Lagrange found for x_1 an expression of the following form:

$$\begin{aligned} x_1 = {} & \epsilon_1\cos(M_1 t + \omega_1) & (9)\\ & - K_{21}\cos(\mu_2 - \mu_1)t - K_{22}\cos 2(\mu_2 - \mu_1)t\ldots\\ & - K_{31}\cos(\mu_3 - \mu_1)t - \ldots\\ & - K_{41}\cos(\mu_4 - \mu_1)t - \ldots\\ & - K_5\cos 2(m - \mu_1)t. \end{aligned}$$

Here the first term, multiplied by na, gives the changing part of the first satellite's radius vector due to its elliptical motion, and the remaining terms give the perturbations due to the actions of the three other satellites and the Sun. The Ks are constants, and μ_1, μ_2, μ_3, μ_4, and m are the mean motions of the satellites and the Sun about Jupiter. In the first approximation, M_1 proves to be equal to μ_1, which means that the line of apsides, in this approximation, remains fixed.

The corresponding expression for y_1 is

$$y_1 = -\frac{2\mu_1\epsilon_1}{M_1}\sin(M_1 t + \omega_1) \qquad (10)$$
$$+ k_{21}\sin(\mu_2 - \mu_1)t - k_{22}\sin 2(\mu_2 + \mu_1)t \ldots$$
$$+ k_{31}\sin(\mu_3 - \mu_1)t - \ldots$$
$$+ k_{41}\sin(\mu_4 - \mu_1)t - \ldots$$
$$+ k_5\sin 2(m - \mu_1)t.$$

The first term, multiplied by n, gives the main term of the equation of centre, and the remaining terms represent perturbations in longitude.

The solution of the third equation proved to be

$$p_1 = nz_1 = n\lambda_1\sin(N_1 t + \eta_1), \qquad (11)$$

where λ_1, and η_1, are arbitrary constants, and in the first approximation $N_1 = \mu_1$, so that the nodes are fixed.

These first-approximation solutions proved to be all that was needed to predict the anomalies so far detected in the satellite motions. The Swedish astronomer Pehr Wilhelm Wargentin (1717–83), on the basis of a long series of careful timings of the eclipses of the satellites as they passed through Jupiter's shadow, and comparisons between his own observations and those of earlier astronomers, had found that the uniform circular motions previously ascribed to the first three satellites did not quite accurately hold, but were modified in each case by a single anomaly. Thus in the case of the first satellite Wargentin found a sinusoidal anomaly with a period of $437^d\ 19^h\ 41^m$ and a maximum value of 3.5 minutes of time. For this inequality to be accounted for by the theory, it had to be derivable from the formula for y_1.

Now in this formula we have first of all the term giving the equation of centre, but its coefficient contains an arbitrary constant, so that its size can only be determined by observation. Of the other terms, there is one that dwarfs all the others, the term with argument $2(\mu_2 - \mu_1)t$. This term is especially large because (a) the coefficients of the terms in the solution have denominators of the form $p^2 - M^2$, p being the coefficient of t and M

being here identical with μ_1; and (b) the mean motion of the first satellite is very nearly twice that of the second (more precisely, $\mu_1{:}\mu_2 = 2.0075{:}1$). Thus the denominator of the term with argument $2(\mu_2 - \mu_1)t$ is especially small:

$$2^2(\mu_2 - \mu_1)^2 - \mu_1{}^2 = \mu_1{}^2\left[4\left(1 - \frac{1}{2.0075}\right)^2 - 1\right]$$
$$= 0.007\,49\mu_1{}^2.$$

Therefore the coefficient is especially large.

The period of the inequality with argument $2(\mu_2 - \mu_1)$ is 1.762 731 77 days, while the synodic period of the first satellite with respect to the Sun is 1.769 860 526 days. (These values are derived from Wargentin's determinations of the synodic periods of the satellites, that is, their periods from one eclipse to the next. For convenience of calculation we have put them in decimal form.) Hence the period P in which the inequality returns to the same phase in eclipses will be given by

$$\frac{1}{P} = \frac{1}{1.762\,731\,77} - \frac{1}{1.769\,860\,526},$$

whence $P = 437.634\,47$ days, very nearly the Q period found by Wargentin.

In the case of the second satellite, Lagrange's theory showed two perturbational terms far surpassing the others in size, the terms with arguments $2(\mu_3 - \mu_2)t$ and $(\mu_1 - \mu_2)$. The ratio $\mu_2{:}\mu_3$, like the ratio $\mu_1{:}\mu_2$, is just slightly in excess of 2:1, and as a result the coefficients of both terms are made large by small denominators. The two inequalities have very nearly the same period, and the period in which they return to the same phase with respect to eclipses of the second satellite is about 437.67 days, very nearly the same as that found for the first satellite. This result agreed with Wargentin's observational finding. In effect the theory yields a single inequality for the second satellite, for as Lagrange deduced from the observations, $2(\mu_3 - \mu_2)t = 180° - (\mu_1 - \mu_2)t$, so that the sines of the two arguments are equal. Two decades later Laplace would undertake to explain *why* this special relation, first pointed out by Lagrange, must hold.

The theory of the third satellite contained a single sizeable perturbational term, with the argument $(\mu_2 - \mu_3)t$, leading to an inequality in eclipses with a period, once again, of some 437.67 days. Wargentin had detected its presence, but had been unable to satisfy himself with regard to its quantity.

Having accounted for the inequalities detected by Wargentin, Lagrange turned in the remainder of his memoir to the further development of the theory. In the second stage of the method of absolute perturbations, the first-stage solutions are substituted back into the differential equations in the terms multiplied by the small coefficient n, and the differential equations are then solved once more. This process, Lagrange showed, results necessarily in terms in which the factor n does not appear – terms therefore belonging to the first-order solution.

Thus in the differential equation for x_1 there is a term of the form $nA_1x_2\cos(\mu_2-\mu_1)t$, where A_1 is a constant and x_2 in the first approximation is given by $e_2\cos(M_2t+\omega_2)$. For the second-stage approximation the constant M_2 takes the form $\mu_2(1-n\beta_2/2)$. With these substitutions, there emerges in the differential equation a term of the form

$$\frac{n}{2}A_1e_2\cos\left[\left(\mu_1-\frac{n\beta_2\mu_2}{2}\right)t+\omega_2\right],$$

and in the solution this leads to the term

$$\frac{\dfrac{n}{2}A_1e_2\cos\left[\left(\mu_1-\dfrac{n\beta_2\mu_2}{2}\right)t+\omega_2\right]}{\left[\mu_1-\dfrac{n\beta_2\mu_2}{2}\right]^2-\left[\mu_1-\dfrac{n\beta_1\mu_1}{2}\right]^2}$$

$$\approx\frac{A_1e_2\cos\left[\left(\mu_1-\dfrac{n\beta_2\mu_2}{2}\right)t+\omega_2\right]}{2\mu_1(\beta_1\mu_1-\beta_2\mu_2)},$$

in which n does not appear, and which therefore belongs to the first order in x_1.

There is worse to come. An entirely similar term is found in x_2, and when this is substituted for x_2 in $nA_1x_2\cos(\mu_2-\mu_1)t$, there emerges in the solution a term with zero for denominator, implying that x_1 contains a term proportional to t; indeed, an infinity of such terms lurks ready to emerge. The same is true for y_1 and z_1. Thus, the procedure of successive approximations in the method of absolute perturbations leads unavoidably to the proliferation of *arcs de cercle*, angles increasing proportionally to the time. Lagrange regarded this as the most important discovery of the present memoir. As he wrote to d'Alembert, "[it] appears to me entirely new and of a very great importance in the theory of the planets".

In fact, the discovery was not new; the same insight lay at the basis of Euler's memoir of 1752 on the inequalities of Jupiter and Saturn, which was published only in 1769. Euler, too, had seen that the *arcs de cercle* emerging from the successive approximations of the method of absolute perturbations did not really belong to the solution, and that a different method was necessary to gather into a single finite expression all the variations of a given order, without permitting their exfoliation into an infinite power series in t, the time.

To the achieving of such a solution, Lagrange devoted a long algebraic development, in which his favourite device of integration by parts plays a prominent role. He introduced new variables, in which the reciprocal relations whereby terms contribute to one another through integration became symbolically explicit; six new differential equations emerged; and an integration was effected by means of integrating factors. The solution was similar in form to Euler's, but free of the algebraic errors by which Euler's had been bedevilled. Thus as the first-order result for x_1 Lagrange obtained

$$x_1=e_{11}\cos\left[\left(\mu_1-\frac{n\rho_1}{2}\right)t+\omega_1\right]$$
$$+e_{12}\cos\left[\left(\mu_1-\frac{n\rho_2}{2}\right)t+\omega_2\right]$$
$$+e_{13}\cos\left[\left(\mu_1-\frac{n\rho_3}{2}\right)t+\omega_3\right]$$
$$+e_{14}\cos\left[\left(\mu_1-\frac{n\rho_4}{2}\right)t+\omega_4\right].$$

Similarly for y_1 he found

$$y_1=-2e_{11}\sin\left[\left(\mu_1-\frac{n\rho_1}{2}\right)t+\omega_1\right]$$
$$-2e_{12}\sin\left[\left(\mu_1-\frac{n\rho_2}{2}\right)t+\omega_2\right]$$
$$-2e_{13}\sin\left[\left(\mu_1-\frac{n\rho_3}{2}\right)t+\omega_3\right]$$
$$-2e_{14}\sin\left[\left(\mu_1-\frac{n\rho_4}{2}\right)t+\omega_4\right].$$

Here the constants ρ_1, ρ_2, ρ_3, and ρ_4 are the solutions of a biquadratic equation. The expressions for x_1 and y_1 when multiplied by n give the variations in the radius vector and the equation of centre of the first satellite. As in Euler's formulation, we can distinguish a 'proper' eccentricity and equation of centre and 'adventitious' eccentricities and equations of centre; the 'proper' being represented by the first term in each of the two expressions, and the 'adventitious' by the last three terms, which can be regarded as having their source in the perturbing satellites. The expressions for x and y in the cases of the second, third, and fourth satellites have an exactly parallel form.

Entirely analogous results emerge for z. They show that the first-order variation in the tangent of the latitude of any of the satellites may be determined by imagining four planes passing through the centre of Jupiter, of which the first retrogrades with constant inclination on the plane of the orbit of Jupiter, the second moves in the same manner on the first, the third on the second, and the fourth, which will be the plane of the orbit of the satellite, on the third.

In the equations for x, y, and z are contained implicitly the secular equations of the eccentricities, aphelia, nodes, and inclinations of the satellites. To find expressions for these, it would only be necessary to combine the four terms of each expression into a single term, in such a way that x and z take the forms

$$x = D \cos(Mt - w),$$
$$z = G \sin(Lt - k).$$

Here D and G will be found to involve a slow sinusoidal change, and w and k to involve a slow, steady motion with a sinusoidal variation superimposed. In other words, the eccentricities and inclinations of the orbits oscillate about mean values; the apses progress and the nodes regress, but with added oscillatory motions.

Lagrange did not carry out the calculation of these secular variations in his memoir on the satellites of Jupiter, but immediately after completing this memoir he applied the same theory to the interactions of Jupiter and Saturn, carrying the approximation to terms of the order of n^2, and deriving the secular variations of the eccentricities, aphelia, inclinations, and nodes that each of these planets causes in the other. His analysis was published in the midst of a long memoir, "Solution de différents problèmes de calcul intégral", in Vol. 3 of the *Miscellanea Taurinensia* (1762–65). His results for the secular variations of these elements were very close to those that Laplace would give in his *Mécanique céleste*, except that Laplace found some-

what smaller values for Jupiter, because of using a smaller value for the mass of Saturn (Lagrange used the Newtonian value, $\frac{1}{3021}$ of the mass of the Sun, while Laplace reduced this to $\frac{1}{3512}$, which is close to the value now accepted).

In another respect, Lagrange obtained a result with which Laplace was soon to take issue: he found secular variations in the mean motions of the two planets. In effect, the differential equation for y (where ny represents what is variable in the longitudinal motion) contains terms proportional to x^2 and z^2, and from these can be derived, in the solution for y, terms proportional to the square of the time. For the increasing inequality in Jupiter's motion in longitude, Lagrange found $+2''.7402 N^2$, where N is the number of revolutions since epoch. For Saturn the result was $-14''.2218 N^2$. These results were empirically plausible, since they showed an acceleration for Jupiter and a deceleration for Saturn, in agreement with what astronomers had been finding from Kepler's time to the 1760s. It was Laplace's discovery (to be discussed in the next chapter) that the mean motions, as well as the mean *radii vectores*, were immune to secular variation; the second-order approximation that had led to Lagrange's result was inadequate.

The prize essays of 1772 and 1774: the problem of three bodies and the secular equation of the Moon

For its contest of 1768, the Paris Academy set the problem of "perfecting the methods on which the lunar theory is founded, in order to determine precisely the equations of this planet that are as yet uncertain, and in particular to examine whether a reason can be derived from this theory for the secular equation of the Moon". D'Alembert early in 1766 encouraged Lagrange to compete, but Lagrange, who was just then entering upon his new duties as head of the mathematical section of the Berlin Academy, declined. No prize was awarded in 1768, and the same problem was set for the contest of 1770, with the offer of a double prize.

In March 1768 Lagrange wrote to d'Alembert that he was at work on the three-body problem, and might have an essay ready for submission in the contest of 1770. In the preceding year, stimulated by Euler's memoirs on the subject, he had written two memoirs on the restricted problem, in which a body is attracted by two fixed centres; these were published in Vol. 4 of the *Miscellanea Taurinensia* (1766–69). Once more, however, the deadline for entry into the contest passed without his submitting an essay. Half the prize was awarded to Leonhard Euler and his son Johann Albrecht Euler, for the lunar theory with rotating coordinates described in the preceding chapter; the remaining half was reserved to be joined to the prize of 1772, when the same problem was to be posed for a third time. The prize commission was evidently not altogether satisfied with Euler's theory, which, despite its innovative method, revealed nothing strikingly new.

The *Essai sur le problème des trois corps* which Lagrange submitted in the contest of 1772, and which won half the prize (the other half being awarded to Euler for a more thorough development of his lunar theory), attacked the three-body problem in a way never before tried. It set out to determine the motions of the bodies relative to each other solely in terms of the distances between them. By means of the integral of *forces vives* and the three integrals expressing the conservation of angular momentum, Lagrange reduced the problem to the solution of seven differential equations: three giving the second derivatives of the squares of the three distances in terms of (a) these distances, (b) the masses of the bodies, and (c) certain auxiliary variables defined in terms of the distances and masses, and the remaining four equations giving relations between the auxiliary variables. Assuming these equations solved, so that the three mutual distances are expressed as functions of the time, Lagrange showed that the coordinates of the bodies with respect to a fixed frame of reference could be found without further integration.

Lagrange's derivation reduced the system of differential equations from the twelfth to the seventh order; later investigators carried the reduction to the sixth order by elimination of the nodes. (It has been claimed, however, that such a reduction is implicit in Lagrange's analysis.) Lagrange failed to give his equations the symmetry they could have had. He defined the three relative velocities as those of bodies B and C with respect to A, and of body C with respect to body B, instead of choosing the velocity of B with respect to A, that of C with respect to B, and that of A with respect to C,

in rotational order. This asymmetry in his equations resulted from his being preoccupied with their application to the case of the Moon.

In the second chapter of his essay, Lagrange dealt with the cases in which his equations admitted of closed solutions, without approximation. He found two such cases, in both of which the distances between the bodies were required to be either constant or in fixed ratios to each other. The bodies might fall in a straight line, with one of the distances equal to the sum of the other two; it was the solution that Euler had earlier found for the restricted problem, in which two of the bodies are in Keplerian motion while the third has negligible mass. Or the three distances between the bodies could be equal, so as to form an equilateral triangle. In both cases the motions were confined to a single plane. If the distances were supposed variable but in fixed ratios, the orbits of any two about the third proved to be conic sections.

Lagrange believed these solutions to be inapplicable in "the system of the world", but in the present century one of the cases has been found to be exemplified in the Trojan asteroids, two groups of asteroids with the same period about the Sun as Jupiter, and occupying positions to the east and west of Jupiter so as to form equilateral triangles with the Sun and Jupiter at the other vertices. The first of these asteroids was discovered photographically by Max Wolf in 1906.

In the last two chapters of his essay, Lagrange showed how his theory could be applied to the motion of the Moon. In an initial step he introduced the supposition that the distances of A (the Earth) and B (the Moon) from C (the Sun) were much larger than the distance between A and B; the two larger distances he set equal to R/i and R'/i, where i is a very small coefficient. He then obtained an approximative solution of the differential equations good to terms of the ith order. He did not, however, attempt a numerical calculation of the coefficients of the terms thus found.

For its contest of 1774, the Paris Academy posed a double question: first, how one might be assured, in the calculation of the lunar motions, that no detectable error results from the quantities neglected; second, whether, taking account if necessary of the actions of the planets and the non-sphericity of the Earth and the Moon, one could explain, on the basis of the theory of gravitation alone, why the Moon appears to have a secular equation in its longitudinal motion. In his prize-winning essay, Lagrange dealt only with the second question, arguing that neither the non-sphericity of the Earth nor that of the Moon could account for a secular equation of the magnitude claimed by the astronomers.

The existence of a continual acceleration in the Moon's motion had first been suggested by Halley in the *Philosophical Transactions* for 1693; in a comparison of modern eclipses of the Sun or Moon with Babylonian observations of such eclipses on the one hand, and Arab observations on the other, he had found that a single value of the mean motion would not serve. In the *Philosophical Transactions* for 1749 Richard Dunthorne examined a number of ancient observations of eclipses and found an acceleration such as to yield an increment of 10 arc-seconds in the first century. Tobias Mayer in his first lunar tables of 1753 gave the increment as 7″, but in his second set of tables published posthumously in London in 1770 this number was increased to 9″. Lalande in a memoir of 1757 put the number at 10″.

In effect, the astronomers were assuming the Moon's mean motion to be given by an expression of the form $Z + iZ^2$, where Z is proportional to the time and i is a very small coefficient. As Lagrange pointed out, an apparent secular acceleration could also result from a sinusoidal term of the form $\sin(A + \mu Z)$, where A is a constant and μ is a very small coefficient. Indeed, as Lagrange's analysis showed, the only way in which the theory of gravitation could account for the supposed secular acceleration would be by means of such a sinusoidal term. And in this case an approximation of the form $Z + iZ^2$ would eventually fail. The question to be considered, then, was whether the longitudinal component of the perturbational forces acting on the Moon could produce a sinusoidal term of the form $\sin(A + \mu Z)$.

A noteworthy feature of Lagrange's analysis in the memoir of 1774 was his use of a potential function in deriving the perturbative forces in given directions. Let F be the force between two bodies A and B, and s the distance between them; then, Lagrange pointed out, the component of the force in the coordinate direction x would be $-F.\partial s/\partial x$. Suppose the point-masses of which the Earth is made up to be dm, dm', dm'',...., at distances s, s', s'',...., respectively, from the centre of the Moon. Then the gravitational forces exerted by these point-masses on the Moon will be dm/s^2, dm'/s'^2, dm''/s''^2, etc., which may be expressed by $-\partial(dm/s)/\partial s$, $-\partial(dm'/s')/\partial s'$, $-\partial(dm''/s'')/\partial s''$, etc., and the total force exerted by all the point-masses will be the sum of these terms, or $-(\partial/\partial s)\int(dm/s)$. Let $\int dm/s = V$. It follows that the total force on the Moon in the coordinate direction x will be $(\partial V/\partial s)(\partial s/\partial x) = \partial V/\partial x$. This result holds for any coordinate, whatever the coordinate system. We have here Lagrange's earliest utilization of the potential function in celestial mechanics.

In searching for a term of the form $\sin(A + \mu Z)$, Lagrange first observed that all the perturbational terms in the longitude could be expressed in the form

$$K \sin(m\xi + ns + p\eta + qz + r\psi),$$

where m, n, p, q, and r are positive or negative integers or zero, and

ξ = the anomaly of the Sun,
s = the anomaly of the Moon,
η = the angular distance of the Moon from the Sun,
z = the longitude of the Moon counted from the equinox,
ψ = the angular distance of the prime meridian of the Earth from the equinoctial colure (the great circle through the equinoxes and the celestial poles).

The problem thus reduced to finding a linear combination of the foregoing variables yielding a nearly constant angle. According to Lagrange, the only combination meeting the requirement was $z - \xi - \eta$, along with its multiples; this combination being equal to the longitude of the Sun's apogee, which moves only about 1° 50′ per century. Lagrange therefore went on to identify, among the terms arising from the solution of the differential equations, all those containing a sine or cosine of this angle or its double or triple. The coefficients of these terms were in part dependent on constants expressing the shape and mass-distribution of the Earth.

The final step of the analysis consisted in evaluating these coefficients. If the northern and southern hemispheres of the Earth were assumed to be exactly alike, the coefficients either vanished or yielded a result far too small to account for the supposed secular equation of the Moon. If the northern and southern hemispheres were assumed to be dissimilar, an apparent secular equation of the magnitude Mayer had assigned could only be derived if the dissimilarity were supposed very large – larger than would be compatible with empirical determinations of the Earth's shape or with the theory of the precession of the equinoxes and nutation of the Earth's axis.

As for the non-spherical shape of the Moon, Lagrange claimed without further calculation that this also was incapable of producing a secular equation in the longitude of the size required. For if the quantities in his formulas pertaining to the Earth were reinterpreted as applying to the Moon, no term containing a sine of the form $\sin(A + \mu Z)$ emerged: no combination of angular arguments would produce a nearly constant angle. Six years later, at the end of his second memoir on the libration of the Moon, Lagrange carried out the calculation in detail and found a term of the kind required, namely a term proportional to the physical libration of the Moon; but its coefficient turned out to be too small by several orders of magnitude to account for Mayer's secular equation.

Could the resistance of an interplanetary medium or aether be responsible for the Moon's secular equation? Euler in his prize-winning memoir of 1772 had pronounced it to be the only possible cause, but Lagrange thought the existence of a detectable aethereal resistance doubtful: it was not confirmed in the case of most of the planets, and was even contradicted by Saturn's apparent secular deceleration (the effect of such resistance being to cause the planet to lose potential energy, hence to come closer to the Sun, hence to have a shorter period).

Lagrange devoted the final sections of his memoir of 1774 to a discussion of the seven observations, from 720 BC to AD 1478, from which Dunthorne had derived his value for the secular acceleration. Although in a general way these observations fitted a curve of acceleration, Lagrange regarded them as too uncertain or vaguely reported to give assurance of the existence of a secular acceleration, in particular one proportional to the square of the time. The best course for the time being, he urged, would be to reject the secular equation entirely, while retaining the mean movement established by Mayer, which agreed with observations since the time of Tycho

without any adjustment for secular acceleration.

This expression of scepticism brought a protest from the Parisian astronomers, particularly Pierre-Charles Le Monnier. But Lagrange, as he explained in a letter to d'Alembert of May 1774, wanted to provoke the astronomers into providing better proofs for their secular equations. There were several responses to the challenge, one in the *Philosophical transactions* for 1780, another in the Berlin *Mémoires* for 1782. The observations by Ibn Yūnus of the eclipses of AD 977 and 978 provided a strong verification, which Laplace, among others, regarded as settling the question.

The source of the secular change appeared to have been found in 1787, when Laplace announced that the apparent secular acceleration of the Moon was caused by the secular diminution in the eccentricity of the Earth's orbit. The truth was, however, that Laplace's explanation gave only part of the cause, the other part being a backward drag on the Earth's rotation due to the Moon's gravitational action on the Earth's tides. But this second factor came to be recognized only in the 1860s.

Lagrange had already, in 1783, noticed that secular changes in the eccentricities and inclinations of a perturbing planet could produce a secular change in the mean motion of a perturbed planet, and he had carried out the calculation of this effect for the mutual perturbations of Jupiter and Saturn, but found it negligible. In 1792 he applied the same formulas to the calculation of the secular equation of the Moon, and like Laplace found it to be a little over '10″ for the first century. But Lagrange's derivation, like Laplace's, contained an error – the omission of negative terms reducing the coefficient – which would first be pointed out by John Couch Adams in 1854. (See on this topic the penultimate section of Chapter 28 below.)

A new approach to the secular equations of the nodes and inclinations, eccentricities and aphelia: 1774–75

In June 1774 Lagrange wrote to d'Alembert:

I have been occupied of late with the solution of this problem: Given different planes that pass through the same fixed point, and each of which moves at the same time on each of the others while maintaining the same inclination, but causing the line of nodes to retrograde with a given uniform movement, to find the position of the planes at the end of any time. *What gave me the idea was the quarrel between MM. [Joseph-Jérôme Lefrançais] de la Lande and [Jean-Sylvain] Bailly as to the discovery of the cause of the variations in the inclinations of the satellites of Jupiter. It seemed to me that it was necessary to consider the question from a more exact point of view than anyone had yet done, and I have found that it presented difficulties that made it worthy of the attention of geometers, independently of the use it can have in astronomy. When there are only two mobile planes, I can give the complete solution of the problem; but if I suppose a greater number, I fall into formulas that are absolutely intractable. However, I have found a particular method for treating the case of as many planes as one wishes, the only restriction being that the mutual inclinations must be very small, and also the movements of the nodes, which is the case with the planetary orbits. If you find this matter interesting enough, I could compose a memoir on it for your Academy, provided there would not be any indiscretion on my part in imposing on it too often my feeble productions.*

Euler, we recall, was the first to derive differential equations for the variations of the node and inclination of the orbit of one planet perturbed by another. In 1755 he had used these equations to deduce the variation in the position of the plane of the Earth's orbit due to perturbation by the other planets. Here he deduced the differential effect produced by each perturbing planet singly, then added the results together, without attempting an integration. Lalande in 1757 applied the same procedure to the five known planets other than the Earth, and Bailly then applied it to the Galilean satellites of Jupiter. The results would presumably be valid for hundreds of years. Lagrange, by contrast, was seeking formulas valid for all time.

In a memoir presented to the Berlin Academy in June 1774, Lagrange showed that when the problem is confined to two planets the inclinations of the two orbits with respect to a fixed plane oscillate between determinable bounds, and the nodes of the orbits on this same plane either retrograde steadily or oscillate between bounds, depending on the relative value of certain constants. If three or more planets were interacting, the solution turned out to depend on elliptic integrals, and so was not obtainable by known methods. However, if the mutual inclinations of the orbits taken two by two were very small, it became possible to represent the problem by a set of linear differential equations of the first order. The application of such a set of differential equations to the determination of the secular variations of the nodes and inclinations of the six then known planets was the subject of a memoir which Lagrange sent to Paris in October 1774.

Lagrange's derivation of the differential equations turned on a substitution of new variables whereby the equations were freed from sines and cosines. For a given planet let the integrals of area in rectangular coordinates be

$$R = \frac{x\,dy - y\,dx}{dt},$$

$$Q = \frac{x\,dz - z\,dx}{dt}, \qquad (12)$$

$$P = \frac{y\,dz - z\,dy}{dt}.$$

These expressions will be variable or constant, depending on whether perturbation is present or absent. The line of nodes in the xy plane will form an angle χ with the x-axis such that $\tan \chi = P/Q$, and the tangent λ of the instantaneous inclination of the orbit to the xy plane will be $[P^2 + Q^2]^{1/2}/R$. The new variables that Lagrange introduced were

$$s = \lambda \sin \chi = \frac{P}{R}, \qquad (13)$$

$$u = \lambda \cos \chi = \frac{Q}{R}.$$

Thus

$$\tan \chi = \frac{s}{u}, \qquad \lambda = \sqrt{s^2 + u^2}.$$

If the orbits are assumed to be and remain circular, and all terms that depend explicitly on the positions of the planets in their orbits are discarded, the differential equations take the form

$$0 = \frac{ds}{dt} + (0,1)(u - u') + (0,2)(u - u'') + \dots, \quad (14)$$

$$0 = \frac{du}{dt} - (0,1)(s - s') - (0,2)(s - s'') - \dots,$$

$$0 = \frac{ds'}{dt} + (1,0)(u' - u) + (1,2)(u' - u'') + \dots,$$

$$0 = \frac{du'}{dt} - (1,0)(s' - s) - (1,2)(s' - s'') - \dots,$$

$$0 = \dots.$$

Here prime marks distinguish the variables referring to different planets, and the expressions (0,1), (0,2),..., (1,0), (1,2),..., etc. are constants depending on the masses and *radii vectores* of the planets, and on the coefficients of the trigonometric series representing the inverse cubes of the distances between perturbing and perturbed planets.

The solution of the system, Lagrange showed, is

$$s = A \sin \alpha + B \sin(bt + \beta) + C \sin(ct + \gamma) + ..., (15)$$
$$u = A \cos \alpha + B \cos(bt + \beta) + C \cos(ct + \gamma) + ...,$$
$$s' = A \sin \alpha + B' \sin(bt + \beta) + C' \sin(ct + \gamma) + ...,$$
$$u' = A \cos \alpha + B' \cos(bt + \beta) + C' \cos(ct + \gamma) + ...,$$
....

Here t is the only variable, and the values of the other letters can be determined on the basis of observation, although the process increases in complication with the number of planets.

In applying this solution to the six known planets, Lagrange observed that the constants (0,1) and (1,0), where '0' refers to Jupiter and '1' to Saturn, are – because of the larger masses of these two planets – larger by several orders of magnitude than the corresponding constants referring to other pairs of planets. He therefore neglected the effects of the other planets on these two, and took account only of their effects on each other. So for Jupiter he found

$$\tan \chi = \frac{s}{u} = \frac{A \sin \alpha + B \sin(bt + \beta)}{A \cos \alpha + B \cos(bt + \beta)}, \quad (16)$$

$$\lambda = \sqrt{s^2 + u^2} = \sqrt{A^2 + B^2 + 2AB \cos(bt + \beta - \alpha)}, \quad (17)$$

and the solution for Saturn is the same except that B is replaced by B'. On evaluating the various constants, Lagrange found that λ and λ', the tangents of the inclinations of the two orbits to the ecliptic, undergo oscillations with a period of 51 150 years; Jupiter's orbit is at maximum inclination when Saturn's is at a minimum, and *vice versa*; the ranges of oscillation are 45' 3" for Jupiter and 1° 45' 51" for Saturn. The motion of the nodes is also libratory with the same period of 51 150 years; the mean positions of the nodes of the two orbits coincide and are fixed with respect to the stars, and the range of libration is 26° 7' for Jupiter and 64° 8' for Saturn.

For each of the remaining four planets, Lagrange went on to derive the secular variations of the variables s_i and u_i, but, because of the complexity of the equations, did not attempt to determine whether for these planets the nodes and inclinations are confined to bounded ranges.

The procedure just indicated for determining the secular variations of the nodes and inclinations is applicable, *mutatis mutandis*, to the secular variations of the aphelia and eccentricities. Laplace read Lagrange's memoir immediately it was received by the Paris Academy, and recognized this possibility; in mid-December 1774 he registered with the Academy a memoir developing the idea. It was published in 1775 in Part I of the Paris *Mémoires* for 1772, some three years before the publication of the memoir Lagrange had submitted in October. But Laplace appended to his memoir an extract of a letter from Lagrange of April 1775, which shows that Lagrange had had precisely the same idea as Laplace.

In the quoted extract, Lagrange started from the solution of the three-body problem given in Clairaut's *Théorie de la lune*:

$$\frac{f^2}{Mr} = 1 - \sin u \, [g - \int \Omega \cdot \cos u \cdot du]$$
$$- \cos u \, [c + \int \Omega \cdot \sin u \cdot du],$$

where f, M, g, and c are constants, r is the planet's radius vector and u its longitude, and Ω is a function of the perturbing forces. He then set

$$g - \int \Omega \cdot \cos u \cdot du = e \sin I = x,$$
$$c + \int \Omega \cdot \sin u \cdot du = e \cos I = y,$$

so that the original equation became

$$\frac{f^2}{Mr} = 1 - e \cos(u - I)$$
$$= 1 - x \sin u - y \cos u,$$

which is the equation of an ellipse of which e is the eccentricity and I the longitude of the aphelion. The equations

$$dx = - \Omega \cdot \cos u \cdot du,$$
$$dy = + \Omega \cdot \sin u \cdot du,$$

will then yield the secular variations of the variables x and y, if Ω and u are expressed in terms of x, y, x', y',..., and t, and only those terms are retained in which x, y, x', y', etc. are linear and multiplied by constant coefficients. Since

$$e^2 = x^2 + y^2$$

and

$$\tan I = \frac{x}{y},$$

the secular variations of e and I are obtainable from those for x and y.

As we have seen, Lagrange's new way of determining the secular variations of the nodes and inclinations, aphelia and eccentricities, showed that these variations could be libratory and periodic. Lagrange continued to classify the perturbational variations as either 'secular' or 'periodic', but the true basis of the distinction could no longer be periodicity. The variations called 'secular', besides being slow, were expressible by formulas in which the orbital positions of the planets did not appear, whereas the variations called 'periodic', besides being short-term, depended explicitly on the configuration of the planets in their orbits.

That the nodes and aphelia of the planets move slowly and unidirectionally had been a premiss of Johannes Kepler's *Rudolphine Tables*, and despite the attempt of Thomas Streete (1622–89) and Nicolaus Mercator (*c.* 1619–87) to challenge this assumption had come to be generally accepted by eighteenth-century astronomers. Steady motion in these orbital parameters did not imperil the stability of the solar system. With the inclinations and eccentricities the case was different: if their variations were not periodic and confined within narrow bounds, the solar system would collapse. The question of the system's stability would soon be addressed, first by Lagrange and then by Laplace.

Lagrange on planetary perturbations, 1775–85

Lagrange, in his letter of April 1775, while thanking Laplace for sending manuscript copies of his memoirs on the secular variations, appeared to relinquish to him the entire subject:

Long ago I proposed to myself to take up once more my old work on the theory of Jupiter and Saturn, to push it further and to apply it to the other planets . . . . But as I see that you have yourself undertaken this research, I willingly renounce it, and I am very happy that you have relieved me of the necessity of undertaking this work, for I am persuaded that the sciences can only gain much thereby.

In thus proposing to cede the topic to his younger rival, Lagrange was probably moved by a sense of Laplace's ambition, as well as by a recognition of the importance of his discoveries, in particular his demonstration that the mean *radii vectores* were immune to secular variation.

But a month and a half later, writing to d'Alembert, Lagrange reclaimed the topic:

I am now at the task of giving a complete theory of the variations of the elements of the planets due to their mutual action. What M. de la Place has done on this matter has pleased me a great deal, and I flatter myself that he will not be displeased with me for not keeping the sort of promise I made to abandon it entirely to him. I have not been able to resist the desire to occupy myself with it again, but I am not the less delighted that he should also work at it on his side. I am even very eager to read his further researches on this subject, but I beg him to send me nothing in manuscript, only the printed memoirs. I beg you to tell him this . . . .

During the course of 1775 Lagrange completed and read to the Berlin Academy two long memoirs on the secular variations of the eccentricities, aphelia, inclinations, and nodes, but as he told Laplace in a letter of May 1776, he did not expect to have them published, "above all as I know that you are occupied with the subject . . . . I have always proposed to make this application, but I expect that your work will now make it unnecessary . . . ,"

If this was a challenge, Laplace did not rise to it. In the years from 1775 to 1785, his mathematical investigations, insofar as they had to do with the planets, were without direct bearing on the secular inequalities. They dealt with gravitational attraction of spheroids, tidal oscillations of the oceans and atmosphere, and the precession of the equinoxes. As he explained in February 1783 – he was responding to Lagrange's announcement that he was now revising his two memoirs for publication –

Our uncertainty as to the masses of the planets and the derangements they undergo due to comets, made me renounce work on the memoir that I was preparing on the variations of the eccentricities and aphelia . . . . I do not doubt that you will shed light on a matter so interesting.

The two parts of Lagrange's revised treatise on the secular variations were published in 1783 and 1784, in the Berlin *Mémoires* for the years 1781 and 1782. The first of these parts contains a classic presentation of the solution of the equations of motion by the variation of orbital parameters. A major feature of it was the use of a potential function, from which the components of the total perturbing force in the coordinate directions were derived by partial differentiation. This function, later dubbed by Laplace *the perturbing function*, had made its appearance in two earlier memoirs, the first presented to the Berlin Academy in October 1776, the other in October 1777; both were published in 1779. Before turning to the treatise on the secular variations, we review what these earlier memoirs accomplished.

The first of the memoirs was entitled "Sur l'altération des moyen mouvements des planètes", and was no doubt a response to Laplace's memoir on the same subject, which had been presented to the Paris Academy in 1773 and published in the *Savants étrangers* in 1776. Laplace demonstrated that the mean motions of the planets were immune to secular variation by showing that the constants determining the secular variation of a planet's mean radius vector cancelled one another out, up to the third dimensions of the eccentricities and inclinations. Lagrange wanted to avoid the latter restriction, and to obtain a proof that was *a priori* in the sense of showing not just *that* the result followed from the theory but *why*. His demonstration, like Laplace's, neglects the squares and products of the perturbing forces.

Let the equations of motion for a single planet in rectangular coordinates be

$$0 = \frac{d^2x}{dt^2} + \frac{Fx}{r^3} + X,$$

$$0 = \frac{d^2y}{dt^2} + \frac{Fy}{r^3} + Y \qquad (18)$$

$$0 = \frac{d^2z}{dt^2} + \frac{Fz}{r^3} + Z,$$

where F/r^2 is the force on the planet due to the Sun, and X, Y, Z are the components of the total perturbative force in the coordinate directions. If X, Y, and Z were zero, the equations would become

$$0 = \frac{d^2x}{dt^2} + \frac{Fx}{r^3},$$

$$0 = \frac{d^2y}{dt^2} + \frac{Fy}{r^3}, \qquad (19)$$

$$0 = \frac{d^2z}{dt^2} + \frac{Fz}{r^3},$$

which are integrable, their complete integrals giving the values of x, y, z in terms of the time and six arbitrary constants of integration. The arbitrary constants determine the size, shape, and orientation of the conic-section orbit in which the planet moves.

Let each of the three integral equations obtained from (19) be differentiated with respect to time; three first-order differential equations result, and these in combination with the three integral equations yield six differential equations of the first order, each containing but one constant of integration. Let any one of these equations be $V = k$, where k is the constant, and V is a function of x, y, z, dx/dt, dy/dt, and dz/dt. Differentiating, we have $dV = 0$, which must be identical with one of the original differential equations (18). If in dV we put in place of the differentials $d(dx/dt)$, $d(dy/dt)$, $d(dz/dt)$ their values – namely $-(Fx/r^3)dt$, $-(Fy/r^3)dt$, $-(Fz/r^3)dt$ – the expression becomes identically zero.

But now let us take account of the perturbative forces X, Y, Z in equations (18). The values of the differentials $d(dx/dt)$, $d(dy/dt)$, $d(dz/dt)$ will now be $-(Fx/r^3 + X)dt$, $-(Fy/r^3 + Y)dt$, $-(Fz/r^3 + Z)dt$, and dV will no longer be equal to zero. Thus k will no longer be a constant, and as Lagrange showed, we shall have

$$dk = -\left[\frac{\partial V}{\partial(dx/dt)}X + \frac{\partial V}{\partial(dy/dt)}Y + \frac{\partial V}{\partial(dz/dt)}Z \right] dt.$$
$$(20)$$

Equation (20) is a recipe for finding the variation of the orbital parameter k.

To find the variation of the orbit's major axis, $2a$, Lagrange made use of an equation obtainable from equations (19) by integration, called the integral of *forces vives*:

$$\frac{1}{\sqrt{x^2 + y^2 + z^2}} - \frac{dx^2 + dy^2 + dz^2}{2F\,dt^2} = \frac{1}{2a}. \quad (21)$$

Here V is given by the left-hand member, and $k = 1/2a$. Using equation (20) to determine dk, Lagrange found

$$d\left(\frac{1}{2a}\right) = \frac{X\,dx + Y\,dy + Z\,dz}{F}. \qquad (22)$$

It is at this point that Lagrange introduced the perturbing function, to which he gave the symbol Ω. (Clairaut, we recall, had given this symbol to a function similar in gathering together all the perturbing forces.) Let there be perturbing planets with masses T', T'', etc., coordinates (x', y', z'), $(x''$,

y'', z''), etc., and distances from the perturbed planet v', v'', etc. Further, let

$$\Omega = T'\left[\frac{xx' + yy' + zz'}{r'^3} - \frac{1}{v'}\right] \qquad (23)$$
$$+ T''\left[\frac{xx'' + yy'' + zz''}{r''^3} - \frac{1}{v''}\right]$$
$$+ \dots.$$

If we treat the Sun as at rest, transferring to the perturbed planet the actions on the Sun of T', T'', etc., but with changed sign, then it will follow that the components X, Y, Z of the perturbing force are given by $\partial\Omega/\partial x$, $\partial\Omega/\partial y$, $\partial\Omega/\partial z$, and the right-hand member of (22) becomes $\mathbf{d}\Omega/F = \mathbf{d}\Omega/(S+T)$, where the boldface '$\mathbf{d}$' means differentiation with respect only to the variables pertaining to the perturbed planet, and S and T are the masses of the Sun and perturbed planet, respectively.

Let θ, θ', θ'', ... be the mean motions of the planets T, T', T'', ... during the time t. Because the eccentricities of the planetary orbits are small, the variables $x, y, z; x', y', z'; x'', y'', z''$; etc. can all be expressed by means of series of sines and cosines of the angles θ, θ', θ'', ..., and their multiples. The distances between T and the perturbing planets can be developed as series of sines and cosines of the angles $(\theta - \theta')$, $(\theta - \theta'')$, etc. Hence the expression $\Omega/(S+T)$ can be reduced to a series of terms of the form

$$M \begin{Bmatrix} \sin \\ \cos \end{Bmatrix} (m\theta + n\theta' + p\theta'' + \dots),$$

where M is a quantity depending on the orbital elements, and m, n, p, ..., are positive or negative whole numbers or zero.

To obtain the derivative $\mathbf{d}\Omega/(S+T)$ it is only necessary to differentiate with respect to θ; the result will be a series of terms of the form

$$\pm mM \begin{Bmatrix} \sin \\ \cos \end{Bmatrix} (m\theta + n\theta' + p\theta'' + \dots),$$

Finally, to obtain the integral of this series, and hence the value of $1/2a$, Lagrange made, in accordance with Kepler's third law, the substitutions

$$\theta' = \theta\left[\frac{a}{a'}\right]^{3/2},$$
$$\theta'' = \theta\left[\frac{a}{a''}\right]^{3/2}, \dots,$$

where a', a'', ... are the semi-major axes of the orbits of the perturbing planets. The required integral is then given by terms of the form

$$\pm mM \frac{\begin{Bmatrix} \sin \\ \cos \end{Bmatrix}[(m + n(a/a')^{3/2} + p(a/a'')^{3/2} + \dots]\theta}{m + n(a/a')^{3/2} + p(a/a'')^{3/2} + \dots} \qquad (24)$$

Lagrange's conclusion was that $1/2a$, being given by a series of terms of form (24), would be subject only to periodic variations, provided that the denominator of (24) is never zero for any particular set of integers m, n, p, ... "But it is easy to be convinced that this case cannot occur in our system, where the values of $a^{3/2}$, $a'^{3/2}$, $a''^{3/2}$, ... are incommensurable with one another."

Lagrange's confidence in his conclusion seems rather astonishing: how could he claim to know that numbers, to which the only access was empirical, were incommensurable? But lack of acuity as to the meaning of incommensurability appears to have been endemic among the mathematicians. Laplace, equally ready with Lagrange to assume the incommensurability of the mean motions, was enthusiastic about Lagrange's argument: "The felicitous application of the beautiful method that you have explained at the beginning of your memoir," he wrote, "... joined to the elegance and the simplicity of your analysis, has given me a pleasure that I cannot put into words." And he went on to infer (erroneously, as he himself would later discover) that the apparent secular inequalities in the motions of Jupiter and Saturn could not be due to their mutual interaction.

A few years later Lagrange discovered that his argument required modification: the coefficient M in (24) could be subject to secular change owing to secular change in orbital elements other than the major axes.

The second memoir published in 1779 used the perturbing function to derive the integrals of motion for a system of bodies interacting in accordance with the inverse-square law of gravitation. Here Lagrange took the centre of gravity of the system as the origin of coordinates, and treated all bodies of the system equivalently. As a consequence the perturbing function took the simple form of a series of terms each having a product of two masses for numerator, and the distance between the two masses for denominator. The proofs that the centre of gravity of the system moves uniformly, and that the angular momentum and *forces vives* of the system are conserved, are remarkably simple and straightforward.

We saw earlier that Lagrange, in his prize-winning memoir of 1774 on the secular equation of the Moon, had used a potential function in deriving the gravitational forces on the Moon due to the non-spherical Earth. Using a potential function in determining the interactions of a system of point-masses was no doubt an obvious extension of the idea. But the importance of this step was nevertheless great. According to Laplace in Vol. 5 of the *Mécanique céleste*, the introduction of this function into celestial mechanics was, because of its utility, "a veritable discovery".

We return now to the treatise of 1783–84 on the secular variations. Its first part is a purely algebraic development, yielding general formulas for the variations of the orbital elements of a planet, and then (by deletion of sinusoidal terms dependent on the planetary positions) arriving at formulas for the variations called secular. Lagrange was here aiming to provide a complete and unified account, using the perturbing function and the procedures he had developed in earlier memoirs.

Once again he showed the mean solar distances to be immune to secular variation. Turning to the two pairs of elements, eccentricity and aphelion, inclination and node, he once again substituted new variables, arriving in both cases at two equations for each planet:

$$x = A \sin(at + \alpha) + B \sin(bt + \beta) + C \sin(ct + \gamma) + \ldots,$$
$$y = A \cos(at + \alpha) + B \cos(bt + \beta) + C \cos(ct + \gamma) + \ldots. \tag{25}$$

From x and y the variations of the inclination and node, or those of the eccentricity and aphelion, can be calculated. Thus the eccentricity is given by the square root of the sum of the squares of two expressions of the form of (25), and the same is true of the inclination. In (25) the letters $A, B, C, \ldots, \alpha, \beta, \gamma, \ldots$ are to be determined from the values of the orbital elements found observationally at a given epoch. The letters $a, b, c, \ldots$ are the roots of an equation of degree n, where n is the number of interacting planets. If the roots of this equation are all real and unequal, the solution will be oscillatory; if any two roots are equal, the solution will contain terms proportional to the time; and if any of the roots contain $\sqrt{(-1)}$ as a factor, the solution

will contain exponential functions of the time. Only in the first case will the variations of the inclinations and eccentricities be bounded.

In the second part of his treatise, Lagrange applied his formulas to the six known planets. Once again, he treated the system of Jupiter and Saturn separately, as being largely uninfluenced by the smaller planets. The solutions for the eccentricities and orbital inclinations of the two planets proved to be oscillatory and confined to narrow bounds. Lagrange concluded:

... it follows that the system of Saturn and Jupiter, insofar as one regards it as independent of the other planets, which is always permitted as we have shown, is of itself in a stable and permanent state, at least if we make abstraction of the action of every foreign cause, such as that of a comet, or of a resistant medium in which the planets move ....

Of the unexplained derangements in the motions of Jupiter and Saturn – Jupiter's apparent secular acceleration and Saturn's apparent secular deceleration – Lagrange was aware. He was asserting the stability of the Sun–Jupiter–Saturn system only in respect to gravitational interactions.

Among the remaining four planets, the masses of Mercury, Venus, and Mars were unknown. Lagrange cleared this obstacle by applying to these planets a relation he found among the densities of the three planets of known mass, Earth, Jupiter, and Saturn (the relation does not in fact apply). For the four planets he obtained eight simultaneous equations of the form of (25), and carried out the detailed process of determining the constants in these equations. The biquadratic equation for the coefficients of the time in the arguments of the sines and cosines proved to have four real and unequal roots. Lagrange concluded:

Thus we are already assured that these expressions cannot contain "arcs de cercle", and that consequently their exactitude will not be limited to a finite time, but will hold for all time.

Uncertainty as to the planetary masses, Lagrange admitted, might lead one to doubt this conclusion; he proposed in the future to seek a way of removing the doubt. (He was to be anticipated by Laplace.) Meanwhile, using his hypothetical masses, he computed upper bounds for the eccentricities and inclinations of the four inner planets, and found them all small. He restated his vision of the solar system at the end of the treatise:

the planets, in virtue of their mutual attraction, change insensibly the form and position of their orbits, but without ever going outside certain limits. The major axes remain unalterable; at least the theory of gravitation implies only alterations that are periodic and dependent on the positions of the planets . . . .

His long treatise on the secular inequalities completed, Lagrange next turned to the periodic inequalities. His treatise on this subject was published in two parts in 1785 and 1786, in the Berlin *Mémoires* for 1783 and 1784. The aim of the work was to develop the periodic inequalities by the same method of variation of orbital parameters used for the secular inequalities. The mathematical analysis in the new treatise was a natural sequel to that in the earlier one. There the secular inequalities were determined by deleting, from the differential equations for the variations of the orbital elements, all sinusoidal terms; the new deduction required that the deleted terms be restored. Given the earlier integrations for the secular variations, the differential equations with the newly included terms proved to be easily integrable, yielding the periodic contributions by themselves.

Given the periodic changes in the orbital parameters, Lagrange then proceeded by means of partial derivatives to deduce the resulting periodic changes in the perturbed planet's longitude, radius vector, and tangent of the latitude. Thus if the planet's true longitude is q, and q depends on orbital parameters p, x, y, s, u, whose values independent of periodic variation are $\bar{p}$, $\bar{x}$, $\bar{y}$, $\bar{s}$, $\bar{u}$, and whose periodic variations are $\tilde{\omega}$, ξ, ψ, σ, and v, then the periodic correction in the longitude will be

$$\left(\frac{\partial q}{\partial \bar{p}}\right)\tilde{\omega} + \left(\frac{\partial q}{\partial \bar{x}}\right)\xi + \left(\frac{\partial q}{\partial \bar{y}}\right)\psi$$
$$+ \left(\frac{\partial q}{\partial \bar{s}}\right)\sigma + \left(\frac{\partial q}{\partial \bar{u}}\right)v.$$

Analogous expressions hold for the radius vector and tangent of the latitude.

In the second part of the treatise, Lagrange applied the formulas thus derived to all the six then known planets, obtaining numerical values for the periodic inequalities of these planets independent of the eccentricities and inclinations. On the whole, the results were not new, but Lagrange's systema-

tic derivation gave them authority. Moreover, the same method could be extended to the derivation of the terms proportional to the eccentricities and inclinations and their powers and products. Lagrange had himself planned to extend his theory to terms proportional to the first powers of the eccentricities and inclinations, but found himself anticipated by Nicolaus Claude Duval-le-Roi (d. 1810), whose derivations and results for perturbations of this order were published in the Berlin *Mémoires* for 1792 and 1793. Earlier, in the Berlin *Mémoires* for 1787, Duval-le-Roi had published the extension of Lagrange's theory to the newly discovered planet Herschel [i.e., Uranus].

More important for later developments, however, were certain theoretical features in the first part of Lagrange's treatise on the periodic variations. At the start, he introduced a change in the analysis whereby, in his earlier treatise on the secular variations, he had obtained the relation between true longitude q and mean longitude p. In the earlier treatise he had derived the differential equation

$$dq + \beta \sin q \cdot dq + \gamma \cos q \cdot dq \qquad (26)$$
$$+ \delta \sin 2q \cdot dq + \epsilon \cos 2q \cdot dq + \ldots = dp,$$

where β and γ are of the first order in the eccentricities and inclinations, δ and ϵ of the second order, and so on. Because of the perturbations, the coefficients β, γ, δ, ϵ, ... had to be regarded as variables. Thus, in the earlier treatise, using integration by parts, Lagrange had derived an approximate integral of (26) in the form

$$q - (\beta)\cos q + (\gamma) \sin q$$
$$- (\delta)\cos 2q + (\epsilon) \sin 2q - \ldots = p, \qquad (27)$$

where (β), (γ), (δ), (ϵ), ... were given in terms of β, γ, δ, ϵ, ... and their derivatives with respect to q. By reversal of series, it was then possible to obtain q as a function of p.

In his new treatment of this topic, Lagrange proposed to keep the left-hand side of (27) in the form it would have if the orbital elements were constant, then apply to the value of p the corrections resulting from the variability of these elements. Thus, setting

$$d\Sigma = -\cos q \cdot d\beta + \sin q \cdot d\gamma$$
$$-\tfrac{1}{2}\cos 2q \cdot d\delta + \tfrac{1}{2}\sin 2q \cdot d\epsilon \qquad (28)$$
$$- \ldots,$$

and adding this equation to (26), he obtained after integration

$$q - \beta \cos q + \gamma \sin q$$
$$- \tfrac{1}{2}\delta \cos 2q + \tfrac{1}{2}\epsilon \sin 2q - \ldots = p + \Sigma. \quad (29)$$

In the differential $d(p + \Sigma)$ of this equation, dp represents the part of the variation of the mean motion that relates to the mean distance; thus we shall have $dp = dt/a^{3/2}$, where a is the semi-major axis of the ellipse, and is subject only to periodic inequalities which are functions of the orbital positions of the perturbed and perturbing planets. On the other hand, $d\Sigma$ is the variation of the mean motion due to variation of the orbital elements other than the mean distance or semi-major axis. It was to be related, Lagrange proposed, to the epoch of the mean movement, "which in invariable orbits contains the sixth arbitrary constant of the integrals, and so constitutes the sixth element of the elliptic movement."

Thus all six of the orbital elements had now been included in Lagrange's theory of the variation of orbital elements. And he had pinpointed a new source of possible apparent secular variation in the mean motion: secular variation in Σ. He proceeded to show that, to the first power of the eccentricities and inclinations and to the first powers of the perturbing forces, Σ is free of secular variation, but that if the approximation is carried to the second dimensions of the eccentricities and inclinations, there emerge terms in $d\Sigma$ that are proportional to the time. Lagrange devoted a special memoir to these secular variations; like his "Théorie des variations périodiques (Première partie)", it appeared in the volume of the Berlin *Mémoires* for 1783, published in 1785. Here he made the application to Jupiter and Saturn, but found the resulting variations far too small to account for the supposed secular variations in the mean motions of these planets. Nearly a decade later he would use the same formulas to deduce, as Laplace had done before him, the secular equation of the Moon.

The fact was that the apparent secular inequalities of Jupiter and Saturn were not really secular, in the sense of being independent of the configuration of the two planets. They were to be found, not in $d\Sigma$, but in dp; and there Laplace found them in 1785, after reading Lagrange's "Théorie des variations périodiques (Première partie)". And this memoir contains a suggestion that may have played a role in Laplace's discovery; it is nothing less than a description of the very procedure Laplace used, but applied to a different quantity.

Lagrange gives an explicit method for determining the terms of dp that are independent of the eccentricities and inclinations – an approximation which, he says, "suffices in most cases". Then he adds: "The calculations will not be more difficult, but only a little longer, when one wishes to take account also of the terms due to the eccentricities and inclinations." In discussing the terms of $d\Sigma$ that are proportional to the eccentricities and inclinations or to their powers and products, he offers the following suggestion about abbreviating the labour:

Because of the smallness of the terms involved, it will suffice to take account of those which will be augmented a great deal by integration; and it is clear that these are the terms which will contain sines and cosines of angles of which the variation is very small in relation to that of the angle p. If, for example, π designates one of these angles, and $d\pi = \nu\, dp$, ν being a very small coefficient, the first integration will introduce into the denominator the coefficient ν, and the second integration will introduce the square of ν. Since the angle π can only be composed of multiples of the angles p, p', p'', … joined together by the signs $+$ or $-$, it is only by the known relations of the mean motions of the planets that one will be able to judge in each case the value of the coefficient ν; there will thus be required a particular examination for each planet of which one wishes to calculate the perturbations.

If in the case of Jupiter and Saturn Lagrange had carried out the particular examination he recommends in relation, not to $d\Sigma$, but to dp, it would have been he rather than Laplace who discovered the source of the unexplained divagations in the mean motions of these planets. But although not uninterested in applying his theories to observations, his gaze was primarily directed toward achieving unity and clarity in the theory. And in the case at hand his focus was upon the inclusion of the sixth orbital element – the epoch – in the theory of variation of orbital elements, and on showing that there could be secular variation in the mean motion owing to secular variation in elements other than the mean distance.

Lagrange on the perturbations of comets

As the subject of its prize contest of 1778 the Paris Academy of Sciences proposed the theory of perturbations of comets, with particular application to the comet that Halley had hypothesized to account for cometary apparitions in 1532 and 1661. (The two apparitions are now recognized to have been due to two different comets.) In his winning essay Lagrange developed the theory of cometary perturbations by the method of variation of orbital elements; Laplace was to follow this method in his *Mécanique céleste*, with evident reliance on the Lagrangian formulations. For configurations in which the solar distance of the comet was either very large or very small relative to the solar distance of the perturbing planet, Lagrange derived analytical approximations permitting determination of the perturbations by formal rather than numerical integration; these formulas, too, would be followed by Laplace. As for the application to the supposed comet of 1532 and 1661, Lagrange indicated how it could be done, but left the calculations for others. He mentioned predecessors without naming them; he was undoubtedly referring to Clairaut in his *Théorie du mouvement des comètes*, and to d'Alembert in his "Théorie des comètes".

Lagrange's *Mécanique analytique*

Lagrange's *Mécanique analytique*, first published in 1788, was his *chef-d'œuvre*, the work from which later students of mechanics imbibed the Lagrangian formulations. Here Lagrange aimed to present the whole of mechanics as a rational science, expounded analytically. The theory and the art of solving mechanical problems were to be reduced to general formulas. Geometrical constructions were to play no part. As Lagrange wrote in the preface to the first edition,

One will find no diagrams in this work. The methods that I expound require neither geometrical nor mechanical constructions or reasonings, but only algebraic operations ordered in a regular and uniform development. Those who love analysis will be pleased to see mechanics become a new branch of it, and will be grateful to me for having thus extended its domain.

The work is divided into two volumes, one on statics and the other on dynamics. In the first edition, the subsection on celestial mechanics presents the Lagrangian theory of perturbation as we have seen it emerge in the earlier memoirs; there is little here that will appear particularly novel to a

reader of the preceding pages. But in the second edition, which Lagrange was preparing during his final years, and of which the first volume was published in 1811, and the second posthumously in 1816, the case is otherwise. Amidst other augmentations and reformulations, volume 2 of this edition contains a totally new section on the variation of arbitrary constants in the solution of dynamical problems. Here appear the concept and the theorem for which Lagrange is chiefly recognized in the later literature on celestial mechanics: the concept of 'Lagrange brackets', and the theorem according to which the arbitrary constants, rendered variable by perturbation, have time-derivatives given by functions of (a) these same constants and (b) the partial derivatives of the perturbing function with respect to these constants.

When Lagrange came to Paris in 1787, it appears that he was mentally exhausted, his interest in celestial mechanics depleted. Aside from his application, in 1792, of a formula he had derived earlier to the derivation of the Moon's secular equation, he wrote nothing on celestial mechanics till 1808. What revived his interest in that year was the presentation to the Institute, on 20 June 1808, of a memoir by Siméon-Denis Poisson (1781–1840), a student of both Laplace and Lagrange. The memoir was a proof that the immunity of the mean solar distances of the planets to secular variation held when the approximation was extended to the squares and products of the planetary masses.

Poisson's demonstration depended on the formulas for elliptical motion. The result, Lagrange conjectured, must be an analytical consequence of the form of the differential equations and of the conditions of variability of the constants, hence a more general truth, not limited in its application to the special solution represented by elliptical motion.

In a memoir read in two segments on 22 August and 12 September 1808, Lagrange proved the correctness of his conjecture. As in his memoir of 1776 on the variation of the major axis, he took the equations of motion of the unperturbed planet to be six differential equations of the first order, each containing a single arbitrary constant together with the coordinates x_i of the planet and the components x_i' of the velocity of the planet in the coordinate directions. Let the arbitrary constants be a, b, c, f, g, h, and let the bracket (a,b) be defined by

$$\sum_{i=1}^{3} \left[\frac{\partial x_i}{\partial a} \frac{\partial x_i'}{\partial b} - \frac{\partial x_i}{\partial b} \frac{\partial x_i'}{\partial a} \right],$$

where the summation extends to all three spatial coordinates. There are parallel definitions for $(a,c),\ldots,(b,c),\ldots,(g,h)$. Once perturbing forces are introduced, there will be a perturbing function Ω, and the constants become parameters subject to variation. Lagrange showed that the partial derivatives of the perturbing function with respect to these parameters are given as linear functions of the time-derivatives of the parameters each multiplied by a bracket. Thus

$$\frac{\partial \Omega}{\partial a}\,dt = (a,b)db + (a,c)dc + (a,f)df$$
$$+ (a,g)dg + (a,h)dh.$$

There are corresponding equations for the partial derivatives of Ω with respect to b, c, f, g, and h. Hence it is possible to solve for the time-derivatives of the parameters, da/dt, db/dt, etc.; they will be given by expressions in terms of the brackets and the partial derivatives of Ω, in which the time t does not explicitly figure.

Now let Ω be developed as a series in sines and cosines of angles proportional to the time. By discarding, in the new expressions for the time-derivatives of the orbital parameters, all terms except those independent of the time, one obtains immediately the equations of the secular variations of these parameters, good to the squares and products of the planetary masses. In the particular case of da/dt, the variation of the major axis, Lagrange found essentially the same formulas as Poisson, showing the immunity of the major axis to secular variation.

The proof of invariability thus obtained by Poisson and Lagrange was limited insofar as it referred only to terms arising from the variation of the parameters of the *perturbed* planet. Would the invariability subsist if account were taken of the variation of the parameters of the *perturbing* planets? To extend the proof in this way, Lagrange used a coordinate system with centre not in the Sun but in the centre of gravity of the solar system; in such a coordinate system, the perturbing function takes a symmetric form with respect to perturbed and perturbing planets, and the extension

of the proof becomes straightforward. Once more, the major axis proved immune to secular variation.

In two further memoirs, read on 13 March 1809 and 19 February 1810, Lagrange generalized his theorem to apply to any mechanical system of interacting bodies, where the equations of motion are integrable when abstraction is made of certain 'perturbing' forces, and account can be taken of these perturbing forces by permitting the constants of integration to vary. He then inserted a unified account of the general theorem in a new subsection, the fifth, of Vol. 2 of the second edition of his *Mécanique analytique*. His working out of this theorem, although marred here and there by minor errors, is a major achievement. So the aged Lagrange, in his mid-seventies, generalizing, completing symmetries, put in place the capstone of his masterwork.

Because of its generality and elegance, Lagrange's variational formulation of the principles of mechanics is still widely used; it is employed, for instance, in the development of relativistic gravitational theory. To Lagrange we owe the introduction of the concept of potential in the formulation of gravitational interactions; the use of this concept, and in particular of the special case of it known as the perturbing function, has been a regular feature of works on celestial mechanics ever since. The expression of the time-rates of change of the orbital elements in terms of 'Lagrange brackets' is also standard in such works. The memoirs of Lagrange's final years form an essential antecedent for the emergence of the Hamiltonian formulation of mechanics.

In Lagrange's own time, a primary result of his work in celestial mechanics was its impact on Laplace. Laplace's first great *coup* in celestial mechanics, his discovery of the immunity of the mean solar distances to secular variation up to the third order in the eccentricities and inclinations, emerged from a comparison of Lagrange's and Euler's differing computations of the secular variations of the orbital elements of Jupiter and Saturn. Lagrange's memoir on the satellites of Jupiter formulated the problem that Laplace resolved in his own study of the interactions of these satellites; and the idea of the pendulum-like oscillation that Laplace showed to be the guarantor of stability in these interactions was probably inspired by Lagrange's establishment of the reality of the physical libration of the Moon. Laplace's discovery in 1785 of the great inequality of Jupiter and Saturn may well have been triggered by an assiduous reading of the memoirs of Lagrange that were published in the same year. Lagrange's vision of a stable solar system, subject only to minor oscillations in form, was adopted and promulgated by Laplace, who made it the theme of his *Exposition du système du monde*.

Neither Lagrange nor Laplace can be looked to today for the kind of mathematical rigour that would be developed after them, by such mathematicians as A.-L. Cauchy and K. T. W. Weierstrass. Lagrange's concern for symmetric form may appear to us to be at times almost a fetish; Laplace's sharp eye for results may seem opportunistic. But in the interactive relation between the two men, opposing qualities complemented each other, and accounted for the main advances of celestial mechanics during the last forty years of the eighteenth century.

Further reading

Oeuvres de Lagrange (ed. J. A. Serret; 14 vols., Paris, 1867–92)

Alfred Gautier, *Essai historique sur le problème des trois corps* (Paris, 1817)

Curtis Wilson, Perturbations and solar tables from Lacaille to Delambre, . . . , Part II, *Archive for History of Exact Sciences*, vol. 22 (1980), 189–243

Curtis Wilson, The great inequality of Jupiter and Saturn: from Kepler to Laplace, *Archive for History of Exact Sciences*, vol. 33 (1985), 15–290

22

Laplace

BRUNO MORANDO

Pierre-Simon Laplace was born on 23 March 1749 at Beaumont-en-Auge in Lower Normandy, a region that produces apple cider and cheeses, among them the famous Camembert. Of Laplace's family little is known. In his later life Laplace, having become Count and then Marquis de Laplace, rich and celebrated, scarcely ever mentioned relatives that he no doubt regarded as unworthy of the glory he had acquired. His father was a cultivator of the land, but not as poor as has sometimes been suggested; his mother, born Marie-Anne Sochon, was the daughter of a rich farmer. The young Pierre-Simon, for his primary and secondary education, attended the school run by the Benedictines of Saint-Maur at Beaumont, where one of his uncles, the Abbé Louis Laplace, had taught. Destined for the priesthood, in 1766 he entered upon the required preparatory studies at the University of Caen. But one of his professors, Pierre Le Canu, interested him in mathematics, and in 1768 sent him to Paris with a letter of recommendation to Jean le Rond d'Alembert (1717–83), doyen of French mathematicians and a powerful influence within the Paris Academy of Sciences. D'Alembert, it is said, was unimpressed by the letter of recommendation; but happily a mathematical essay that the young man brought him a few days later won his admiring approval, and he arranged for Laplace to be given the post of professor of mathematics at the École Militaire.

Thus began Laplace's scientific career; it would be a long and brilliant one, lasting nearly six decades. After initially seeking, without success, for a position at the Academy of Berlin, in 1773 he was named an associate member for mechanics in the Paris Academy of Sciences. In 1784 he became examiner for the Royal Artillery Corps, succeeding the illustrious Étienne Bezout. And in 1785 he became a full member of the Academy of Sciences,

hence recipient of a pension. In March 1788 he married Charlotte de Courty by whom he had a son, who would study at the École Polytechnique and then become a general in the army, and a daughter who died at the age of 21 while bringing into the world a daughter, the future Marquise de Colbert.

Laplace's work in mathematical astronomy to 1789

The first of the mathematical memoirs that Laplace submitted to the Paris Academy of Sciences was registered by the Academy's secretary on 28 March 1770, five days after Laplace's twenty-first birthday. More than a hundred such would follow, and the last would be entered in the register on 8 September 1823, when Laplace was 74. The memoirs of the first few years dealt with a variety of mathematical topics: application of the integral calculus to the study of finite differences, use of finite differences in the study of probability, recurrent series as applied to probability, investigation of particular solutions of differential equations, the probability of causes (Bayes's rule). Astronomical questions also made their appearance early: the motion of the nodes of the planetary orbits, variation of the plane of the ecliptic, determination of the lunar orbit, perturbations of the planets caused by the motions of their satellites, the secular equations of the planets, etc.

Laplace's first published memoir on astronomy
In 1776, in the volume of the Paris Academy's *Savants étrangers* for the year 1773, there appeared Laplace's "Sur le principe de la gravitation universelle, et sur les inégalités séculaires des planètes qui en dépendent" – the first of his astronomical memoirs to be published. It began with a presentation of "the general equations of motion of a body

22.1. Pierre-Simon de Laplace.

of any figure, acted upon by any forces"; here Laplace showed himself in full command of the theory of the motion of rigid bodies as developed by d'Alembert and Leonhard Euler (see Chapter 17 above).

The memoir next turned to "an examination of the principle of universal gravitation". According to Laplace, Newton in his application of this principle to the celestial bodies made four assumptions, which were now generally accepted: (1) the attraction exerted on a body varies directly as its mass and inversely as the square of the distance from the attracting body; (2) the attractive force of a body is the sum of the attractions of each of its parts; (3) the force is propagated instantaneously; and (4) it acts in the same way on bodies at rest and in movement.

In the discussion that followed Laplace examined these assumptions both speculatively and critically. He suggested a possible metaphysical justification for the inverse-square law in the fact that only this law permits the planetary paths to remain geometrically similar to themselves under proportional shrinkage or magnification of all distances and velocities. He examined the evidence for the second assumption, urging that the phenomena of precession and nutation required its

acceptance. He also showed that if, contrary to the third and fourth assumptions, the speed of propagation of gravitation was finite, and its effect on bodies dependent on their relative speeds, then there were mathematical consequences to be reckoned with. Daniel Bernoulli, in his prize-winning memoir of 1740 on the tides, had imagined that as much as two days might be required for the attractive force to travel from the Moon to the Earth. Laplace showed that the hypothesis of a finite speed would account for the secular acceleration of the Moon, provided that the velocity of propagation were supposed more than six million times the speed of light. The planets also, under this hypothesis, would be subject to secular accelerations. The idea is presented as a conjecture for the consideration of philosophers; it would be acceptable, Laplace says, only if it were known that the Moon's secular acceleration could not be explained by more ordinary assumptions. With other mathematicians of the time he saw this acceleration as a problem demanding solution. He examined the suggestion that the friction of the trade winds might slow the rate of the Earth's diurnal rotation, and so make the Moon appear to accelerate; but concluded that no such effect could result. As possible explanations he also considered, but dismissed, the frictional effect of an interplanetary aether on the Earth's motion, and the Sun's loss of mass due to the emission of light, causing the Earth to recede into a larger orbit.

In the second half of the memoir, Laplace turned to the derivation of the secular inequalities of the planets, and quickly arrived at a new and important result. He began by pointing out that, of all the inequalities, the secular ones are the most important; the planets in each of their revolutions move very nearly in ellipses, and the perturbational effect most important to determine is the change in the orbital elements of these ellipses. And among the secular inequalities, the most crucial is that of the mean movement; but, Laplace urged, it had not been determined with the precision that its importance demanded. Euler, in his memoir of 1752 on the irregularities of Jupiter and Saturn, had concluded that both Jupiter and Saturn were accelerating at the present epoch. Joseph Louis Lagrange (1736–1813), in a treatise published in 1766 in the *Mémoires de Turin*, had determined Jupiter to be accelerating and Saturn decelerating, a result in agreement with observations, but still, according to Laplace, incorrect: the derivable effects were negligible. In a footnote, Laplace stated that after reading his memoir to the Academy, he had been able to prove that the secular inequalities of the mean movement were identically zero.

Laplace's derivation had certain novel procedural and notational features that were to remain characteristically his in his later work on planetary perturbations. His differential equations for the longitude (ϕ), radius vector (r), and tangent (s) of the latitude of a perturbed planet were essentially Lagrangian. But into them Laplace proceeded to introduce the differential operator δ, borrowed from Lagrange's formulation of the calculus of variations. He initially designated the masses of the perturbed and perturbing planets by δm and $\delta m'$, respectively, thereby indicating their smallness relative to the Sun's mass (denoted by S). He followed Lagrange in distinguishing the variables referring to the second planet by prime marks. Deleting from the differential equations for the longitude and radius vector the quantities involving the factor $\delta m'$, he obtained the following equations:

$$\frac{d\phi}{dt} = \frac{c}{r^2}, \tag{1}$$

$$0 = \frac{d^2 r}{dt^2} - \frac{c^2}{r^3} + \frac{S + \delta m}{r^2}, \tag{2}$$

where c is a constant of integration. These equations when integrated yield

$$\phi = nt + A' - 2ae\sin(nt + \epsilon) \\ + \tfrac{5}{4}a^2 e^2 \sin 2(nt + \epsilon) \\ + \cdots \tag{3}$$

and

$$r = a\left[1 + \frac{a^2 e^2}{2} + ae\cos(nt + \epsilon) \right. \\ \left. - \frac{a^2 e^2}{2}\cos 2(nt + \epsilon) + \cdots \right], \tag{4}$$

which are the equations for motion in an ellipse, in which a is the semi-major axis, ae the eccentricity, n the planet's rate of mean motion, A' the planet's mean longitude when $t = 0$, and ϵ the angle by which the planet has advanced beyond its aphelion at the same moment. (Later, in the *Mécanique céleste*, Laplace would shift the starting-point for anomaly from aphelion to perihelion, a change which is effected simply by changing the sign of e in the preceding equations, and which allows for a uniform treatment of cometary and planetary motions.) In the foregoing equations the tangent s of the latitude of the perturbed planet is assumed to be so small that s^2 can be neglected. The symbol a was the tag Laplace used to keep track systematically of orders of magnitude.

With a view to re-introducing into his equations the terms involving $\delta m'$, Laplace next differentiated (3) and (4) using the operator δ, and taking advantage of the fact that this operator commutes with the standard differential operator 'd'; then he added to the resulting equations the terms from the original equations of motion involving $\delta m'$. He thus obtained

$$\frac{d\delta\phi}{dt} = -\frac{2c}{r^3}\delta r \tag{5}$$

$$-\frac{\delta m'}{r^2}\int r\, dt \sin(\phi' - \phi)\left[\frac{1}{r^2(1+s^2)^{3/2}} - \frac{r'}{v^3} \right],$$

$$0 = \frac{d^2\delta r}{dt^2} + \frac{3c^2}{r^4}\delta r - \frac{2(S + \delta m)}{r^3}\delta r$$
$$+ \text{ terms involving } \delta m'. \tag{6}$$

Here v denotes the distance between perturbing and perturbed planets. The variations of r and ϕ

distinguished by the operator δ are those due to the perturbations caused by $\delta m'$.

In a first step towards solving (5) and (6), Laplace replaced the terms containing $\delta m'$ as a factor by expressions involving undetermined coefficients A, B, C, D, and including only such terms as contributed to the secular variations up to the degree of approximation he was aiming to achieve. Thus the terms of (6) containing $\delta m'$ were replaced by

$$a\frac{\delta m'}{a^3}A + a^2 a\frac{\delta m'}{a^3}Bnt$$
$$+ aa\frac{\delta m'}{a^3}C\cos(nt+\epsilon)$$
$$+ aa\frac{\delta m'}{a^3}D\sin(nt+\epsilon),$$

and the last term of (5) was replaced by

$$-a^2\frac{\delta m'}{a^3}\frac{a^2}{2c}Bnt.$$

For the quantity $(S+\delta m)/a^3$ Laplace substituted n^2, in accordance with Kepler's third law, and for the quantity $\delta m'/S, \delta\mu'$. Then, for $\delta\phi$ and δr in (5) and (6), he substituted the expressions

$$\delta\phi = \delta\mu'gnt + a^2\delta\mu'hn^2t^2, \qquad (7)$$

and

$$\delta r = a\delta\mu'[L + aPnt\cos(nt+\epsilon)$$
$$+ aQnt\sin(nt+\epsilon)$$
$$+ a^2Knt]. \qquad (8)$$

By comparing the coefficients of t, t^2, $\cos(nt+\epsilon)$, and $\sin(nt+\epsilon)$, he then determined g, h, L, P, Q and K in terms of A, B, C, and D. In this way he obtained expressions for the secular increase in the equation of centre, the movement of the aphelion, and the acceleration of the mean movement in terms of A, B, C, and D. Thus the acceleration of the mean movement took the form

$$\frac{3}{2}\pi a^2\delta\mu'(B-eD)\cdot i\cdot 360, \qquad (9)$$

where i is the number of revolutions.

The final step in evaluating the expressions for the secular variations was the determination of the values of A, B, C, and D in terms of the trigonometric series expressing the inverse third, fifth and seventh powers of the distance between perturbing and perturbed planets. Letting $a'/a = z$ be the ratio

of the semimajor axes of the two planets, and θ the heliocentric angle between their *radii vectores*, Laplace set

$$(1 - 2z\cos\theta + z^2)^{-\frac{3}{2}} = b + b_1\cos\theta + b_2\cos2\theta + \dots,$$
$$(1 - 2z\cos\theta + z^2)^{-\frac{5}{2}} = b' + b'_1\cos\theta + b'_2\cos2\theta + \dots,$$
$$(1 - 2z\cos\theta + z^2)^{-\frac{7}{2}} = b'' + b''_1\cos\theta + b''_2\cos2\theta + \dots.$$
$$(10)$$

(We have here an early form of the double indicial notation that he would make standard for the coefficients of the trigonometric series.) The acceleration of the mean motion turned out to involve the factor

$$\{3[b - b' + \tfrac{1}{2}(b'_2 - b_2)]$$
$$+ \tfrac{1}{4}z[7(b'_1 - b'_3) - 5(b''_1 - b''_3)]$$
$$- z^2[3(b' - \tfrac{1}{2}b'_2) - \tfrac{5}{4}(5b'' - 2b''_2 - \tfrac{1}{2}b''_4)]$$
$$- \tfrac{5}{4}z^3(b''_1 - b''_3)\}. \qquad (11)$$

At first, in applying his formula to Jupiter and Saturn, Laplace substituted numerical values for the bs. On finding the result to be nearly zero, he then made use of the relations first developed by Euler whereby all these coefficients can be expressed in terms of the first two, b and b_1, and showed that the above factor reduced identically to zero.

The result meant that, at least to the order of $a^4\delta\mu'$, secular accelerations or decelerations in the motions of the planets – that is, terms in the mean motion proportional to t^2, and independent of the positions of the planets – did not occur. Laplace drew the conclusion that the apparent secular acceleration of Jupiter and secular deceleration of Saturn must be due to causes other than their mutual gravitation; they might be caused, he suggested, by the actions of comets. Neither he, nor Lagrange when he learned of Laplace's new result, considered the possibility that the observed alterations in the mean motions of Jupiter and Saturn might be produced by long-term *periodic* inequalities – oscillations in longitude depending on the positions of the planets, but with periods much longer than those of the planets themselves.

Our sketch of the course of Laplace's derivation, besides exhibiting notational features that were to remain distinctively Laplacian, illustrates characteristics of his mathematical style: while gifted with a keen sense for subtle consequences of both mathematical assumptions and empirical data, he almost always set his sights on some final practical or numerical result to be obtained, and frequently took the short way through a tangle without fully explaining his processes. In contrast to Lagrange, he was little moved by the 'beauty' of a theory, as a quality independent of the theory's power to yield practical results. To the determination of order of approximation in the calculation of approximate numerical results, on the other hand, he brought a new rigour.

Secular inequalities of the eccentricities, aphelia, inclinations, and nodes

As reported in the preceding chapter, Lagrange in October 1774 presented to the Paris Academy a memoir giving a new method for determination of the secular variations of the nodes and inclinations of the planetary orbits; it would be published in 1778. (Memoirs published in the *Mémoires de l'Académie Royale des Sciences de Paris* will herein-after be designated by *MARS*, followed first by the year of the volume in question and then by the year of publication; and similarly for other publication series. Thus the memoir of Lagrange just cited appeared in *MARS* 1774/78.) The stroke of genius in Lagrange's memoir consisted in the introduction of new variables $s = \theta \sin \omega$, $u = \theta \cos \omega$, where θ is the tangent of the inclination and ω is the angle between the line of nodes and some chosen reference line; there were corresponding variables for each of the planets. The differential equations for the secular part of the variables s and u of the several planets, independent of short-term oscillations, then proved to be linear and of the first order in s and u. Laplace read this memoir immediately it reached the Paris Academy, and realized that the same procedure was applicable to the aphelia and eccentricities. He proceeded to register his elaboration of this idea with the Academy's secretary, and it was published in Pt I of *MARS* 1772/75, three years before Lagrange's essay appeared. Before it was printed, however, Laplace received from Lagrange a letter of 10 April 1775 outlining precisely the same application, and this he appended to the end of his memoir. Laplace's new variables were of the form $x = e \sin L$ and $y = e \cos L$, where e is the eccentricity of the planetary orbit and L the longitude of the aphelion; and in analogy

with the case of the nodes and inclinations he derived first-order linear differential equations for the secular variations of x and y.

Further consideration of this Lagrangian procedure led Laplace to a new method for integrating differential equations by approximation — a method of eliminating the *arcs de cercle* or terms proportional to the time which, as Lagrange had shown in his prize-winning memoir of 1766 on the satellites of Jupiter (see the preceding chapter), emerge inevitably in the solution of the differential equations when they are solved by the ordinary process of successive approximation. Laplace presented this method in Pt II of *MARS* 1772/76. The gist of the method consisted in replacing the eccentricity, aphelion, inclination, and node by the new-style variables, and then showing that, wherever in the solution of the differential equations the time t appeared outside the arguments of sines and cosines, it could be eliminated by taking into account the secular variations of the new variables. From the latter it was then a matter of simple algebra to deduce the secular variations of the eccentricites, aphelia, inclinations, and nodes.

Having explained a general method for the determination of all perturbations, both periodic and secular, Laplace concluded: "It would now remain to apply the preceding theory to the different planets; but the already excessive length of this memoir obliges me to put off these applications to another time …". As reported in the previous chapter, Lagrange, who was expecting this application to be soon forthcoming, in his letter to Laplace of 10 April 1775 went so far as to say that he was ceding the entire subject to his younger rival. A month and a half later, however, he withdrew this offer, suggesting that in their further work on planetary perturbations he and Laplace engage in a friendly rivalry, exchanging only the published memoirs. Lagrange himself proceeded to develop his own general treatment of perturbations by the method of variation of orbital parameters, with application to all the known planets; the resulting memoirs would finally begin to appear in published form in 1783. Laplace, by contrast, published nothing on this subject for a decade; as he explained in a letter to Lagrange of 10 February 1783, he was deterred from pursuing the topic by uncertainties as to the masses of the planets and as to the extent of the derangements they underwent due to the gravitational action of comets.

Tides, precession of the equinoxes, nutation

During the decade from 1775 to 1785, Laplace turned his attention to the gravitational action of *extended* bodies: the attraction of spheroids, the tidal motions of the oceans and atmosphere, the precession of the equinoxes and the nutation of the Earth's axis. One concern was undoubtedly an unresolved difficulty concerning the shape of the Earth: Isaac Newton had deduced, for a homogeneous fluid spheroid rotating with the speed of the Earth's diurnal rotation, an ellipticity of $\frac{1}{230}$; the several geodetic measurements made in Lapland, Peru, France, and the Cape of Good Hope implied an ellipticity greater than $\frac{1}{230}$, while universal gravitation together with plausible assumptions appeared to imply that the ellipticity should be less (see Chapter 15). But this difficulty was as yet unresolvable, and Laplace saw that there were prior problems that needed to be addressed, and first of all the theory of the tides and the role of the fluid parts of the Earth in relation to the actions causing precession and nutation.

Beginning with Newton, the earlier analysts investigating the tides had left the Earth's axial rotation out of account; they had imagined a stationary Earth covered by a thin layer of fluid and subject to the attraction of a single celestial body (Moon or Sun); the fluid would assume an equilibrium shape under the attraction, and if the Moon or Sun were then set into a real or apparent motion of revolution about the Earth, the resulting bulge of fluid would be dragged round the Earth without change of form. This analysis implied that the two full tides occurring each day would differ considerably in height; observations indicated, on the contrary, that the difference was very slight. The traditional analysis failed to take into account another fact: in the Earth's diurnal rotation the linear velocities eastward were different at different latitudes, and the resulting relative displacements of ocean water were of the same order of magnitude as those caused by the attractions of the Moon and the Sun. Moreover, given the inclinations of the Moon's and Sun's paths to the Earth's equator, the tidal distributions in the southern and northern hemispheres must differ, and the effect of the asymmetry on the precession and nutation of the Earth's axis needed to be evaluated.

In a three-part memoir that appeared in *MARS* 1775/78 and 1776/79 and ran to over 200 pages, Laplace grappled with these issues. He showed that, when all factors of the same order of magnitude were taken into account, consecutive high tides would be of nearly the same height if and only if the sea were assumed to be of uniform depth, and further that the slight difference between consecutive high tides was proportional to the influence of the tides on the ratio of nutation to precession. In a further memoir "Sur la précession des équinoxes" (*MARS* 1777/80), he corroborated the result for the precession by a simplified analysis using d'Arcy's principle (conservation of angular momentum); here he announced the theorem:

If the Earth be supposed an ellipsoid of revolution covered by the sea, the fluidity of the water in no way interferes with the attraction of the Sun and the Moon on precession and nutation, so that this effect is just the same as if the sea formed a solid mass with the Earth.

Because d'Arcy's principle applies to all interactions including those involving friction, shock, and turbulence, Laplace now claimed that his analysis could be extended to "the case in nature", in which the shape of the Earth and the depth of the sea are very irregular.

Determination of cometary orbits

Another topic that Laplace concerned himself with in this period was the determination of cometary orbits. In 1771 R. J. Bošković had presented to the Paris Academy a memoir on the determination of a cometary orbit from three observations close to one another in time; he assumed that, during the interval of the observations, the motions of the comet and the Earth were rectilinear and uniform. The memoir appeared in the *Savants étrangers* in 1774, and on reading it Laplace saw at once that the method neglected quantities of the same order as those entering into the supposed solution. He delivered a devastating critique from the floor of the Academy; Bošković, believing himself to have been personally affronted, made a reply. A commission appointed to adjudicate the controversy concluded in its report of June 1776 that Laplace's critique was analytically correct, but that the manner in which it had been couched was unnecessarily abrasive.

In 1781 Laplace returned to the problem of determining cometary orbits and undertook to deal with it in a new way. The motion of a body in the solar system, perturbations aside, is entirely determined if one knows six initial conditions of motion, for instance the three spatial coordinates and the three components of velocity in relation to the Sun at a given 'initial' instant of the motion. Three observations of a planet or comet on the celestial sphere at three different instants give six quantities (right ascensions and declinations or ecliptic longitudes and latitudes), from which the elements of the planet's or comet's heliocentric orbit can be determined. The problem is a difficult one. Newton and others, particularly Lagrange and A.-P. Dionis du Séjour, had contributed to a partial solution of it.

To begin with, Laplace proposed deriving, for the mean instant of the observations, a geocentric longitude and latitude, together with their first and second time-derivatives, from multiple observations (preferably many) by way of the calculus of finite differences. Euler had suggested a similar method for comparing lunar observations with theory in a memoir published in *Mémoires de Berlin* 1763/70. The advantage was that the further analysis could be made rigorous, while less reliance was placed on individual observations, which were in any case necessarily imprecise. For the mean instant of the observations the comet's heliocentric coordinates and components of velocity in relation to the Sun could then be deduced, provided that one knew the geocentric coordinates of the Sun (as given by ephemerides) and the distance of the comet from the Earth at the given instant. Laplace determined the latter quantity as the root of an equation of the seventh degree (or of the sixth degree if the orbit was supposed parabolic). To resolve this equation he used an iterative method, beginning with an approximate value of the distance sought.

In 1784 the Abbé A.-G. Pingré in his *Cométographie* pronounced Laplace's method to be the best available. Together with the later methods of H. W. M. Olbers and C. F. Gauss it has continued in favour with astronomers (see Chapter 25). It has been improved by Henri Poincaré and A. O. Leuschner.

Laplace played a role in the determination of the orbit of the planet Uranus. This planet, discovered by William Herschel on 13 March 1781, was at first taken for a comet (see Chapter 24), but the efforts of P.-F.-A. Méchain, A. J. Lexell, Bošković, and in fact of Laplace himself to determine parabolic elements proved unsuccessful. J.-B.-G. Bochart de Saron, Laplace's friend and protector, appears to have been the first to show that the heliocentric distance of the newly discovered 'star' was at least 14 AU, and hence that one had to do with a planet. On 22 January 1783 Laplace presented a memoir to the Academy giving orbital elements for the planet. In Vol. 3 of the *Mécanique céleste* he would give a set of more precise elements, and calculate the perturbations due to Jupiter and Saturn.

The attractions of spheroids

In late 1782 or early 1783 Laplace was appointed (along with Étienne Bezout and d'Alembert) to a committee for the review of a memoir by Adrien-Marie Legendre (1752–1833) on the attraction of homogeneous spheroids of revolution. Legendre's aim was to prove the theorem:

If the attraction of a solid of revolution is known for every external point on the prolongation of its axis, it is known for every external point. With the origin of coordinates placed at the centre of the spheroid, the radial component of the force of attraction on an external point is given by an integral over all the points within the spheroid. If r is the radius vector of the external point, r' the radius vector of an arbitrary internal point, and γ the angle between them, then the integrand can be expressed as a function of r'/r and of cos γ.

Legendre showed how this function could be expanded as an infinite series in even powers of r'/r, with coefficients P_2, P_4, ... (now called Legendre polynomials) that were rational integral functions of cos γ. At a suggestion from Laplace, he went on to obtain the potential function for the spheroid in terms of Legendre polynomials, and with its aid evaluated the component of the attraction at right angles to the radius vector.

Stimulated by Legendre's work, Laplace took up the problem on his own account, generalizing it and making some applications to the figure of the planets, in Pt II of *Théorie du mouvement et de la figure elliptique des planètes* (1784), and in *MARS* 1782/85 and 1783/86. To know the attraction exerted by an spheroid (taking this to be a body bounded by any surface expressible by a single equation in spherical coordinates), it was only necessary to know the corresponding potential function, and this, Laplace now asserted, was subject to a fundamental partial differential equation – the equation now known as Laplace's equation or the potential equation. In rectangular coordinates, it is

$$\frac{\partial^2 V}{\partial x^2} + \frac{\partial^2 V}{\partial y^2} + \frac{\partial^2 V}{\partial z^2} = 0. \qquad (12)$$

In his early memoirs on the equation, however, Laplace gave it in the more complicated form it has in spherical coordinates, without informing the reader whence it was derived. To solve it he proposed an expansion in series of the form

$$V = \frac{U_0}{r} + \frac{U_1}{r^2} + \frac{U_2}{r^3} + \ldots; \qquad (13)$$

the U_n, he then showed, can be expressed as integrals in terms of the Legendre polynomials. This result enabled him to obtain a series expression for the potential of any spheroid differing little from a sphere. Armed with this general theory, Laplace returned to the problem of tidal oscillations in *MARS* 1776/79, and showed that equilibrium conditions entailed periodicity, and further that the equilibrium would be stable only if the layer of fluid covering the spheroid were less dense than the underlying spheroid. In *MARS* 1783/86 he brought the theory into relation with the available data on the Earth's figure, and in *MARS* 1787/89 he made a tentative application to the rings of Saturn.

The observations of the London optician and instrument-maker James Short (1710–68) had shown that the two rings of Saturn distinguished by G. D. Cassini (Cassini I) were actually multiple, each consisting of many rings. Laplace initially treated each of these many rings as a thin stratum of liquid, its cross-section by any plane through Saturn's axis being a narrow ellipse; he then sought the conditions of equilibrium of this stratum. He took the duration of the rotation of each ring to be equal to that of a satellite at the same distance from Saturn's centre. Finally, he showed that for the equilibrium to be stable the separate rings must be irregular in shape, their centres of gravity differing from their centres of figure. Proof that the rings cannot be fluid, but must be ensembles of tiny satellites, would have to await James Clerk Maxwell's essay of 1856 "On the stability of the motion of Saturn's rings".

A new attack on planetary perturbations

On 23 November 1785 Laplace read to the Academy a memoir "Sur les inégalités séculaires des planètes"; it marks his return, after the lapse of a decade, to the problem of the mutual perturbations of the planets. The memoir as finally printed (in *MARS* 1784/87) contained a sketch of a new theory concerning the mutual perturbations of the satellites of Jupiter, and to the title was added the phrase "et des satellites"; but the presence of this section in the original memoir may be doubted, for in its original form the memoir may have dealt with only two subjects: the anomalies in the mean motions of Jupiter and Saturn, and a proof of the stability of the solar system.

Laplace's new attack on the perturbational problem was manifoldly influenced by memoirs of Lagrange published between 1779 and 1785. In these Lagrange had introduced the perturbing function as applied to the planets conceived as mass-points; had derived with its aid all the known integrals of motion; had articulated an argument for the stability of the solar system; and in the first part of his "Théorie des variations périodiques", published in 1785, had made a suggestion about locating perturbational terms that would be made large by small divisors, owing to the near-commensurability of the mean motions of the planets concerned (see Chapter 21). All these features played a role in Laplace's new assault; and in fact Lagrange's suggestion about small divisors may have triggered Laplace's discovery of the source of the puzzling anomalies in the mean motions of Jupiter and Saturn.

At the start Laplace reviewed what he and Lagrange had proved with regard to the immunity of the mean motions to secular change from mutual gravitational action; his own proof was valid up to the third powers of the eccentricities and inclinations, while Lagrange's proof was valid independent of the eccentricities and inclinations, on the assumption that the mean motions of the planets were incommensurable, "as is the case for the planets in our system". But, Laplace went on, there were perceptible variations in the revolutions of Jupiter and Saturn; Edmond Halley had found, for a period of 2000 years, 3° 49' of accelerational advance in Jupiter's motion, and 9° 16' of retardation for Saturn.

These alterations, Laplace next argued, were "very probably" an effect of the mutual action of Jupiter and Saturn. For from the integral of *forces vives*, by neglecting all terms of the third order in the masses (m^3), and terms of the order m^2 that were either periodic or constant, he had deduced the relation

$$\frac{m}{a} + \frac{m'}{a'} + \frac{m''}{a''} + \ldots = \text{const}, \qquad (14)$$

where $m, m', m'', \ldots$ are the planetary masses and a, a', a'' the semi-major axes of the orbits. Because the masses of Jupiter and Saturn were very large relative to the masses of the other planets, all terms in the foregoing relation other than those pertaining to these two planets could be neglected, and it followed that if Jupiter's mean solar distance (a) diminished, Saturn's mean solar distance (a') would increase. From Kepler's third law one could derive that

$$mn^{\frac{2}{3}} + m'n'^{\frac{2}{3}} = \text{const}, \qquad (15)$$

where n and n' are Jupiter's and Saturn's mean motions. In mutual action affecting the mean motions it followed that

$$\delta n' = -\frac{m}{m'}\left(\frac{a}{a'}\right)^{1/2}\delta n, \qquad (16)$$

or with the values Laplace accepted for the masses and mean distances, $\delta n' = -2.33\,\delta n$. The ratio of Halley's values for the secular changes, $-9°\ 16'/3°\ 49'$, is -2.43, nearly the same.

It is therefore very probable that the observed variations in the movements of Jupiter and Saturn are an effect of their mutual action, and since it is established that this action can produce no inequality that either increases constantly or is of very long period and independent of the situation of the planets, and since it can only cause inequalities dependent on their mutual configuration, it is natural to think that there exists in their theory a considerable inequality of this kind, of which the period is very long.

The mean motions of Jupiter and Saturn, Laplace now observed, were nearly as 5 to 2; "whence I concluded that the terms which, in the differential equations of motion of these planets, have for argument five times the mean longitude of Saturn minus two times the mean longitude of Jupiter, could become sizeable because of the integrations, though they are multiplied by the cubes and products of three dimensions of the eccentricities and inclinations of the orbits." If λ is the mean longitude of Jupiter and λ' the mean longitude of Saturn, then the quantity

$$V = 5\lambda' - 2\lambda \qquad (17)$$

varies very little with time and the terms in $\sin V$ in the longitudes of Jupiter and Saturn cause inequalities having a period of about 850 years and large amplitudes which account for the accelerations and retardations observed in the motions of these two planets. In a preliminary formulation Laplace gave the inequality in the longitude of Jupiter as

$$20'\sin(5n't - 2nt + 49°\ 8'\ 40''),$$

and that in the longitude of Saturn as

$$46'\ 50''\ \sin(5n't - 2nt + 49°\ 8'\ 40''),$$

where the zero of time is taken as the beginning of 1700; a later, more accurate derivation produced somewhat larger coefficients, 21.1' for Jupiter and 49.0' for Saturn.

Laplace's discovery of this inequality not only resolved an enigma but demonstrated the importance of resonant terms in celestial mechanics (that is, terms involving near-commensurabilities in the mean motions). C. G. J. Jacobi would later give an elegant method for isolating these resonant terms in the perturbing function, which was published in *Astronomische Nachrichten* in 1848.

It is a remarkable fact that the "great inequality of Jupiter and Saturn", as it is called, reached its null point toward the end of the sixteenth century, so that in the observations of Tycho Brahe employed by Johannes Kepler the displacements of Jupiter and Saturn due to this inequality were close to zero. Kepler's values for the mean motions of these planets, obtained from a comparison of Tycho's observations with ancient observations, were therefore very close to the correct values. Throughout the seventeenth and most of the eighteenth centuries Jupiter's average motion was found to be more rapid and Saturn's slower than Kepler's numbers implied. Then in 1773 Johann Heinrich Lambert showed that with respect to the seventeenth-century observations of Johannes

Hevelius the motion of Jupiter was now slowing and that of Saturn accelerating: the apparent acceleration of Jupiter and deceleration of Saturn were rounding off. The displacements in longitude caused by this inequality reached their maxima toward the end of the eighteenth century, then commenced to decrease.

In the second part of the memoir Laplace undertook to provide what he believed to be a definitive answer to the question of the stability of the solar system. Given the immunity of the mean distances to secular change (the Laplacian and Lagrangian proofs had in fact established this immunity only to the order of the first powers of the perturbing masses), there remained the problem of secular changes in the eccentricities and inclinations; if an orbital eccentricity were to increase beyond the value 1, the motion would cease to be periodic. Laplace now gave an argument to show that neither the orbital eccentricities nor the tangents of the orbital inclinations contained *arcs de cercle* or exponential terms that could lead to indefinite increase; ergo the solar system was stable. Given the fact that all the planets turn in the same direction round the Sun, he derived from the conservation of angular momentum, on neglecting terms of the order of m^2, the two equations

$$\text{const} = m(a)^{\frac{1}{2}}(e)^2 + m'(a')^{\frac{1}{2}}(e')^2 + \dots, \quad (18)$$

$$\text{const} = m(a)^{\frac{1}{2}}(\theta)^2 + m'(a')^{\frac{1}{2}}(\theta')^2 + \dots, \quad (19)$$

where $e, e', \dots$ are the eccentricities, and $\theta, \theta', \dots$ are the tangents of the inclinations. These equations, he showed, could not hold if the es and θs contained terms that would cause them to increase indefinitely with time.

In 1808 Siméon-Denis Poisson, a protégé of Lagrange and Laplace, would demonstrate that the immunity of the mean distances to secular change held to the second order in the perturbing masses. Yet the general problem of stability is a more difficult one than Lagrange, Laplace, and Poisson envisaged; it remains today a subject of active research (see the appendix to Chapter 28).

Laplace next devoted himself to a systematic and thorough development of the mutual perturbations of Jupiter and Saturn; the resulting "Théorie de Jupiter et de Saturne" was read in two parts on 10 May and 15 July 1786, and published in *MARS* 1785/88.

In an extended initial section Laplace developed the equations of motion into forms convenient for the calculations that he now knew to be necessary. Thus for the perturbations of the radius vector he obtained

$$\delta r = \frac{\left\{ \begin{matrix} + a \cdot \cos v \cdot \int n dt \cdot r \cdot \sin v \cdot \{2 \int DR + r\, \partial R / \partial r\} \\ - a \cdot \sin v \cdot \int n dt \cdot r \cdot \cos v \cdot \{2 \int DR + r\, \partial R / \partial r\} \end{matrix} \right\}}{\sqrt{1 - e^2}}, (20)$$

where v is the true anomaly, R is the perturbing function, and 'D' means differentiation with respect only to the variables pertaining to the perturbed planet. Similarly, for the perturbations in longitude he obtained

$$\delta v = \frac{\left\{ \begin{matrix} (2r \cdot d\delta r + \delta r \cdot dr)/a^2 n dt \\ + 3a \int n dt \int DR + 2a \int n dt \cdot r\, \partial R / \partial r \end{matrix} \right\}}{\sqrt{1 - e^2}}. \quad (21)$$

The first of these equations is closely parallel to the perturbative part of Clairaut's equation for the radius vector, and like Clairaut's equation has the merit of permitting expeditious application. but the perturbing function R is now understood as being developable systematically as an infinite series in accordance with increasing powers of the eccentricity and tangent of the orbital inclination; Laplace reduces this development to an essentially mechanical decision procedure, allowing him to calculate the value of any term in the expansion. For the inequalities independent of the squares and higher powers of the eccentricities and inclinations, a complete derivation was feasible, and Laplace proceeded to supply it; but for the higher powers such a derivation would have been costly in time and labour, and here he resorted to a species of sharpshooting, picking out the terms in R farther down the line that would become large enough to be detectable on integration, and deriving the inequalities resulting from these terms by themselves. In this way he obtained not only the "great inequality" of Jupiter and Saturn, proportional to $\sin(5n't - 2nt + \text{const})$, with coefficients for the two planets dependent on the third order in the eccentricities and inclinations, but also several terms proportional to the squares and products of two dimensions in the eccentricities, for instance, in the longitude of Saturn the term $-13' \ 16''$ $\sin(2nt - 4n't - 2° \ 27' \ 4'')$; this and other oscillations of the same order in the longitude had probably caused as much perplexity for his predecessors as the great inequality itself.

The derivation of these inequalities presupposed values for the eccentricities, aphelia, inclinations, nodes, and mean longitudes at epoch; all which must be determined from observations. But there is a logical circle here, of which Laplace was clearly aware: to determine the orbital elements, it was necessary to have good observations from which the perturbations had been subtracted. Thus "the determination of the inequalities of Jupiter and Saturn and that of the elements of their orbits depend reciprocally on one another, and we can come to know them well only by successive approximation." Laplace proceeded to use equations of condition in the manner of Tobias Mayer to determine differential corrections to the orbital elements; for a description of his procedure see Chapter 26 below.

On 10 May 1786, the day on which Laplace presented the first part of his "Théorie de Jupiter et de Saturne", Jean-Baptiste Joseph Delambre (1749–1822), an astronomer who had recently made a name for himself in revising the solar tables of N.-L. de Lacaille and in detecting an error in the accepted tables of Mercury, offered to Laplace his services in the correction and refinement of the data – the observed oppositions – on which any theory of Jupiter and Saturn must rest. All the original unreduced observations had to be re-examined and assessed, and corrected for aberration, nutation, and faulty assumptions as to the positions of stars. Then, on the basis of Laplace's theory and making use of his method of equations of condition, the orbital elements had to be re-determined, and the results embodied in new tables of the two planets. This work, Delambre tells us, "required nine months of the most constant and assiduous application". Delambre would similarly undertake the observational refinement of Laplace's theories of Uranus and of the Galilean satellites of Jupiter.

Long-term inequalities of the satellites of Jupiter
Lagrange, in his memoir of 1766 on the inequalities of the Galilean satellites of Jupiter, had found certain sinusoidal terms in the longitudes of the first three satellites to be especially large, because of near-commensurabilities between the mean motions of these satellites. Thus if the mean motions for the first, second, and third satellites are

n, n', and n'', then $n:n' = 2.0073:1$ and $n':n'' = 2.0147:1$. For the second satellite two large terms resulted, one with the argument $(n - n')t$ and the other with the argument $2(n' - n'')t$, and the curious fact was that these two inequalities appeared to have precisely the same period; for Pehr Wargentin in his observational study of the satellite had found only one inequality. Therefore, to within the precision of the observations, the *difference* between the mean motions of the first and second satellites was exactly *twice the difference* between the mean motions of the second and the third:

$$n - n' = 2(n' - n''), \quad \text{or } n - 3n' + 2n'' = 0. \quad (22)$$

Lagrange had further noticed that (to use Laplace's symbols)

$$(n - 3n' + 2n'')t + \epsilon - 3\epsilon' + 2\epsilon'' = 180°, \quad (23)$$

where ϵ, ϵ', and ϵ'' are the longitudes of the three satellites at epoch. At some point in the mid-1780s Laplace set out to try to explain these equations on the basis of gravitation. Inspiration for his inquiry came from Lagrange's work on the libration of the Moon.

The explanation that he found was sketched in MARS 1784/87, then given with full detail in MARS 1788/91. Let $s = n - 3n' + 2n''$, and $V = (n - 3n' + 2n'')t + \epsilon - 3\epsilon' + 2\epsilon''$. From an examination of the mutual perturbations dependent on the squares and products of the masses Laplace obtained two differential equations having to do with variations in the mean motions, namely

$$\frac{ds}{dt} = an^2 \sin V, \quad (24)$$

where a represents a complicated function, and

$$\frac{dV}{dt} = s. \quad (25)$$

From (24) and (25) it followed that

$$\frac{d^2V}{dt^2} = an^2 \sin V, \quad (26)$$

whence

$$\pm \frac{dV}{dt} = \sqrt{\lambda - 2an^2 \cos V}. \quad (27)$$

Here λ is a constant of integration. The case in which a is positive and $\lambda < 2an^2$ required that V be periodic with a value oscillating around $180°$, and this appeared to be the case in nature, the oscillations being very small; their amplitude and zero-point could be determined, if at all, only observationally.

A practical consequence was that, in the construction of tables, equations (22) and (23) could and should be taken as exact. On a more theoretical note, one could conclude that, even if originally the distances of the satellites from the centre of Jupiter had not been in the ratios now observed, the pendular oscillations implied by (27) would have brought them into that relation. It was a case entirely analogous to that which Newton had conjectured and Lagrange had derived for the physical libration of the Moon (see Chapter 21).

The secular equation of the Moon

Besides the anomalies in the mean motions of Jupiter and Saturn, so triumphantly accounted for on the basis of Newtonian theory in Laplace's "Théorie de Jupiter et de Saturne" (*MARS* 1785/88), there was another major unexplained anomaly in the solar system: the secular equation of the Moon. It had been a puzzle since Halley first drew attention to it in 1693. In 1749 Richard Dunthorne, from an examination of ancient and modern lunar and solar eclipses, had found the rate of acceleration to be 10″ per century per century; Johann Tobias Mayer in his first lunar tables of 1753 put it at 7″, but in the revised version of these tables published in London in 1770 it was raised to 9″. D'Alembert and Euler had shown that this acceleration could not arise from direct perturbations due to the Sun and planets. Lagrange, in a prize-winning memoir of 1774, had proved that it could not be caused by the spheroidal shapes of the Earth and Moon, and had raised doubts about the reality of the acceleration. (In his *Mécanique céleste* Laplace would claim that it was placed beyond doubt by ancient eclipses, and those observed in AD 977 and 978 by Ibn Yūnus.) As we have seen, Laplace in his first published astronomical memoir pointed out that the assumption of a finite speed of propagation of the gravitational force would account for it; but he regarded this assumption as doubtful. After completing his "Théorie de Jupiter et de Saturne" in August 1786, he set as his next challenge the answering of this question: could the anomalous acceleration of the Moon be accounted for on the basis of a strictly Newtonian theory?

On 19 December 1787 Laplace read to the Academy a first draft of his resolution of the question; the completed memoir "Sur l'équation séculaire de la lune" was published in *MARS* 1786/88. The acceleration, he argued, was the result of an indirect perturbation. Owing to planetary perturbations, the eccentricity of the Earth's orbit is subject to an oscillation of very long period; in the present age of the world (as he had proved in an earlier memoir) it is diminishing. The diminution causes a slight decrease in the average radial component of the Sun's perturbing force on the Moon (the perturbing force being the difference between the Sun's force on the Earth and its force on the Moon); as a consequence the Moon is less attracted away from the Earth, and the increased net force toward the Earth causes it to revolve more rapidly. After a few hundred thousand years the eccentricity will cease to diminish and begin to increase, and then the Moon's motion will be subject to a corresponding deceleration.

Laplace's explanation, however, does not completely resolve the problem. As John Couch Adams was to point out in 1854, the Laplacian explanation accounts for only about half the Moon's observed secular acceleration; the effect of the diminishing eccentricity on the tangential components of the Sun's perturbing force, which Laplace failed to include in his calculation, subtracts from the total effect. The remaining acceleration is due to the fact that the scale of mean time, in accordance with which all the celestial motions were timed in Laplace's day, is not uniform because of the slowing of the Earth's rotation caused by the friction of the tides. Laplace, in his earliest astronomical memoir, we recall, had considered and then dismissed the possibility that the secular acceleration of the Moon could be an illusion resulting from frictional slowing of the Earth's rotation. The idea would not be taken up again until the 1860s, and an atomic clock capable of establishing its validity would not be available till the twentieth century.

Revolution, Consulate, Empire, and Restoration

The Revolution, and more especially the Terror, turned upside down the lives of Frenchmen generally, including scientists. Initially the revolutionaries took an antipathetic view toward science, especially physical and mathematical science; the academies, seen as elitist, were suppressed by a decree of the Convention of 8 August 1793. A number of scientists, friends of Laplace like A.-L. Lavoisier, J.-S. Bailly, and Bochart de Saron who had been president of the Parlement and one of Laplace's first protectors in the Academy, were guillotined. Laplace had been appointed to the Commission on Weights and Measures which was charged with the establishment of the metric system; but he was removed from this commission by the decree of 3 nivôse An II (23 December 1793), on suspicion of lacking "republican virtues and the hatred of kings". He took refuge at Melun, southeast of Paris and, there pursued his work, writing his *Exposition du système du monde* and the first volumes of his *Mécanique céleste*. He was assisted by the young Alexis Bouvard (1767–1843), who became his faithful collaborator and verified the calculations of the *Mécanique céleste*.

Often quoted is the stupid remark of a member of the Convention on the occasion of Lavoisier's death: "The Republic has no need of *savants*". Anti-scientific rhetoric was abandoned, however, with the Thermidorean reaction and rise of the first Directory, for the new leaders recognized the importance of scientific research for a great nation – a nation, moreover, engaged in a long and difficult war with half of Europe. The law of 7 messidor An III (25 June 1795) created the Bureau des Longitudes, charged with calculating and publishing ephemerides, directing the Paris Observatory, and perfecting the theories of celestial mechanics; among the members originally appointed to it were Lagrange and Laplace. The École Polytechnique, created in December 1794, received on 1 September 1795 the name by which it has been known ever since; the École Normale had its beginnings in January 1795, but then underwent temporary eclipse till 1812. The academies were re-created in the form of the five classes of the Institut de France by the law of 3 brumaire An IV (25 October 1795), and on 29 brumaire An IV (20 November 1795) Laplace and Lagrange were named members of the mathematical section of the class of physical and mathematical sciences.

The Republican Calendar had been adopted by the Convention in 1793, with 22 September 1792 designated as the beginning of its first year. The group that had drawn it up, under the direction of the Committee of Public Instruction, had included Lagrange but not Laplace; on one occasion, however, the latter had served as a consultant. On 22 fructidor An XIII (9 September 1805) it was Laplace, now a senator, who presented the report advising a return to the Gregorian Calendar, to be effective on 1 January 1806. In the first two editions of the *Exposition du système du monde* (1796 and 1799), Laplace presented and explained the Republican Calendar; in the third edition (1808) he did not mention it, but confined himself to praising the Gregorian Calendar.

After the *coup d'état* of 18 brumaire An VIII (9 November 1799), by which Napoleon came to power, Laplace was named Minister of the Interior, only to be replaced six weeks later by Lucien Bonaparte. In 1803 he became Chancellor of the Senate. Honours accumulated: Mme Laplace became Dame of Honour in attendance on the Princess Elisa, Napoleon's sister, and Laplace became Grand Officer of the Legion of Honour and Count of the Empire in 1808.

In 1814, with France assaulted by enemies on all frontiers, the Senate, Laplace included, voted the end of the Empire and Napoleonic power. During the Hundred Days (from Napoleon's return in March 1815 to Waterloo, 18 June 1815), Laplace avoided rallying to the Bonaparte cause; thus, after Waterloo and the definitive fall of the Empire, he remained in favour with Louis XVIII. He received the Grand Cross of the Legion of Honour, and was given the title Marquis. In 1816 he was made a member of the Académie Française, and here he is said to have played an ultraconservative role, refusing even to support freedom of the press. For this attitude – the docile acceptance of the regimes that succeeded one another in France during this troubled time, without any sign of having made critical distinction among them – Laplace has been much blamed. He has been defended, on the other hand, as the scientist totally preoccupied with science; could one not argue that it was proper for him to accept the honours that successive governments, whatever their character might be, owed him as the greatest scientist of his nation and time?

Laplace always showed much solicitude for the young scientists who came to him for counsel and support. In 1806 he acquired a country house at Arcueil near Paris, and with the chemist Claude Berthollet, a colleague in the Academy of Sciences who had a house nearby, he created the "Société d'Arcueil". It aimed to foster the association of illustrious elder statesmen of science, like Laplace and Berthollet themselves, with younger scientists who could benefit from the knowledge, experience, and counsel of such elders. Among those who profited from the society were J. A. Chaptal, J. L. Gay-Lussac, Alexander von Humboldt, Jean-Baptiste Biot, François Arago, and S.-D. Poisson.

Laplace's work in celestial mechanics after 1789

In 1793, while Laplace was taking refuge from the Revolution at Melun, he began his *Exposition du système du monde*; the first edition appeared in 1796, and the sixth, which was posthumous but in large part revised by the author before his death, in 1835. In the interim there had appeared four other editions, in 1799, 1808, 1813, and 1824. This work is a very special one among Laplace's writings: it is a general exposition of all the astronomical knowledge of the time in which there appears not a single mathematical formula, and – what is perhaps stranger yet – not a single geometrical figure. The exposition is nevertheless perfectly clear. According to Arago in his "Notice" on Laplace, "This work, written with a noble simplicity, an exquisite propriety of expression, a scrupulous correctness, is classed today, by universal consent, among the beautiful monuments of the French language."

The work contains five books. The first concerns the apparent movements of the celestial bodies (the diurnal motions, the apparent motions of planets and satellites, etc.). The second is entitled "On the real motions of the celestial bodies", and describes the heliocentric orbits of the planets and the laws of planetary motion. The third is dedicated to mechanics (forces, the motion of a mass-point under the action of forces, etc.). In Bk IV, entitled "On the theory of universal gravitation", we are given an exposition of Laplace's own discoveries, for instance those having to do with the secular inequalities and the stability of the solar system. Book V gives a history of astronomy from Antiquity to Newton, and concludes with a chapter entitled "Considerations on the system of the world and on the future progress of astronomy", in which Laplace draws attention to the structure of the solar system, to the fact that the orbits are nearly circular and little inclined to one another, and to the stability of the observed motions, which he

believes he has definitely demonstrated. One finds here a scarcely veiled rejection of direct action by the Deity as an explanation for astronomical phenomena. "These phenomena and some others similarly explained, lead us to believe that everything depends on these laws [the primordial laws of nature] by relations more or less hidden, but of which it is wiser to admit ignorance, rather than to substitute imaginary causes solely in order to quieten our uneasiness about the origin of the things that interest us."

Note 7 at the very end of Bk V has made the *Exposition du système du monde* especially celebrated. It contains the hypothesis on the formation of the solar system known as Laplace's nebular hypothesis. Georges-Louis Leclerc, Comte de Buffon, here cited by Laplace, had suggested that the matter that would later form the planets had been dragged out of the Sun by the attraction of a passing comet. A very similar idea was to be suggested in 1920 by Sir James Jeans, who proposed that matter had been dragged out of the Sun as the result of tidal action due to the near passage of a star.

Laplace rejected Buffon's hypothesis because it failed to explain why the direction of revolution of the satellites around their primary planets is the same as that of the planets around the Sun, or why the direction of rotation of the planets about their axes is again the same, or above all why the planetary orbits are so little eccentric. He proposed instead that a vast nebula in rotation had cooled and condensed to form the Sun at its centre and, by the agglomeration of the remaining matter at different distances from the Sun, the planets. (Immanuel Kant had proposed a similar hypothesis in 1755, but Laplace probably did not know of it.) In the successive editions of the *Exposition du système du monde* Laplace revised his formulation of the hypothesis and presented it with increasing confidence. He was aware that William Herschel, in his examination of the nebulae, had persuaded himself that they were stars in process of formation through condensation. We know today that this is true for a gaseous nebula like the Orion nebula, but false for other nebulae such as the galaxies and the globular clusters. Remarkably, Laplace cited the example of the Pleiades, suggesting that this group had been created in the middle of a "nebula with several nuclei" – an hypothesis completely in

TRAITÉ

Hh II – 64

DE

MÉCANIQUE CÉLESTE,

PAR P. S. LAPLACE,

Membre de l'Institut national de France, et du Bureau
des Longitudes.

TOME PREMIER.

———

DE L'IMPRIMERIE DE CRAPELET.

A PARIS,

Chez **J. B. M. DUPRAT**, Libraire pour les Mathématiques,
quai des Augustins.

———

AN VII.

22.2. The title-page of the first volume of Laplace's *Traité de mécanique céleste*.

accord with twentieth-century astrophysics. The hypotheses of Buffon and Jeans have been abandoned, while Laplace's nebular hypothesis retains the interest of modern astronomers who strive to establish the theory on a more scientific basis.

The *Traité de mécanique céleste* (Figure 22.2), commonly known simply as the *Mécanique céleste*, is Laplace's most famous work, contributing most to his renown both in France and abroad. For the astronomers of the nineteenth century, it was the chief source from which they derived the foundations of the theories by which they made their century "the golden age of celestial mechanics". If

Lagrange's *Mécanique analytique* (1788) had, in some of its theoretical formulations, an equal importance, it was the *Mécanique céleste* that combined theoretical formulation with detailed application in such a way as to serve most effectively as guide and spur to further research; as such it was undoubtedly the most influential work of celestial mechanics since Newton's *Principia*. Numerous translations were made almost immediately, but the most celebrated – the bedside book of American astronomers like Simon Newcomb and G. W. Hill – was the English translation by Nathaniel Bowditch, published in Boston between 1829 and

1839. Bowditch added explanatory notes and re-did all the calculations.

For a twentieth-century astronomer the reading of the *Mécanique céleste* is not easy; nor was it easy for Laplace's contemporaries. Bowditch and Biot, who read page-proofs, complained of the absence of numerous intermediary calculations, replaced by "it is easy to see"; to bridge these gaps they were forced to devote long hours of reflection and calculation. F. F. Tisserand, in the preface to the first volume of his *Traité de mécanique céleste* (1889), proposed his own work as an easy introduction to an arduous science, but urged those who wished to penetrate more deeply to turn to Laplace's treatise, "of which all the chapters present, still today, and even to the most accomplished astronomers, varied subjects for fruitful meditation". A century later, the interest of Laplace's text must be admitted to be largely historical.

The work contains five volumes of which the first four appeared between 1798 and 1805 and the fifth in 1825. Laplace's purpose, as formulated in a brief preface, was to bring together into a coherent whole all the theories of celestial mechanics, both his own and those of his predecessors and contemporaries. For celestial mechanics thus unified he had an explicit goal:

Astronomy, considered in the most general manner, is a grand problem in mechanics, in which the elements of the celestial motions are arbitrary constants; its solution depends both on the accuracy of the observations and on the perfection of Analysis, and it is very important to banish all empiricism and to borrow nothing from observation except indispensable data. It is the aim of this work to achieve, as much as may be in my power, this interesting result.

The first volume consists of two books and the second volume of three. These two volumes contain the general theory of the motions and figures of the celestial bodies. Book I presents a general course in mechanics, based on the laws of equilibrium and motion. In Bk II Laplace introduces universal gravitation, with its consequences for the elliptical motion of the centres of gravity of the planets about the Sun and the perturbations of this motion due to mutual planetary attraction. Book III deals with the figures of the heavenly bodies, and includes a comparison of theoretical predictions and observational data relative to the figure

of the Earth. Book IV concerns the oscillations of the sea and atmosphere. In Bk V Laplace treats of the rotations of the heavenly bodies about their centres of gravity, and deals in particular with the precession and nutation of the Earth's axis, possible changes that the Earth's rotation might undergo, the libration of the Moon, and the rotation of the rings of Saturn.

The third volume consists of two books: Bk VI which gives the detailed theories of the motions of the several planets, and Bk VII which contains the detailed theory of the Moon's motions. The fourth volume contains three books: Bk VIII deals with the motions of the satellites of Jupiter, Saturn, and Uranus; Bk IX with the theory of comets; and Bk X with miscellaneous topics including astronomical refraction and the possible effect of a resisting medium on planetary motion.

The fifth and last volume is divided into six books, each concerned with one of the major topics dealt with in the earlier books, but giving results that Laplace had published more recently in the *Connaissance des temps* or in the *Mémoires* of the Institute. Each of these books begins with an historical notice, describing the relevant work of Laplace's predecessors along with his own contributions.

As we might expect, the *Mécanique céleste* is to a considerable extent a presentation and elaboration of the results of Laplace's earlier work – on the secular equations and the stability of the solar system; the attraction of spheroids with its consequences for the figure of the Earth, the tides, and the precession and nutation of the Earth's axis; the effect of near-commensurabilities in the motions of planets and satellites in producing inequalities of long period; the libration in the motion of the first three Galilean satellites of Jupiter; and the secular equation of the Moon. But the *Mécanique céleste* also incorporated a number of discoveries and developments dating from after 1789; the following deserve special mention.

The invariable plane
In the late 1790s Laplace made the discovery that there exists in the solar system an invariable plane, about which the planetary orbits, owing to mutual planetary interactions, perpetually oscillate through small angles to either side. In the study of the motions of the planets astronomers ordinarily

use the ecliptic – the plane of the Earth's orbit about the Sun – as a plane of reference; but this choice has the disadvantage that because of perturbations the ecliptic does not remain parallel to a fixed plane in a system of absolute coordinates. Laplace defined his invariable plane as follows: it passes through the centre of gravity of the solar system, and is such that if the areal velocity of the projection of each planet onto the plane is multiplied by the planet's mass, then the sum of these products is a maximum. Using the later language of vectors we may say that the invariable plane is the plane perpendicular to the vector sum of the angular momenta of the planets. For the year 1750, Laplace found the ecliptic to be inclined to the invariable plane by an angle of 1° 35′ 31″. Changes in the inclination and position of the node of the ecliptic with respect to the invariable plane are very slow; it is important to determine them, however, whenever long-term variations in the orientations of the planetary orbits are considered.

The figure of the Earth

Laplace had early recognized the disagreement between the gravitational theory of the Earth's figure, as it had been developed by A.-C. Clairaut, and the several measures of a degree of the meridian in different latitudes; the ellipticity apparently implied by the latter was a good deal greater than that implied by the theory. In Bks III and XI of the Mécanique céleste Laplace carried the development of the theory further than Clairaut, but with general corroboration of Clairaut's results: the Earth should be an ellipsoid of revolution, very nearly, with layers of increasing density toward the centre, and with an ellipticity considerably less than $\frac{1}{230}$ (the precession and nutation set limits between which the ellipticity must fall: $\frac{1}{304}$ and $\frac{1}{578}$ according to Laplace, $\frac{1}{279}$ and $\frac{1}{578}$ according to Bowditch). How were these conclusions to be reconciled with the observational results? Laplace was undoubtedly one of those who had supported the basing of the metric system on a new measurement of the meridian of France: the surreptitious aim was to resolve the enigma of the Earth's shape. From the re-measurement carried out by Delambre and Méchain during the years from 1792 to 1799, compared with the measured length of the degree in Peru, Laplace in Bk III of the Mécanique céleste derived an ellipticity of $\frac{1}{334}$, and in Bk XI an

ellipticity of $\frac{1}{308}$. Similar low values were supported by the variation of the length of the seconds-pendulum with latitude, and by two inequalities in the Moon's longitude that were dependent on the Earth's oblateness. Only the measure of the degree of the meridian in Lapland appeared to belie these conclusions; but the re-measurement of that degree by Jons Svanberg in 1802–03 showed that the original result obtained by Maupertuis and his colleagues in the 1730s had erred greatly in excess. (For further details see Chapter 15 above.)

The tides

As we have seen, already by the early 1730s, Laplace had achieved the first truly dynamic theory of the tides; a systematic presentation of the theory is to be found in Bk IV of the Mécanique céleste. At the beginning of the twentieth century G. H. Darwin paid homage to Laplace as the first to put in evidence the whole difficulty of the problem, and to demonstrate the fundamental role that the Earth's rotation plays in producing the phenomena of the tides.

Laplace based his theory, he tells us, on the result that "the state of a system of bodies in which the initial conditions of motion have disappeared owing to resistance is periodic, like the forces to which it is subject." At the end of Chapter 3 of Bk IV, he established a formula giving the height of the tide at a given place as a function of the hourly coordinates of the Moon and the Sun and of certain constants characteristic of the place. In the next chapter he proceeded to compare this formula with a series of measurements of the tides made at the port of Brest during the years 1711–16, and in Bk XIII he carried out a further comparison with new measurements made at Brest during the years 1807–22. Laplace's formula is still used to calculate the Brest-Reference Tide, which serves as an intermediary in the prediction of tides in all ports situated on the coast of France.

Development of the perturbing function of the planets

In 1774 Lagrange had introduced a potential function in the calculation of the attraction exerted by an extended body, and in memoirs published in Mémoires de Berlin 1776/79 and 1777/79, he had introduced a potential function, later called by Laplace "the perturbing function", for the determination of the mutual perturbations of the planets considered as point-masses. The introduction of the perturbing function, according to Laplace, constituted "a veritable discovery". In his "Théorie de Jupiter et de Saturne" he proceeded to employ this function to obtain formulas for the perturbations in radius vector, true anomaly, and latitude; these formulas reappear as formulas (X), (Y), and (Z) of Bk II, Chap. 6 of the *Mécanique céleste* (equations (X) and (Y) are identical with our equations (20) and (21) above). Of them he says:

The formulas (X), (Y), (Z) have the advantage of presenting the perturbations under a finite form. This is very useful in the theory of comets, in which those perturbations cannot be found, except by the [mechanical] quadrature of curves. But the smallness of the eccentricities, and the inclinations of the orbits of the planets to each other, enables us to develop their perturbations in converging series of sines and cosines of angles, increasing in proportion to the time, and we can then arrange them in tables which will answer for an indefinite time.

The development of the perturbations in series entailed that of the perturbing function, which entailed in turn the development in series of the odd inverse powers of the distance between the perturbed and perturbing planets, or

$$(1 - 2a\cos\theta + a^2)^{-s} = \frac{1}{2} \sum_{-\infty}^{+\infty} b_s^{(j)}\cos j\theta, \quad (28)$$

where a is the ratio of the mean solar distances of the perturbed and perturbing planets, $s = \frac{1}{2}, \frac{3}{2}, \frac{5}{2}, \ldots$, and

$$b_s^{(-j)} = b_s^{(j)}.$$

The bs, here given in the double indicial notation that Laplace employed in the *Mécanique céleste*, are today called Laplace coefficients. They are functions solely of a. In deriving them, Laplace made use of a device introduced by Lagrange in 1766, whereby the left-hand member of equation (28) is factored into two complex factors, each of which is then expanded by the binomial theorem; the two resulting series are then multiplied together. The resulting product gives the bs as hypergeometric series. Laplace also exploited the Eulerian relations among the bs and their derivatives with respect to a, thus reducing the number of necessary summations of series to two.

Laplace then proceeded to develop the perturb-

ing function to the third order in the eccentricities and inclinations. This development was much utilized by the astronomers of the early nineteenth century, for instance by Bouvard in his theory of Uranus. Some fifty years were to elapse before the development was carried to the sixth order in the eccentricities and inclinations by Philippe de Pontécoulant and Benjamin Peirce, and then to the seventh order by U. J. J. Le Verrier.

The theory of comets

In Bk IX Laplace took up the problem of determining the changes resulting in cometary orbits owing to perturbation by planets. Here, in contrast to the case of planets perturbed by planets, no analytic formulas could be derived that would give the perturbations for all time; it was necessary to have recourse to mechanical quadratures, and starting at a given time, proceed step by step round the cometary orbit. Following Lagrange's prize-winning essay on cometary perturbations (see the preceding chapter), Laplace employed the eccentric anomaly as independent variable, and proposed calculating the variations of the orbital elements – eccentricity, perihelion, mean solar distance and mean motion, epoch, orbital inclination and node of the cometary orbit on the ecliptic – for each degree or two degrees round the orbit. For the required numerical integrations, he developed the appropriate approximating formulas by means of a generating function. Like Clairaut (see Chapter 19 above), in the superior half of the orbit he separated the calculation into two parts, one concerned with the (very small) direct effect of the planet on the comet, the other with the effect of the planet on the Sun; in both cases formal integration could be invoked. And again like Clairaut, he proposed calculating backwards in the last quadrant, from 360° to 270° of eccentric anomaly, when the date corresponding to 360° was known. Clairaut had used a single cometary reference ellipse for the determination of planet–comet distances; Laplace proposed rectifying the reference ellipse after every fourth of a quadrant of eccentric anomaly in the first and fourth quadrants, and at least once in the superior part of the orbit. These procedures were applied by the Baron de Damoiseau, J. K. Burckhardt, O. A. Rosenberger, Pontécoulant and J. W. Lubbock to the calculation of the return of Halley's Comet in 1835, and led to predictions considerably more accurate than the prediction Clairaut had been able to achieve in 1759.

In a second chapter on the cometary perturbation, Laplace considered the case of a comet coming so close to a planet that for a time it could be considered as moving in an unperturbed ellipse round the planet; later, when it had passed out of the 'sphere of influence' of the planet, it could be considered as moving in an ellipse about the Sun. The formulas derived from this conception were applied by Burckhardt, at Laplace's request, to Comet Lexell. This comet had been seen only once, in 1770; from the observations Lexell and Burckhardt concluded that it had a period of 5.6 years, and a solar distance at perihelion of about 0.68 AU. Applying Laplace's formulas to the near approach of this comet to Jupiter in 1767, Burckhardt showed that the comet had previously had a mean solar distance of 13.29 AU, and a solar distance at perihelion of 5.08 AU, so that it would have been invisible from the Earth. Applying the formulas again to the near approach of the comet to Jupiter in 1779, Burckhardt showed that the orbital elements must have been so altered as to make the mean solar distance 6.39 AU, and the solar distance at perihelion 3.33 AU; thus the comet once more became invisible from the Earth.

Laplace concluded Bk IX by assembling evidence to show that the masses of comets are not sufficiently great to perturb the planets detectably; the consideration that had deterred him from pursuing the analysis of mutual planetary perturbations during the decade 1775–85 was thus disposed of.

Astronomical refraction

Light rays from a star on entering the Earth's atmosphere are refracted in such a way that the apparent zenith distance of the star is less than its true zenith distance. The estimation of astronomical refractions had been a concern of astronomers since Tycho's time; again and again the tables drawn up for this purpose had to be refined, to keep pace with the increasing precision of observational instruments. Delambre in 1797 made observational determinations of the refractions from 70° to 90° of zenith distance, and Jean-Charles Borda undertook similar studies. In Chap. 1 of Bk X of the *Mécanique céleste* Laplace derived a formula, which now carries his name, and which gives the astronomical refraction to the third order of the tangent of the zenith distance of the star.

The scientific legacy of Laplace

The work of Laplace largely defined the problems with which celestial mechanics would be concerned during the nineteenth century, and provided many of the methods that the celestial mechanicians of that century would use in dealing with them. The heirs and appropriators of the Laplacian legacy, both in France and abroad, were numerous and brilliant. The works of Bouvard, Damoiseau, Pontécoulant, Le Verrier, C.-E. Delaunay, and A.J.-B. Gaillot in France, of Newcomb and Hill in the United States, of Adams in England and of Peter Andreas Hansen in Germany, were direct developments from the foundation laid in the *Mécanique céleste* (see Chapter 28 below for further detail). We have already insisted on the continuing importance of Laplace's cosmogonic hypothesis as presented in his *Exposition du système du monde*.

Laplace was a proponent of philosophical determinism, and no-one has been more resolute in championing the doctrine, as the following oft-quoted passage testifies:

We must therefore envisage the present state of the universe as the effect of its anterior state and as the cause of its future state. An intelligence that knew, for a given instant, all the forces acting in nature and the respective situations of the beings that made it up, if it were in addition vast enough to submit these data to analysis, would embrace in a single formula the motions of the largest bodies of the universe and those of the smallest atom: nothing would be uncertain for it and the future like the past would be present to its eyes.

The tranquil assurance here evident could hardly have survived unshaken in face of such later developments in science as the emergence of quantum mechanics. Recently it has been shown that, because of sensitivity to tiny errors in the initial conditions, predictions of the future state of the solar system are valid only for durations of a hundred million years or so (see the appendix to Chapter 28). Celestial mechanics as Laplace conceived it nevertheless remains the fundamental tool of the astronomers for calculating ephemerides of the bodies of the solar system.

Almost one hundred years after Newton, on 5 March 1827, Laplace died in Paris, at 108 rue du

Bac, a house which still stands. The comparison with Newton is an interesting one. Newton stood at the origin of a scientific revolution without precedent. He discovered a law applying to the least particle wherever it is found in the universe; at one blow he provided explanations of the most diverse phenomena, and with apparently unrestricted validity. The success of his derivations encouraged the hope that the employment of mathematics as a tool would suffice for deriving all the consequences of universal gravitation. The genius of Laplace, on the other hand, consisted in wielding mathematical devices in such a way as to discover and demonstrate a very considerable number of these consequences. He definitively placed the law of universal gravitation in the rank of the great scientific truths by showing that the phenomena that appeared to contradict it or limit its domain

(e.g. the great inequality of Jupiter and Saturn, the secular acceleration of the Moon, the motions of the satellites of Jupiter) were, on the contrary, explained by it.

Despite his achievement, Laplace is relatively little known to the general public. Is it, as has suggested Jacques Merleau-Ponty, because Laplace was not the originator of a spectacular revolution but the hero of "normal science"?

Further reading

Article on Laplace, in *Dictionary of Scientific Biography*, vol. 15 (New York, 1978), 273–403

H. Andoyer, *L'oeuvre scientifique de Laplace* (Paris, 1929)

B. A. Vorontsov-Veliaminov, *Laplace* (Moscow, 1985; in Russian)

François Arago, Laplace, in J. A. Barral (ed.), *Oeuvres de François Arago*, Vol. 3 (Paris, 1859)

Oeuvres complètes de Laplace (14 vols, Paris, 1878–1912)

PART VII

Observational astronomy and the application of theory in the late eighteenth and early nineteenth centuries

23

Measuring solar parallax: the Venus transits of 1761 and 1769 and their nineteenth-century sequels

ALBERT VAN HELDEN

Since the time of Copernicus, the astronomical unit – the mean distance between the Earth and the Sun – has been the fundamental unit of our solar system. Kepler's third law, published in 1619, made it an easy matter to express with great precision the distances of the planets from the Sun in terms of this astronomical unit. It was, however, extremely difficult accurately to express the distance from the Earth to the Sun in terms of precisely defined earthly distance units. In order to express the sizes and distances of all bodies within the solar system in terms of common terrestrial measures, astronomers had to bridge this gap between cosmic and earthly dimensions. For this, at least one distance in the heavens had to be measured in earthly terms.

Important as this task was for cosmology, measuring at least one celestial distance was crucial to predictive astronomy because precise knowledge of the astronomical unit, or of the angle from which it is determined (the solar parallax, or the angle subtended by the Earth's radius as seen from the centre of the Sun*) is fundamental to planetary theory. In the Copernican universe, the motions of the planets must be referred to the Sun, and therefore the Earth's motion around the Sun must be known accurately. Determining this motion requires accurate measurements of the Sun's positions, which are obtained by correcting the raw measurements downwards for atmospheric refraction and upwards for solar parallax. Since historically these two corrections were interdependent, it was impossible to construct reliable refraction tables until the solar parallax was known with acceptable accuracy. The uncertainties in the Sun's altitudes resulting from inadequate refrac-

tion and parallax corrections introduced errors in the obliquity of the ecliptic and the eccentricity of the Earth's orbit. Errors in the Earth's orbit led to systematic errors in the calculated orbits of the other planets, so that their calculated positions deviated from their observed positions. Moreover, in the physics of Newton, without an accurate knowledge of solar parallax, it was impossible precisely to determine the Earth's mass relative to the Sun's, a ratio essential in determining the perturbations of the Earth on the other planets. Finally, there is a connection between solar parallax and the parallactic inequality of the Moon, an essential component in lunar theory: one is needed to determine the other.

Now much of this was academic at the time of Copernicus: planetary theory had not progressed much beyond Ptolemy and measuring instruments perhaps not at all. Over the next two centuries, however, both theory and measurement increased immeasurably in sophistication and precision, and by the beginning of the eighteenth century, after the elimination of many other errors and uncertainties, the accurate determination of solar parallax had become a central problem in technical astronomy.

The quest for the astronomical unit had begun in the middle of the sixteenth century, and it was to take several centuries before the problem would reach a reasonably satisfactory conclusion. By the end of the seventeenth century, telescopic observations had resulted in a consensus figure of 10 to 12 arc-seconds for solar parallax, corresponding to a solar distance of some 20 500 Earth radii or about 80 million miles. This figure was, however, the result not of direct measurements of a parallax but rather of a combination of factors. The assumption that solar parallax was less than about 12″ led to better refraction tables. That figure also made the

*Unless otherwise noted, in this chapter 'solar parallax' stands for the mean horizontal solar parallax.

Earth, a planet with a moon, larger than Venus, a planet without a moon, while making Mercury, a primary planet, larger than the Moon, a secondary planet (see Chapter 7 in Volume 2A).

The centre piece in this new consensus was the measurement of the parallax of Mars made by French and English astronomers in 1672, when a favourable opposition brought Mars unusually close to the Earth. A close examination of these measurements shows, however, that at that time Mars's parallax still lay within the error margin of the measuring instruments. In retrospect, therefore, the consensus was not based on a positive measurement.

Although Gian Domenico Cassini (Cassini I) and John Flamsteed, the architects of the new consensus, at times cautioned their audiences against assigning too great a precision to their measurements, they firmly believed that they had measured Mars's parallax in 1672 and had confirmed it by subsequent measurements. For several generations measurements of Mars's parallax during favourable oppositions remained the preferred method of determining solar parallax. In 1685 Francesco Bianchini used it to arrive at a solar parallax of about 15"; in 1704 and 1719 Cassini's nephew Giacomo Filippo Maraldi concluded that solar parallax was about 10"; in 1719 James Pound and his nephew James Bradley (1693–1762) reported a value of between 9" and 12"; in 1736 Cassini's son Jacques (Cassini II) obtained a result of 11" to 15"; and in 1751 the abbé Nicolas-Louis de Lacaille (1713–62) obtained a value of 10" from his measurements of the diurnal parallaxes of Mars and Venus, made at the Cape of Good Hope. In view of the improvements in instruments and observing techniques between 1670 and 1750, if Cassini, Flamsteed, and their successors had actually measured Mars's parallax, one might have expected that during those years the Sun's parallax would have been confined to increasingly narrow limits around the actual value of 8".8. Since, however, the values reported above show the opposite tendency, we can see that Mars's parallax was still lost in the error margin of the instruments.

From the beginning, one astronomer, Edmond Halley (1656–1742) refused to lend his name to the consensus. Acting on a suggestion by James Gregory, Halley had observed Mercury's transit of 1677 in St Helena, where he was mapping the southern skies. He was the first to observe both the planet's ingress onto the solar disk and its egress, measuring their times as accurately as he could. Only one observer in Europe – Jean Charles Gallet in Avignon – had made similar measurements: a comparison of the two observations yielded a solar parallax of 45", a value already considered much too large by influential astronomers. Although Halley did not put much stock in this result because Mercury's parallax is only slightly greater than the Sun's, he did think the method to be useful, especially if applied to transits of Venus. His subsequent role as an assistant to Flamsteed in several measurements of Mars's parallax led him to conclude that this method was useless. In his *Catalogus stellarum australium* of 1678 he wrote:

There remains but one observation by which one can resolve the problem of the Sun's distance from the Earth, and that advantage is reserved for the astronomers of the following century, to wit, when Venus will pass across the disk of the Sun, which will occur only in the year 1761 on 26 May [Julian]. For if the parallax of Venus on the Sun is then observed by this method I have just explained, it will be almost three times greater than the Sun's, and the observations required for this are the easiest of all, so that through this phenomenon men can instruct themselves of all they could wish on that occasion.

Over the next decade, one of Halley's research projects was the improvement of the elements of Venus and Mercury, building on the work of Johannes Kepler, Jeremiah Horrocks, and Thomas Streete. His own daylight measurements of Mercury's positions showed that Streete's elements were more accurate than those of Kepler. In 1691 Halley published his results in a paper in the *Philosophical Transactions* entitled "An Astronomical dissertation on the Visible Conjunctions of the Inferior Planets with the Sun," the first paper ever devoted entirely to the subject of transits. He gave the inclination of Venus's orbit to the ecliptic as 3° 23′ (modern value for 1700, 3° 23′ 28″) and the location of the ascending node as 14° 18′ in the sign of Gemini (modern value 13° 59′ Gemini). He calculated that in eight sidereal years Venus moved 1° 30′ 28¼″ more than thirteen revolutions, in 235 years 42′ 21″ less than 382 revolutions, and in 243 years 48′ 8″ more than 395 revolu-

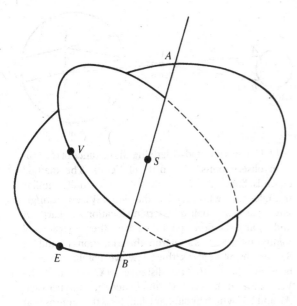

23.1. Diagram showing the orbits around the Sun (*S*) of Venus (*V*) and the Earth (*E*). Transits of Venus are observed when Venus and the Earth are simultaneously in direction *SA* or simultaneously in direction *SB*.

tions. Transits of Venus happen when the planet is at or very near one of its nodes and the line from the centre of the Sun through that node passes through the Earth (Figure 23.1). Since the Sun's radius is about 16', the planet may be a little distance removed from its node and still appear on the Sun's disk. The greatest duration of a Venus transit occurs when the planet passes over the centre of the Sun's disk: Halley gave this time as 7 hours 56 minutes. Transits at the ascending node occur early in December and those at the descending node early in June.

With this and similar information for Mercury, the dates of past and future transits could easily be calculated. Halley gave a list of 29 transits of Mercury between 1615 and 1789, and seventeen transits of Venus between 918 and 2004. He correctly gave the June transits of Venus of 1518, 1526, 1761 and 1769 and the December transits of 1631 and 1639 (the last of which had been observed by Horrocks and William Crabtree). For those farther removed from his time period, his predictions were not as good. His list therefore does not show the rigid sequence of alternating pairs of June and December transits, the transits of each pair being separated by 8 years, and the successive

pairs by 235 years. (Thus, June transits occurred in 1518 and 1526, 1761 and 1769, and will occur in 2004 and 2012; December transits occurred in 1631 and 1639, and 1874 and 1882.) The important thing was, however, that he was correct in his prediction of the next set of Venus transits: they were to happen in June 1761 (already predicted by Kepler) and 1769 (not predicted by Kepler).

Halley concluded his article with the statement that the chief use of Venus transits was the determination of solar parallax, a quantity astronomers had investigated in vain by various methods and with the most delicate instruments. During a transit, one could measure the time lapse between ingress and egress (using the moments of internal contact) accurately to a second of time, or 15 arcseconds, by means of even a mediocre telescope and a pendulum clock that runs accurately for six or eight hours. How, from two such observations, made simultaneously at different locations, the solar parallax and distances could be determined to one part in 500 he promised to explain in a future paper.

Although Halley's more senior colleagues, Newton and Flamsteed, did not react publicly to this paper, some of his younger ones did. In his *Astronomiae physicae & geometricae elementa* of 1702, David Gregory repeated his uncle James Gregory's remarks about the use of transits in determining solar parallax, and stated that the world would have to wait until the Venus transit of 1761 for an accurate determination. In his *Praelectiones astronomicae* of 1707 William Whiston drew on Halley's 1691 paper to advocate the use of Venus transits. Both books went through several English and Latin editions.

Halley himself returned to the subject in 1716. In that year he published in the *Philosophical Transactions* a paper entitled "A Singular Method by which the Parallax of the Sun or its Distance from the Earth can be Determined Securely, by means of Observing Venus in the Sun." In this paper, Halley's most elaborate treatment of the subject, he reviewed the history of this problem and gave some harmonic reasons why the Sun's parallax should be taken as about $12\frac{1}{2}''$ (this would make the Earth larger than Venus and Mercury larger than the Moon) until the observations he was about to advocate could be made. The measurements that had been made up to then, he insisted, were useless because they often produced a negative solar parallax.

Halley related how in 1677, in St Helena, he had been able to observe the ingress and egress of Mercury: he had been able to time the transit of Mercury "without an error of one second" by taking the internal contacts as his reference points. The reason was that the moments when a thin filament of the Sun's light first appeared between the Sun's limb and the dark body of the planet at ingress (first internal contact), and disappeared at egress (last internal contact) could, in Halley's judgment, be timed very accurately. If Mercury were not so close to the Sun, transits of this planet would be ideal for determining parallax, for they happen frequently. It remained therefore to use the much less frequent transits of Venus, whose parallax is four times greater than the Sun's. Differences in the measured times of a Venus transit as observed from different places on Earth should allow a determination of the Sun's parallax to within a small part of an arc-second. For such observations the only instruments needed were common but good telescopes and clocks, and all that was required of the observers besides trustworthiness and diligence was a modicum of skill in astronomical matters. Nor was it necessary scrupulously to measure the latitude of the locations of the observations but only to record the local times accurately.

The length of Venus's path across the Sun was a function of the planet's observed latitude on the Sun and ranged from a maximum equal to the Sun's diameter to zero when the planet merely grazed the Sun. By measuring the times of ingress and egress, an observer could determine the length of time Venus took to cross the Sun and therefore its path. The angular separation between two paths observed at two observing stations was the measure of the parallactic displacement of Venus on the Sun caused by the separation between the stations (Figure 23.2). Since the ratio of the distances of Venus and the Sun was known, the parallaxes of Venus and the Sun could be calculated.

In London, the transit of 5 June 1761 would begin before sunrise, but observers in more northern latitudes, Norway and perhaps the Shetlands, would be able to observe both ingress and egress. Since at the midpoint of the transit the Sun would

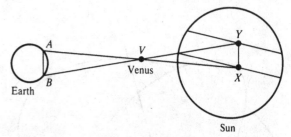

23.2. Halley's method of deriving the distance of the Sun from observations of transits of Venus. The method depends, first, on consideration of two (virtually) similar triangles each with vertex at the planet (V), one triangle having as base a pair of observing locations on Earth (A and B) and the other having as base the projections of Venus on the Sun as seen from these locations (X and Y). Second, from Kepler's third law we obtain the ratio between the Earth–Sun distance (AX, say) and the Venus–Sun distance (VX), and hence the ratio between AV and VX, which by similar triangles is the same as that between AB and XY. Third, we use the known distance in miles between A and B to obtain the distance in miles separating X and Y. Multiplying this by the angular diameter of the Sun divided by the angular separation of X and Y, we obtain the linear diameter of the Sun in miles. By comparing this linear diameter with the Sun's angular diameter, we obtain the Sun's distance from Earth.

The drawing is not to scale, and the distance XY is greatly exaggerated. It is therefore important that terrestrial observating locations A and B are chosen that are very different in latitude; even so, XY is typically only 1/50th of the Sun's disk. For the same reason transits of Venus are chosen in preference to those of Mercury, for Venus is nearer to Earth when in transit and so the separation between X and Y is greater.

In the method that Halley proposed, the angle XY is obtained by observing from both A and B the times taken for the complete transit of the planet across the Sun, for (if we extract from the observations the effects of the Earth's diurnal and annual movements during transit) the times the transit takes are proportional to the lengths of the lines across the Sun traced out by X and Y, and knowledge of these lengths allows us to locate these lines on the Sun.

be directly overhead near the northern shore of the Bay of Bengal, the entire transit would be visible there as well. It would also be useful to have an observation at a point where the midpoint of the transit occurred at midnight, and this would be the case in the Hudson Bay colony, where ingress would happen shortly before sunset on 4 June and egress shortly after sunrise on 5 June. Halley also advocated observations in Madeira and Benkulen (on the western shore of Sumatra). All these were areas controlled by the British. If the French wished to make observations, it would be useful to have observers in Pondicherry on the western shore of the Bay of Bengal; if the Dutch had a desire to advance the science of the heavens in this matter, the observatory of their celebrated city of Batavia in the Dutch East Indies would be a good observing station. Halley wished this phenomenon to be witnessed by many observers in different locations, so that out of a consensus a more trustworthy result would emerge, but also to ensure that the effort would not be foiled by cloudy skies:

Therefore to the curious investigators of the stars to whom (when we shall have ended our lives) these observations are reserved, we recommend again and again that, bearing in mind this warning of ours, they set themselves to carrying out this observation actively and with all their powers. And we wish them luck and pray above all that they be not deprived of the desired spectacle by the untimely gloom of cloudy skies, and that at last the magnitudes of the celestial orbs, confined within narrower limits, will bestow upon them eternal glory and fame.

The paper ended with a geometric analysis of the 1761 transit as seen from London, and a brief mention of the 1769 transit.

Halley's two papers were the basic canon of transits. Since the seventeenth-century attempts to measure the parallax of Mars were, in retrospect, failures, Halley's establishment of the science of transits on a solid footing, educating his fellow astronomers on how parallaxes could be measured by means of Venus transits, is perhaps his most important contribution to astronomy.

If Halley eschewed the use of Mercury transits, concentrating his attention entirely on Venus, others were not so quick to abandon the more frequent transits of the smaller planet. The transit of this planet predicted by Halley for 1723 would be partly visible in western Europe. Whiston published a broadsheet in which he showed the paths of Mercury and Venus across the Sun in their transits from the sixteenth to the end of the eighteenth century, and he urged astronomers to observe the transit of Mercury of 1723 with the hope of determining solar parallax. Indeed, Whis-

ton argued that transits of Mercury were more useful for determining solar parallax than transits of Venus. Observations of this particular transit in England and France, however, produced no useful results, and as a consequence astronomers, especially in England, began to focus more on the upcoming transits of Venus. In France, the astronomical establishment led by the Cassini and Maraldi families clung for some time to their belief in the accuracy of measurements of the parallax of Mars. A new generation of French astronomers, however, led by Joseph-Nicolas Delisle (1688–1768), became champions of the transit method.

After a review of the literature in 1723, Delisle had become convinced that measurements of the parallax of Mars were hopeless and therefore turned his attention to transits. He proposed a slightly different method of observation. Halley's method meant that the times of both ingress and egress had to be measured. But, as Delisle pointed out, local weather conditions might allow only one of these events to be observed. In some places one of the events might occur when the Sun was below the horizon. In such cases, Delisle pointed out, a measurement of the exact moment of ingress or egress would suffice, provided that the observer could accurately determine the longitude of his observing station. The time difference between the same event at different places was a function of the difference in path length, and so would produce the desired result as well. In practice, Delisle's method was more practical for it meant that observations could be made in more locations – more practical that is, provided that observers could make accurate measurements of their longitude. He used this method in his observation of the transit of Mercury of 1723, with disappointing results. On a visit to London, in 1724, however, Delisle visited Halley and was reinforced in his conviction that transits, especially those of Venus, were the only practical way of determining parallax.

During his tenure as director of the observatory of the Russian Imperial Academy of Sciences in St Petersburg (1725–47), Delisle retained his interest in this problem, and through his wide correspondence he kept transit observations before the astronomical community. Delisle gathered information from correspondents on the relationship between the length (i.e., power) of the telescope and the duration of the transit, a knotty technical problem that had to be solved before useful results could be extracted from a comparison of observations. Upon his return to France he became the central figure in the preparations for the transit of Venus of 1761.

Two events were important in those preparations: Lacaille's observations at the Cape of Good Hope in 1751 and the transit of Mercury of 1753. Lacaille went to the tip of Africa to map the southern stars. Before he left he asked European astronomers to measure the positions of Mars and Venus, so that from comparisons with his measurements at the Cape the parallaxes of those planets could be determined. As it turned out, the results were disappointing, but in advocating this method Lacaille pointed out several weaknesses of the transit method. He argued that the accuracy of one part in 500 mentioned by Halley was impossible to attain. First, in those places where ingress and/or egress occurred when the Sun was close to the horizon, the uncertainty of the Sun's limb caused by changing refractions would make it impossible to time internal contacts accurately. Second, in instruments of high magnification, the Sun moves very rapidly through the field of view, and this would make the timing of internal contacts difficult. Since Mercury moves across the Sun $1\frac{1}{2}$ times as fast as Venus, its internal contacts were easier to determine with precision than those of Venus, yet during the Mercury transit of 1743 the most skilled astronomers using excellent instruments differed on the moment of internal contact by as much as 40 seconds. Obviously, one should not put all one's hopes on the upcoming transits of Venus: parallax investigations should be conducted whenever an opportunity arose. Events would prove Lacaille's comments justified.

The Mercury transit of 1753 promised to be the most favourable for parallax investigations, because the planet would pass very near the centre of the Sun. If there was any hope of determining solar parallax by means of transits of this planet, this would be the occasion. In preparation for this transit, Delisle reviewed Halley's planetary tables, published posthumously in 1749, and made several corrections for Mercury. He urged his many correspondents to prepare for the observation, and in the winter preceding the event he published an *avertissement* on this subject. Here he instructed observers on how to make the measurement, specifying his own method instead of Halley's, and

included a figure illustrating the different paths of Mercury across the Sun predicted by the tables of Streete (1661), Philippe de La Hire (1702), Jacques Cassini (1740), and Halley as corrected by Delisle. Delisle saw to it that the little tract reached the far corners of Europe and that instructions reached the New World as well. Through his English colleagues he hoped, too, to enrol observers in the East Indies. As the event drew nearer he also prepared a world map on which the visibility of the transit was depicted for easy reference.

Observations of the Mercury transit of 1753 failed to produce a satisfactory measure of solar parallax, but the event served almost as a dress rehearsal for the impending Venus transit of 1761. Here (and in Lacaille's expedition to South Africa) we see the efforts at marshalling an international body of observers through printed tracts and private correspondences, the correction of tables in order to predict the event more exactly, and the circulation of diagrams and instructions. In fact, a number of the key observers of the Venus transits learned how to observe transits at this time.

In preparing for the Venus transit of 1761, the first in order of business was the correction of the tables. The Mercury transit had confirmed that in spite of the many transits of that planet that had been observed since the first such observation by Gassendi in 1631, there were still important errors in its tables. The best tables were those of Jacques Cassini and Halley, but Delisle's calculations based on corrections of Halley's tables had still predicted the entry of Mercury on the Sun 17 minutes late. There was every reason to believe that the tables of Venus (of which only one transit had been observed, in 1639) contained greater errors. In 1753 the young astronomer Guillaume-Joseph-Hyacinthe-Jean-Baptiste le Gentil de La Galaisière (1725–92) compared the predictions of different tables (chief among them those of Jacques Cassini and Halley), and produced a diagram showing the paths of Venus across the Sun's disk in 1761 and 1769 according to the different tables. It was clear that there was much uncertainty concerning the times and paths of the transits. This uncertainty made it difficult to choose the best locations for observing stations.

The corrections occupied Delisle and his assistants for the next several years, and, finally, early in May 1760 – only thirteen months before the

1761 transit – his world map, showing the visibility of the transit, and his memoir of instructions were published by the Paris Academy of Sciences. Delisle's memoir showed that observations made from Port Nelson in the Hudson Bay Colony, as advocated by Halley, would not be useful for the determination of solar parallax and he therefore advocated a station farther north – at a latitude of at least 65°. The optimum locations for viewing the entire transit were in India and the East Indies. The shortest apparent duration of the total transit would be in Tobolsk, Siberia, while the longest would be in Batavia in the Dutch East Indies. But Delisle also stressed the importance of stations where only ingress or egress would be visible, suggesting Yakoutsk, Kamchatka, the Cape of Good Hope, and the island of St Helena for this purpose. He further urged that observations be made in Peking, Macao, Archangel, Torneo, and St Petersburg. Observers were advised to pay close attention to the characteristics of their telescopes. Copies of Delisle's memoir and map were sent to all regions of Europe, including St Petersburg, Constantinople, and Stockholm.

Interest in the Venus transit had grown in France since 1753, often extending to the popular press, and calls for expeditions were heard frequently. Acting on his own initiative, le Gentil found funding for an expedition to Pondicherry and left early in 1760. Shortly after Delisle presented his memoir to the Academy, late in April of that year, Joseph-Jérôme Lefrançais de Lalande (1732–1807), who was to become the central figure in the Venus transits after Delisle's death in 1768, read a memoir that ended with the following call to arms:

The occasion presented to us by this celebrated phenomenon is one of those precious moments of which the benefit, if we let it escape, will not be compensated later – neither by the efforts of genius nor by dint of hard work, nor by the munificence of the greatest kings. It is a moment that the past century envies us and which in the future will be, I dare say, an injury to the memory of those who will have neglected it.

The Academy now took official action and weighed various possibilities. There were no locations in French hands on the west coast of Africa, and cooperation with the Portuguese and Dutch would take time to arrange. In the meantime, it was decided to send Alexandre-Gui Pingré (1711–96) to the island of Rodrigue in the Indian Ocean, in the shipping lanes of the French East India Company. On the invitation of the St Petersburg Academy, J.-B. Chappe d'Auteroche was to go to Tobolsk in Siberia. Le Gentil was already underway to Pondicherry, and César-François Cassini de Thury (Cassini III, 1714–84) was to observe in Vienna. Efforts to organize a joint expedition with the Dutch to Batavia were scrapped when it became apparent that the British would send an expedition to Benkulen.

Events in England had taken a somewhat different course. Since 1723, transits of Mercury had been assiduously observed by British astronomers, but, now following the lead of Halley, they had used them only to improve the elements of that planet. Although British scientists kept themselves informed of the events in France, preparations for the transit of 1761 did not begin until the Royal Society discussed Delisle's memoir in June 1760. That same month the Council of the Society took charge of the observations and quickly decided to send expeditions to St Helena and Benkulen. Bradley, the Astronomer Royal, supplied the Society with a list of instruments needed: "A Reflecting Telescope of Two Foot with Dollond's Micrometer, Mr Dollond's Refracting Telescope of Ten Feet; a Quadrant of the radius of Eight Inches; and a Clock or time piece".

Because of the lateness of British preparations, the expedition to Benkulen was under time-pressure from the start. Charles Mason (1728–86) and Jeremiah Dixon (1733–79) left Portsmouth early in December 1760, on board the *Seahorse*, a warship provided by the admiralty. Britain and France were engaged in the Seven Years War, and although both Crowns had instructed their navies not to interfere with the expeditions sent out by their respective enemies, such orders were difficult to enforce. Before it could get out of the English Channel, the *Seahorse* was attacked by a French frigate and after a violent battle had to return to port for repairs. The ship did not put out to sea again until 3 February 1761. As had been foreseen by Mason and Dixon, this was too late to reach the East Indies in time, and when the *Seahorse* reached the Cape of Good Hope on 27 April, they decided to make the observation there. The expedition to St Helena under Nevil Maskelyne (1732–1811) had, in the meantime, left England on 17 January 1761 aboard the East Indiaman *Prince Henry* and arrived in St Helena without trouble.

Europeans had thus sent out six expeditions to faraway parts: the French sent le Gentil to Pondicherry, Pingré to Rodrigue (where he arrived on 28 April 1761), and Chappe to Tobolsk (where he arrived on 19 April 1761), while Cassini de Thury went to Vienna; the English sent Maskelyne to St Helena and Mason and Dixon to Benkulen (although as described above they succeeded in getting only as far as the Cape of Good Hope). As the event approached, the Harvard professor John Winthrop (1714–79) organized an expedition to St Johns in Newfoundland, paid for by the government of the Massachusetts colony.

At the island of Rodrigue, it was overcast on 6 June, the day of the transit, and Pingré and his assistant Denis Thuillier had only brief glimpses of the transit, not including ingress and egress. But the astronomers did make some useful measurements of Venus's positions on the Sun, noting the exact times. Chappe was able to observe the entire transit, and his measurements were of the greatest importance. He also observed an eclipse of the Sun on 3 June. Le Gentil had the worst luck. After many complications and delays, he arrived near Pondicherry aboard a ship of the French fleet, only to find that this post had been captured by the British. The fleet turned back toward Île de France (Mauritius), and le Gentil observed the transit from the deck of the ship. He was not able to make useful measurements.

Mason and Dixon, at the Cape of Good Hope, observed the entire transit, and both were able to measure the moments of ingress and egress, differing 4 and 2 seconds respectively in their times. They were also able to make observations of Jupiter's satellites for longitude purposes and determined the latitude of their observatory accurately. Their expedition was, therefore, a great success. Maskelyne and Robert Waddington, on the other hand, had bad luck, for on St Helena the sky was overcast on 6 June. During brief periods of visibility they were able to make some quick measurements of Venus's position on the Sun, but on the whole their expedition was unsuccessful.

At St Johns in Newfoundland, only the egress, shortly after sunrise, was visible. The sky was clear and Winthrop was able to make five measurements

of Venus's positions on the Sun and, most importantly, time the moment of egress exactly. After some difficulties he was able also to make an observation of an occultation from which the longitude of his station could be determined.

Besides these expeditions, the Europeans organized observations in a number of places closer to home. The Swedish and Danish academies were particularly active in this respect, and the Jesuits also observed at a number of stations.

In all, more than 120 observations were made, mostly in Europe, but also in Peking, Calcutta, Madras, and Constantinople. Most of these observations were, of course, made by amateurs with little experience in precise astronomical measurements. The best cadres of observers were to be found in France, Sweden, and Britain.

The data obtained from all these observations were not unproblematic. First, the quality of the successful observations varied greatly, and calculators had to choose which data to favour and which to ignore, choices that were, in retrospect, not always the best. Sometimes supposed errors in the observations were even "corrected". In the absence of accepted statistical methods of dealing with such a large and uneven body of data, there was no standard of data reduction to which practitioners could appeal for unanimity. Second, Halley's prediction of an accuracy of one part in 500 was vitiated by the "black drop" effect. The moment of internal contact was defined as the first or last continuous filament of sunlight between the dark body of the planet and the Sun's limb. In practice, the planet seemed to adhere to the Sun's limb as though attached to it by a sticky substance (see Figure 23.3). The duration of this effect varied from observer to observer and depended on several factors. Third, several observers saw a luminous ring around the planet that made the moment of internal contact even more difficult to time. This was the first evidence that Venus has an atmosphere, a fact confirmed a few years later by William Herschel. Finally, in cases where only the ingress or egress could be observed it was crucial to have exact longitude measurements. In most cases there was an unacceptable uncertainty in these measurements. Lalande admitted that even the longitude difference between the established observatories in Paris and Greenwich was uncertain to 20 arc-seconds.

23.3. One of the problems encountered in observations of transits of Venus: the "black drop" effect when Venus is near internal contact. The black silhouette of the planet then appears to be joined to the limb of the Sun by a ligament, and this makes the moment of internal contact observationally ill-defined.

As a result, calculations of the solar parallax based on the transit observations ranged from 8".28 to 10".60, without any immediate sign of consensus. In this respect, then, the expeditions of 1761 were a disappointment. On the other hand, many valuable lessons had been learned that could be used in observing the transit of 1769, when, it was hoped, there would be peace between the French and British. Moreover, the expeditions did much more than observe the transit: their longitude observations made possible corrections to maps and charts, and their studies of local geology, flora, and fauna, brought back a wealth of important information that is useful to this day (Figure 23.4).

The arguments about the value of the solar parallax kept the importance of Venus transits before a wide audience, and therefore preparations for the 1769 transit began early. Le Gentil, who had been foiled by the war, decided to stay in the East, and proceeded to Manila in 1767. The French Academy, however, advised him to proceed to Pondicherry, now that the war with the British was over. Le Gentil arrived there in March 1768, fourteen months before the transit. An expedition, under the leadership of Pingré, left France on 9 December 1768. Its primary function was to test chronometers for the purpose of finding longitude at sea. This expedition observed the transit from Cap-François, Saint Domingue.

Chappe proposed to go to the South Sea islands, but Spanish cooperation for this project could not be secured. A proposal to make the observation from the California peninsula was, however, accepted by the Spanish Crown, and a joint French–Spanish expedition was organized. Chappe departed from Paris in September 1768, set sail from Cadiz a month later, and arrived in Vera Cruz on 8 March 1769. It took more than two months more to cross Mexico and traverse the sea between the mainland and the peninsula, where they arrived on 16 May, three weeks before the transit. Although the French–Spanish team was successful in their observation, all died from an epidemic, with the exception of Vincente de Doz, who brought the observations back to Europe.

The British, likewise, took the opportunity to send an expedition whose goals were wider than the observation of the Venus transit. The *Endeavour* sailed from Plymouth on 26 August 1768, under the command of James Cook. Charles Green was on board to make the astronomical observations, while Joseph Banks went along to study the natural history of the places visited. Cook was to explore the South Pacific, test the chronometers of John Harrison, and make the observation of the transit from the newly discovered island of Tahiti. Under the guidance of the Royal Society, the British Crown also sent expeditions to the northern part of Norway and County Donegal in northwest Ireland.

In the British colonies in North America, efforts were made to organize expeditions to the West but funds could not be found. Nevertheless, the transit of 1769 was observed at no fewer than nineteen stations in the colonies, the most celebrated of which were those of Winthrop at Harvard, Benjamin West in Providence, Rhode Island, and David Rittenhouse in Norristown, Pennsylvania.

Father Maximilian Hell, SJ, was invited by the Danish/Norwegian Crown to observe from the island of Vardö off the northeastern shore of Norway (Figure 23.5). Having obtained permission from his superiors and the Hapsburg Crown, Hell headed a successful expedition whose results were for some time mistrusted because of a delay in publication. Lalande's suspicion led J. F. Encke, in the 1820s, to reject Hell's observations. In 1835 Karl Ludwig von Littrow published what he thought was direct proof that Hell had falsified his data. Hell's results were rehabilitated by Simon

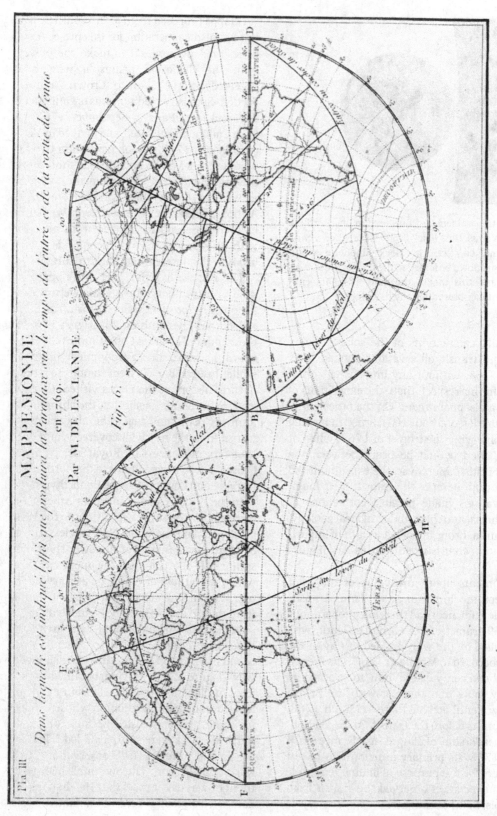

23.4. Lalande's world map for the 1769 transit of Venus, showing the times when the planet will be seen entering and leaving the Sun.

23.5. Maximilian Hell observing the transit of 1769 on the island of Vardö.

Newcomb (1835–1909) in 1883 after an examination of his journal of observations.

In Russia, with the enthusiastic backing of Catherine the Great, the St Petersburg Academy coordinated a number of expeditions, equipped with instruments ordered from James Short in England. Observing stations were established as far east as Yakutsk (east longitude 130°), in the Urals (east longitude 55° and 60°), near the Caspian sea, and on the Kola peninsula between the Arctic Ocean and the White Sea.

As in the previous transit, the Swedes organized a large and coherent effort, making certain that observations were made from a number of locations in Swedish territory.

Altogether, this transit was prepared for by 151 observers at 77 stations, from Yakutsk, Manila, Peking, and Batavia in the East, to Baja California and Tahiti in the West. The British counted no fewer than 69 observers, the French 34, and the Russians 13. Of all the experiences, Le Gentil's is the most poignant. Having been foiled by the war in 1761, and having spent the intervening eight years in the East, in 1769 he was foiled by the weather.

Calculations of the resulting solar parallax began immediately but were not definitive until the results from the expeditions to faraway places arrived. Thus, in his memoir of 1770 Pingré calculated solar parallax to be $8''.88 \pm 0''.05$ but upon further consideration concluded (in 1772) "that the horizontal parallax of the Sun at its median distances is very nearly eight seconds and eight tenths." Lalande, who on the basis of the 1761 transits had used $9''.0$ in his influential *Astronomie* of 1764, put the value between $8''.5$ and $8''.75$ in his memoir of 1770. The next year he narrowed the range to $8''.55 - 8''.63$: "thus, in taking in round numbers $8''.6$ for the middle latitudes, like the one of Paris, one cannot differ sensibly from the truth; the observations made in 1769 cannot suffice for removing that small degree of incertitude of a twelfth of a second." From the English observations Maskelyne found a parallax of $8''.8$; Anders Planmann of the Swedish Academy arrived at a figure of $8''.43$, but upon reconsidering agreed with Lalande on a value of $8''.50$. At the St Petersburg Academy, Anders Johan Lexell based his calculations on Leonhard Euler's theoretical treatment of the determination of solar parallax by means of transits, and found a value of $8''.68$ in his memoir of 1771 but reduced that value to $8''.63$ the next year.

There was, therefore, an agreement considerably closer than in the case of the transit of 1761. Yet, nagging doubts remained. The two transits had clearly shown Halley's claim, that transits of Venus could reveal solar parallax accurate to one part in 500, to be a mirage: the best observations still yielded a range of some $0''.4$, or about one part in 20. The limitation was due to several sorts of problem. First, the longitudes of many of the far-flung places were not known with a sufficient degree of accuracy. This was a problem that might, over time, disappear. Second, there was the problem of the black drop effect and the luminous rings around Venus. Here one had to choose between the moment of internal tangency or the first and last appearance of the bright filament between the dark body of the planet and the limb of the Sun. But all these estimates depended on the condition of the atmosphere as well as the power

and quality of the telescope used. This problem was much less tractable. Last, there was the problem of various inaccuracies introduced by human factors: the skills of the observers varied greatly. Those who assembled the observations and calculated the parallax had to decide how much weight to give to each observation, a difficult task in the absence of a formal theory of error.

The rough consensus of a value around $8''.6$ held steady for about eighty years. In his *Exposition du système du monde* (1796) Pierre-Simon Laplace derived a solar parallax of $8''.6$ from the parallactic inequality of the Moon, thus confirming Lalande's findings. Laplace concluded: "It is very remarkable that an astronomer without leaving his observatory, by merely comparing his observations with analysis, has been enabled to determine with accuracy the magnitude and figure of the Earth and its distances from the Sun and Moon, elements, the knowledge of which has been the fruit of long and troublesome voyages in both hemispheres." And in his influential *Histoire de l'astronomie du XVIIIe siècle* (1827) Jean-Baptiste Joseph Delambre, upon reviewing the results of the Venus transits, concluded that the solar parallax could be held to be $8''.6$ in round numbers.

As better statistical methods of evaluating data became available, astronomers reviewed the results of the Venus transits. In memoirs dated 1822 and 1824 Encke subjected the data to C. F. Gauss's method of least squares (1809) and concluded that the mean horizontal solar parallax was $8''.5776 \pm 0''.0370$. In 1835 Encke further refined his figure to $8''.57116 \pm 0''.0371$. It seemed that the accuracy of the measure, now one part in 200, was approaching Halley's prediction after all. Encke's results were very influential: astronomers felt supremely confident of the accuracy of this, and other, fundamental constants. The consensus fell apart, however, in the middle of the century.

In his researches on lunar theory, Peter Andreas Hansen derived from the parallactic equation a solar parallax of $8''.916$ (published in 1857 and 1863), a value considerably larger than Encke's. At about the same time Urbain Jean Joseph Le Verrier studied the Earth's perturbations on Mars and Venus and from the Earth's mass found a solar parallax of $8''.95$. These values were supported by the results of Jean Bernard Léon Foucault's laboratory determinations of the speed of light, which, by way of the aberration of light, produced a solar parallax of 8".86. In the meantime, the greater accuracy of nineteenth-century measuring instruments promised that perhaps the parallax of Mars could now be measured more accurately. At the favorable opposition of 1862, a number of observers in widely separated observatories measured the planet's declination, and their results yielded values of solar parallax averaging around 8".95.

These developments led to new reappraisals of the observations of 1761 and 1769, with a view to improving on them in the upcoming Venus transits of 1874 and 1882. In his doctoral dissertation published in 1864, Karl Rudolph Powalky went back to the data and by eliminating a large number of observations teased out a value of 8".83. Four years later Edward James Stone took the observations of five stations where both ingress and egress were visible in 1769 and produced a value of 8".91. But neither of these analyses was entirely satisfying.

In preparing for the transit of 1874, several different approaches were taken. Astronomers were generally in agreement that new techniques and better measuring instruments held out great hope of improving on the results of the observations of the previous century. The Germans, under the leadership of Hansen and A. J. G. F. von Auwers, pinned their hopes on heliometers, with which they equipped all ten of their expeditions. The English expeditions, coordinated by the Astronomer Royal, George Biddell Airy, used both visual and photographic techniques (Figure 23.6). Because heliometers were not available for this purpose in the United States, the Americans, under Newcomb, used traditional instruments to observe contacts but also equipped each expedition with a photoheliostat (a forty-foot horizontal instrument in which the image of the Sun was projected by a rotating mirror). The French also availed themselves of a version of this instrument. Expeditions were also organized by Russia and other European countries.

The entire course of the transit could be observed only in East Asia and the Indian and Pacific Oceans. Information was gathered on weather conditions before the choices of observing stations were made. Altogether the Europeans sent over fifty expeditions to these areas. Travel by ship was now relatively easy and, since the European powers were at peace with each other, safe. Further, equipment had been standardized, especially by the Germans and Americans, and observers and photographers had been trained. (The Americans went so far as to construct an artificial transit machine on which their observers could practise.) In view of all these preparations, astronomers were optimistic about the results.

Although the full reports took decades to complete, preliminary results began to appear within a few years. They were very disappointing. Airy found 8".76 for the solar parallax, Stone 8".88, G. L. Tupman 8".81, Auwers 8".88 (heliometer measurements), and D. D. Todd 8".88 (photoheliostat measurements). This range was much larger than had been hoped. As a result, astronomers now realized that transit observations were inadequate for this purpose, and there was decidedly less enthusiasm for observing the transit of 1882 (although a number of observations were made).

There was now a large body of data from four transits, and these were subjected to thorough analyses. Newcomb (no great enthusiast for this method) reviewed the eighteenth-century observations, incorporating up-to-date determinations of the longitudes of the observing stations. His effort, published in 1891, produced a solar parallax of 8".79 with a mean error of 0".051 and a probable error of 0".034. In *The Elements of the Four Inner Planets and the Fundamental Constants of Astronomy* of 1895, Newcomb incorporated the results of the 1874 and 1882 observations. From the measurements of Venus's distance from the centre of the Sun, made by means of heliometers by German observers and by means of photographs by American observers, Newcomb derived a solar parallax of 8".857 ± 0".016. He then compared the observed contacts of Venus with the Sun's limb from all four Venus transits, arriving at a parallax of 8".794 ± 0".023.

Astronomers in the eighteenth century may have been very satisfied with a solar parallax accurate to one part in several hundred, but their successors a century later were not, and they looked for alternative means of determining that important constant. Measurements of Mars at opposition held out promise, and David Gill (1843–1914) even went to Ascension Island to determine the diurnal parallax of the planet by means of a

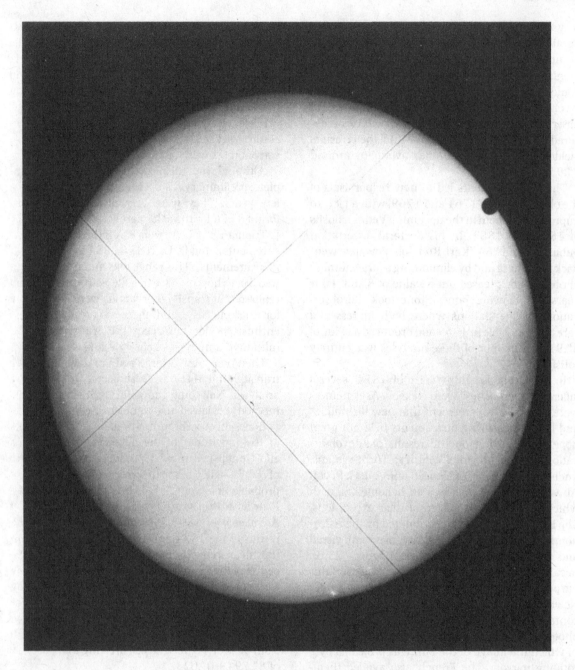

23.6. A photoheliograph of Venus near the solar limb during the transit of 1874, taken by the Greenwich Observatory team on Honolulu.

heliometer during the perihelion opposition of 1877, deriving a solar parallax of 8″.78. Although Gill's result was judged reasonably accurate, it was felt that the use of Mars for such measurements was of limited value because of the inevitable inaccuracy involved in centring the image of a star on the planet's disk in a heliometer or centring the threads on the disk in a transit instrument. Newcomb said of this method that "we have here a cause so mixed up with personal error in making the observations that the objective and subjective effects cannot be completely separated."

There were, perhaps, alternatives to Mars. In 1872 Johann Gottfried Galle had proposed that instead of Mars a planetary body so small that it showed no disk be used. The minor planets proposed were Iris, Victoria, and Sappho, whose nearest approaches to the Earth were all about 0.85 astronomical units – much greater than Mars's distance of 0.38 AU at its perihelion oppositions. In 1888–89 observatories in Europe and the southern hemisphere made measurements of the positions of these bodies, and from them Gill derived a solar parallax of 8″.802, a value that agreed very well with the parallax figure of $8″.80 \pm 0″.01$, derived from the new measurements of the speed of light by Newcomb and Albert Abraham Michelson around the same time.

The minor-planet method received an unexpected boost when, in 1898, Karl Gustav Witt discovered asteroid no. 433, later named Eros. Computations of its orbit showed that its perihelion lay well within the orbit of Mars and approached the Earth's orbit closely. At its opposition of 1900–01 it would approach to within 0.27 AU from the Earth, and during its opposition of 1930–31 to within 0.17 AU. Both visual and photographic measurements were made during the first opposition, resulting in solar parallax values of $8″.806 \pm 0″.004$ and $8″.807 \pm 0″.003$. The reduction of the data of the 1930–31 opposition, published by Harold Spencer Jones in 1942, yielded the value $8″.790 \pm 0″.001$. The confidence limits of these measurements were thus almost an order of magnitude better than those of the value derived by Newcomb from the observations of Venus transits. Yet, the difference between the 1900–01 and 1930–31 measurements is considerably greater than the calculated error margin of either one.

Several conclusions can be drawn from the astronomers' great struggle to determine the astronomical unit to within acceptable error margins. An analysis of seventeenth-century measurements of the parallax of Mars shows that this method of determining solar parallax was not satisfactory. A not uncharitable interpretation of these measurements is that Cassini I and Flamsteed did little more than determine the error margin of their instruments. Yet the resulting partial agreement that solar parallax was 10″–12″ fitted in rather well with other considerations such as refraction. From 1672 to 1761 this rough consensus was confirmed time and again by measurements of the parallax of Mars and transits of Mercury.

The observations made during the Venus transits of 1761 and 1769 were of widely varying quality and could support solar parallaxes anywhere from 8″.5 to over 9″.0. The trend from 1770 to 1850 was toward a lower solar parallax, from Lalande's 8″.6 to Encke's 8″.57. Then, because values for the solar parallax derived by means of other methods tended to argue for a larger value, astronomers returned to the same body of data and duly teased out a slightly larger value. In so doing, they made ever better corrections for the longitudes of the observing stations and subjected the thus improved data to ever more sophisticated statistical analyses. Yet, even in the work of Newcomb, there is a residual degree of arbitrariness that serves to bring the final value closer to the value desired at that particular time. In retrospect it is difficult to escape the conclusion that, in spite of the increased sophistication of the analyses from Lalande to Encke and Newcomb, the values of solar parallax obtained from the transit observations of 1761 and 1769 have an accuracy no better than $\pm 0″.2$.

Measurements of important constants of nature are not made in isolation. The very importance of these constants ensures that there will be converging pressures toward a value that is consistent with the limits imposed by several related approaches to the same problem. Thus, Encke's value was rejected when Hansen, Foucault, and Le Verrier arrived at values fully 0″.3 larger by three different methods. Astronomers promptly derived such larger values from the 1761 and 1769 data.

Observations in astronomy are, therefore, not one-time affairs that serve for a while as the best and are then replaced by better ones. Observations have a rather long life and their raw data are analysed time and again in the hope that they will yield results that are more in agreement with other and more recent observations. The observations of the transits of Venus of 1761 and 1769 provide a good example of this phenomenon.

In 1976 the International Astronomical Union (IAU) adopted 8″.794148 as its value of the mean horizontal solar parallax, corresponding to a mean solar distance of $1.49597870 \times 10^{11}$ metres and an equatorial radius of the Earth of 6.378140×10^6 metres. The range of error is $\pm 0''.000007$. Since 1984 the IAU astronomical constants have been the basis for all the national ephemerides.

The modern system of sizes and distances was first arrived at in rough outline by astronomers at the end of the seventeenth century. It was then fine-tuned by subsequent measurements in which the observations of the Venus transits of 1761 and 1769 were central. These measurements were made at enormous cost, measured in terms of money as well as manpower. The Venus expeditions were a great achievement in international scientific cooperation, something that had never been tried on this scale. Since then such cooperative projects among astronomers have become routine. In the eighteenth century scientific expeditions were anything but routine: astronomers underwent great hardships and some lost their lives. Le Gentil spent more than eleven years of his life thousands of miles away from home and missed both transits. When he finally returned, his heirs were dividing his estate among themselves in the mistaken belief that he was dead. Chappe and all his assistants save one died of an epidemic on the Lower California peninsula. The magnitude of such sacrifices and the great financial cost of such expeditions give us a measure of the power of the spirit of the Enlightenment.

Further reading

R. d'E. Atkinson, The Eros parallax, 1930–31, *Journal for the History of Astronomy*, vol. 13 (1982), 77–83

A. J. Meadows, The transit of Venus in 1874, *Nature*, vol. 250 (1974), 749–52

Simon Newcomb, Discussion of observations of the transits of Venus in 1761 and 1769, *Astronomical Papers prepared for the Use of the American Ephemeris and Nautical Almanac*, vol. 2 (1890), 295–405

Simon Newcomb, *The Elements of the Four Inner Planets and the Fundamental Constants of Astronomy* (Washington, 1895)

Simon Newcomb, On Hell's alleged falsification of his observations of the transit of Venus in 1769, *Monthly Notices of the Royal Astronomical Society*, vol. 43 (1883), 371–81

Albert Van Helden, *Measuring the Universe: Cosmic Dimensions from Aristarchus to Halley* (Chicago, 1985)

Harry Woolf, *The Transits of Venus: A Study of Eighteenth-Century Science* (Princeton, NJ, 1959)

24

The discovery of Uranus, the Titius–Bode law, and the asteroids

MICHAEL HOSKIN

The Mars–Jupiter 'gap' in the planetary system

The traditional Ptolemaic astronomy that the Middle Ages inherited from Antiquity placed the Earth at the centre of the cosmos. As a result not only the distances, but even the order of the planets was uncertain; for observers on a supposedly central Earth were (it seemed) handicapped when it came to determining cosmic dimensions.

All this changed with the publication in 1543 of Nicholas Copernicus's *De revolutionibus*, which located the Sun at the centre of the cosmos. If Copernicus was right, then the Earth-based observer was in fact on a planet circling the Sun (at a mean distance later termed 'the astronomical unit' or AU); and earlier astronomers had been misled by the Earth's changes of position when interpreting their observations of the apparent movements of other planets. It now seemed that the circular movements with a period of one year that occurred in the traditional geometric models of the planets were no more than the reflection of the terrestrial motion; if so, then the radius of each of these circles should now be equated with the astronomical unit, which thus became a yardstick for a true planetary system.

The system is portrayed in outline for the first time in the famous diagram in Bk I of *De revolutionibus*. But Bk I is cosmological and essentially qualitative, and Copernicus's diagram is not drawn to scale. Had it been to scale, the reader could not have failed to note the disparately large gap between the (mean) orbits of Mars and Jupiter. This anomaly was to jar the Platonic sensibilities of the young Johannes Kepler (1571–1630), who in the closing years of the sixteenth century set himself to understand the motives that had led the Divine Geometer to construct the universe in the way he had chosen. In the Preface to his *Mysterium cosmographicum* (1596) Kepler tells us that he tried an approach

of striking boldness. *Between Jupiter and Mars I placed a new planet, and likewise another between Venus and Mercury, two new planets that perhaps we would not see on account of their tiny size; and I assigned periodic times to them. For I reckoned that in this way I should produce some equality between the ratios, as the ratios between the pairs would be respectively reduced in the direction of the Sun and increased in that of the fixed stars. . . . Yet the interposition of a single planet was insufficient for the enormous gap between Jupiter and Mars; for the ratio of Jupiter to the new planet remained greater than is the ratio of Saturn to Jupiter.*

Eventually Kepler found suitable motivation for the Divine Geometer in a quite different approach the nesting of spheres and regular solids (see Chapter 5 in Volume 2A). But such a solution did not commend itself to the generations that followed. Isaac Newton (1642–1727) in 1692, when pressed by the theologian Richard Bentley, suggested that God had located the massive planets Jupiter and Saturn at a great distance from the lesser planets so that their gravitational attraction should not disrupt the system:

. . . Jupiter and Saturn, as they are rarer than the rest, so they are vastly greater, and contain a far greater Quantity of Matter, and have many Satellites about them; which Qualifications surely arose not from their being placed at so great a Distance from the Sun, but were rather the Cause why the Creator placed them at a great Distance. For by their gravitational Powers they disturb one another's Motions very sensibly . . . and had they been placed much nearer to the Sun and to one another, they would by the same Powers have caused a considerable Disturbance in the whole System.

Two of the cosmological speculators of the mid-eighteenth century likewise found the explanation of the gap in the sheer size of the outer planets. Immanual Kant (1724–1804) declared in the

Eighth Section of Pt II of his *Allgemeine Naturges-chichte und Theorie des Himmels* (1755) that the gap "is worthy of the greatest of all planets, namely, of that which has more mass than all the others together [i.e. Jupiter]". Johann Heinrich Lambert (1728–77) was as committed as Newton had been to stability in the large-scale structure of the universe, but he was prepared to accept the Mars–Jupiter gap as the consequence of the powerful attraction of Jupiter. In the first of his *Cosmologische Briefe* (1761), he asks: "And who knows whether already planets are missing that have departed from the huge space between Mars and Jupiter? Is it then true of celestial bodies as well as of the Earth, that the stronger harass the weaker, and are Jupiter and Saturn destined to plunder forever?"

Was the gap between Mars and Jupiter real (and therefore requiring an explanation) or merely apparent, and occupied by planets as yet undiscovered? Astronomers had been taught a lesson by the telescopic discoveries of the seventeenth century and were alive to the possibility of discoveries yet to come. William Whiston, for example, Newton's successor at Cambridge and one of the most influential writers of semipopular treatises in astronomy, repeatedly speaks of the 'known' planets, thereby reminding his readers that there may be some as yet unknown. However, most speculations about undiscovered planets concentrated on regions where planets would inevitably be hard to see: a planet within the orbit of Mercury, it was pointed out, might well remain lost to our sight in the glare of sunlight, while a planet beyond Saturn would be only faintly illuminated by the Sun and therefore difficult to detect. Only occasionally did a writer follow the young Kepler and posit a planet to fill the Mars–Jupiter gap. One who was later reputed to have done so in the early decades of the eighteenth century was the Scottish mathematician Colin Maclaurin. Another, from the next generation, was the amateur astronomer and maverick theologian Thomas Wright of Durham (1711–86). A speculation of his in Letter I of the *Second Thoughts* that remained in manuscript until our own day (and so had no influence on the history of astronomy) is nevertheless so remarkable as to merit citation:

That comets are capable of distroying such worlds as may chance to fall in their way, is, from their vast

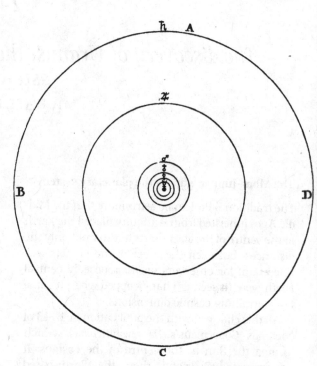

24.1. The orbits of the then-known planets of the solar system (Figure 1 of Plate 1 in David Gregory's *Elements of Astronomy*). The diagram illustrates how disporportionate was the apparent gap between Mars and Jupiter. The planet symbols are ♄ (Saturn), ♃ (Jupiter), ♂ (Mars), ♁ (Earth), ♀ (Venus), and ☿ (Mercury).

magnitude, velocity, firey substance, not at all to be doubted, and it is more than probable from the great and unoccupied distance betwixt ye planet Mars and Jupiter some world may have met with such a final dissolution.

The formulation of the Titius–Bode law

Most writers, however, concerned themselves with the known planets, and astronomical treatises would list their distances from the Sun. David Gregory, professor at Oxford and disciple and confidant of Newton, in his *Astronomiae elementa* (1702) (Figure 24.1) expresses these distances in proportional numbers:

... supposing the distance of the Earth from the Sun to be divided into ten equal Parts, of these the distance of Mercury will be about four, of Venus seven, of Mars fifteen, of Jupiter fifty two, and that of Saturn ninety five.

Gregory's work quickly became a standard text, the original Latin edition being followed by an English translation in 1715 and second editions in both languages in 1726. The sentence quoted is prominently placed in Prop. I of Section 1 of Bk I, and it came to the notice of the German philosophical popularizer Christian Wolff (1679–1754), who incorporated a translation of it in Section 85 of his *Vernünfftige Gedanken von den Absichten der natürlichen Dinge*, which appeared in 1724 and went through several editions. There it was read by Johann Daniel Titius of Wittenberg (1729–96), and when in 1766 Titius put out a German translation of the *Contemplation de la Nature* (Figure 24.2) that had been published two years earlier by the distinguished French natural philosopher Charles Bonnet, he drew on these quantities for an (unsigned) interpolation he chose to make in Chap. 4 of Pt I of Bonnet's text. Titius had evidently noticed that if Gregory's numbers were slightly modified, so that Mars was assigned 16 instead of 15, and Saturn 100 instead of 95, then the resulting sequence fell into an arithmetical pattern, whereby each number was equal to 4 plus a suitable multiple of 3. Bonnet's text was now made to read:

Take note of the distances of the planets from one another, and recognize that almost all are separated from one another in a proportion that matches their bodily magnitudes. Divide the distance from the Sun to Saturn into 100 parts. Then Mercury is separated by 4 such parts from the Sun, Venus by $4 + 3 = 7$ of the same, the Earth by $4 + 6 = 10$, Mars by $4 + 12 = 16$. But note that between Mars and Jupiter there occurs a departure from this very exact progression. After Mars there is a gap of $4 + 24 = 28$ such parts, but thus far no planet or satellite has been sighted there. But should the Lord Architect have left this space empty? Never! Let us therefore confidently assume that this space belongs without doubt to the as yet undiscovered satellites of Mars [!]; let us add that perhaps Jupiter also has several around itself that are not yet visible in any telescope. Next to this for us still unexplored space comes Jupiter's sphere of influence at $4 + 48 = 52$ parts, and Saturn's at $4 + 96 = 100$ parts. What a remarkable relationship!

Titius published a second edition of his translation, with the relationship now more appropriately located in a translator's footnote, in 1772, just as the young Johann Elert Bode (1747–1826) was putting the finishing touches to the second edition of his introduction to astronomy, *Anleitung zur Kenntniss des gestirnten Himmels*, which he had published when he was only nineteen and which was to be in print for nearly a century. Bode rejected the nonsense about the satellites of Mars, but he was attracted by the arithmetical relationship, and inserted it (in this edition, without acknowledgement) as a footnote to his text:

This latter point appears to follow in particular from the remarkable relationship that the six known major planets follow in their distances from the Sun. Call the distance from the Sun to Saturn 100, then Mercury is separated from the Sun by 4 such parts; Venus $4 + 3 = 7$; the Earth $4 + 6 = 10$; and Mars $4 + 12 = 16$. But now comes a gap in this very orderly progression. After Mars there follows a gap of $4 + 24 = 28$ parts, where up to now no planet is seen. Can one believe that the Creator of the Universe has left this position empty? Certainly not. From here we come to the distance of Jupiter by $4 + 48 = 52$ parts and finally to that of Saturn by $4 + 96 = 100$.

Bode was convinced that a primary planet lay undiscovered in the Mars–Jupiter gap, at a distance from the Sun of some 28 such units. Within a few years, his conviction was to receive a wholly unexpected boost.

The discovery of Uranus

On 13 March 1781, in the fashionable English resort of Bath, the Hanoverian-born organist and amateur astronomer William Herschel (1738–1822) was devoting his spare time to what he termed his "second review of the heavens". Self-taught in astronomy, and equipped with home-made reflectors of previously unattained optical quality, Herschel was fascinated by problems of stellar astronomy, in contrast to the professionals of his day whose main concern was the solar system. Herschel was systematically using a 7-ft (2.1-m) reflector of 6.2-inch (16-cm) aperture (Figure 24.3) to examine the stars down to magnitude 8, partly to familiarize himself with the heavens, partly to collect a store of double stars that might be suitable subjects for Galileo's method of measuring annual parallax. That evening he was studying stars in Taurus, when he came across one that – so excellent was his mirror, and so experienced was he as an observer – he instantly recognized as anomalous: "In the quartile near ζ Tauri the lowest of two is a curious either Nebulous Star

nämlich die Nebenplaneten jährlich um ihren Haupt-
planeten, wie Trabanten, herumlaufen.

Venus und die Erde haben jegliche ihren Trabanten. Mit der Zeit wird man ohne Zweifel auch einen
um den Mars entdecken. Jupiter hat ihrer viere, Saturn fünfe, nebst einem Ringe, oder einer leuchtenden
Atmosphäre, welche die Stelle vieler kleinen Monden zu
vertreten scheint. Da er beynahe dreyhundert Millio-
nen Meilen von der Sonne entfernt ist, so würde er ein
schwaches Licht bekommen, wenn nicht seine Trabanten und sein Ring dasselbe zurückwarfen und vermehrten.

Wir kennen siebzehn Planeten, die unser Sonnensy-
stem ausmachen helfen; aber wir sind nicht versichert,
daß ihrer nicht noch mehrere vorhanden sind. Ihre An-
zahl ist seit Erfindung der Fernröhre sehr gewachsen;
vielleicht wird sie noch mehr wachsen, wenn wir noch
vollkommenere Werkzeuge, noch fleißigere und glückli-
chere Bemerker bekommen. Der Trabante der Venus,
der im vorigen Jahrhunderte nur auf einen Augenblick
gesehen, seit kurzem aber aufs neue erblicket worden,
verkündiget der Sternkunde noch manche neue Entde-
ckungen.

Gebet einmal auf die Weiten der Planeten von einander Achtung; und nehmet wahr, daß sie fast alle in
der Proportion von einander entfernt sind, wie ihre kör-
perliche Größen zunehmen. Gebet der Distanz von der
Sonne bis zum Saturn 100 Theile, so ist Mercurius
4 solcher Theile von der Sonne entfernt: Venus
4+3=7 derselben; die Erde 4+6=10; Mars
4+12=16. Aber sehet, vom Mars bis zum Jupiter kömmt eine Abweichung von dieser so genauen
Progression vor. Vom Mars folgt ein Raum von
4+24=28 solcher Theile, darin weder ein Haupt-
noch ein Nebenplanet zur Zeit gesehen wird. Und
der Bauherr sollte diesen Raum ledig gelassen haben?
Nimmer-

X 4

Nimmermehr! lasset uns zuversichtlich setzen, daß dieser
Raum sonder Zweifel den bisher noch unentdeckten Trabanten des Mars zugehöre, lasset uns hinzuthun, daß
vielleicht auch Jupiter noch etliche um sich habe, die bis
ist noch mit keinem Glase gesehen werden. Von diesem,
uns unbekannten Raume erhebt sich Jupiters Wirkungs-
kreis in 4+48=52; und Saturnus seiner, in
4+96=100 solcher Theile. Welches bewunderns-
würdige Verhältniß!

Es war der heutigen Sternkunde vorbehalten, nicht
nur unsern Himmel mit neuen Planeten zu bereichern,
sondern auch die Gränzen unsers Sonnenwirbels viel
weiter hinauszusetzen. Die Kometen, welche, ihres
beträchtlichen Anblickes halber, ihres Schweises, ihres
haarigten Kernes, ihrer den Planeten oft entgegen ge-
setzten und von ihnen verschiedenen Richtung, ihres Er-
scheinens und Verschwindens wegen, für Erscheinungen
gehalten wurden, die eine erzürnte Macht in der Luft
angezündet hatte; diese Kometen sind zu planetischen
Körpern geworden, deren lange Laufbahnen unsre Stern-
kundige berechnen, ihre entfernte Rückkehren vorherfa-
gen, und ihren Ort, ihre Annäherungen und Entfer-
nungen bestimmen. Vierzig dieser Körper erkennen an-
itzt schon die Herrschaft unsrer Sonne, und die Bahnen,
welche einige von ihnen um dieselbige beschreiben, sind
so sehr ausgedehnt, daß sie solche, erst nach einer lan-
gen Reise von Jahren, oder wohl gar in vielen Jahr-
hunderten, einmal durchlaufen.

Gleichergestalt war es ein Vorrecht der neuern Stern-
kenntniß, zu zeigen, daß diese Kometen vermuthlich die-
jenigen Wandelsterne sind, wodurch die unzähligen
Systeme so vieler Sonnen zusammen hängen, und die
das eigentliche Verbindungsglied in der gesammten Kette
der Sterngebäude abgeben. Denn wozu wäre der große
Raum nöthig, der vom Saturn bis zum nächsten Fir-
sterne

24.2. Pages 7 and 8 of Titius's German translation of Bonnet's *Contemplation de la nature*, with the statement of the numerical relationship between the planetary distances interpolated by Titius into Bonnet's text.

24.3. The home-made 7-ft (2-m) reflector with which Herschel discovered Uranus, from the drawing by his friend and ally William Watson.

or perhaps a Comet", he noted (Figure 24.4). Four days later he returned to the object and found that it had moved: it was indeed a member of the solar system.

Interestingly, it did not occur to Herschel that his "curious" object might be a planet, perhaps because while he knew very well that no new planet had been discovered since the dawn of history, he was unaware of the widespread opinion among astronomers that there were nevertheless planets to be found. Nevil Maskelyne, the Astronomer Royal, was within days writing to a mutual friend to tell him of Herschel's "comet or new planet", and on 23 April he wrote to Herschel:

I am to acknowledge my obligations to you for the communication of your discovery of the present Comet, or planet, I don't know which to call it. It is as likely to

be a regular planet moving in an orbit nearly circular round the sun as a Comet moving in a very excentric ellipsis.

Maskelyne in fact had had difficulty in identifying the object to which Herschel was referring; for whereas the organist with his home-made reflector had seen at a glance that this was no ordinary star, the Astronomer Royal with the professionally-built instruments of Greenwich Observatory had been able to find nothing unusual in that region of the sky and had been forced to identify Herschel's object by its movement. Thomas Hornsby, the professor at Oxford, had done no better.

Meanwhile, Herschel himself was running into difficulties. His micrometric observations of the angular diameter of the object in the month following discovery seemed to imply that it had nearly

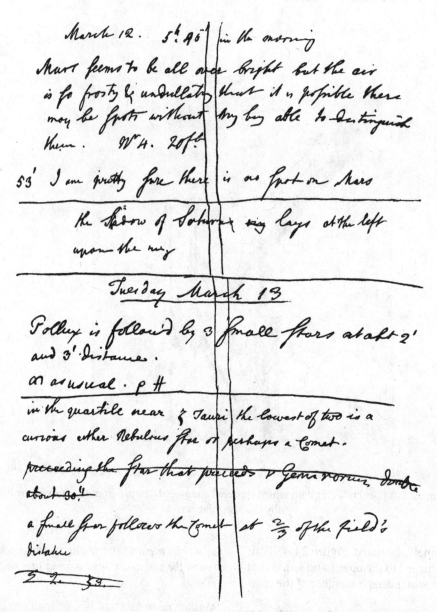

24.4. The page of Herschel's journal with his discovery of Uranus, thought by him at first glance to be "a curious either Nebulous Star or perhaps a Comet".

doubled in apparent diameter and so was rapidly approaching the Earth. In fact these changes were illusory; and their publication was to cast a question-mark over Herschel's reliability. But the unknown amateur's discovery of a planet moving so slowly that its change of position in a single night was almost imperceptible, astonished astronomers throughout Europe and gave notice that an observer of exceptional ability had appeared on the scene.

Meanwhile, the question of the nature of the object remained unresolved. The first orbits were computed on the assumption that the object was a comet. P.-F.-A. Méchain derived a perihelion distance of 0.46 AU and a perihelion date of 23 May 1781. In contrast, Anders Johan Lexell, professor of mathematics at St Petersburg and a specialist in cometary orbits, who chanced to be in England at the time, estimated a perihelion distance of 16 AU, and a perihelion date of 10 April 1789. But within weeks it was clear that no cometary orbit would suffice. Who first gave a satisfactory demonstration that the orbit was nearly circular – that is, planetary – and at a distance about twice that of Saturn is not clear, but Lexell's later claim to priority in the derivation of the elements of a circlar orbit may well have been justified. He found the radius of the orbit to be 18.93 AU and its motion no more than some 4° 20′ per annum, both excellent values: the body was a primary planet of the solar system. The discovery – and some tactful lobbying by his friends – soon earned Herschel royal patronage that would enable him to give up music and devote himself to astronomy, and he was happy in return to name his planet Georgium Sidus in honour of King George III; but astronomers preferred to follow tradition and adopted instead Bode's suggestion of Uranus, since in mythology Uranus was father to Saturn as Saturn was father to Jupiter.

The search for a planet between Mars and Jupiter

Gregory had taken the mean Earth–Sun distance (1 AU) as equal to 10 units, and on this scale had assigned the satisfactory value of 95 units to the distance of Saturn from the Sun. Titius and Bode had forced an arithmetical relationship out of the data by making two modifications to Gregory's numbers, one of which was to increase the 95 to 100; and they had then used this figure of 100 to define the units in which they were now working. But even when this "fudging" is seen for what it is, the fit between the next term in the series, $4 + 192 = 196$, and what was found to be the mean distance of Uranus is remarkable. For within a very few years of the planet's discovery, students of its orbit were taking into account pre-discovery observations of the planet, and even the perturbations by other planets, and nearly all agreed that the mean distance of Uranus was between 19.0 and 19.2 AU, with a gradual convergence towards the upper end of that range – a fit good enough to convince not only Bode but Baron Franz Xaver von Zach (1754–1832), the court astronomer at Gotha, of the validity of the Titius–Bode relationship and so of the existence of a planet between Mars and Jupiter. In 1787 Zach undertook a solo search for the planet, equipping himself for the purpose with a catalogue of zodiacal stars arranged by right ascension; but without success. The autumn of 1799 found him visiting astronomers in

Celle, Bremen and Lilienthal, and (as he later explained to readers of his journal *Monatliche Correspondenz*) it was there that the idea of a cooperative attack on the problem emerged:

It was the opinion of these men of discernment, that to get onto the trail of this so-long-hidden planet, it cannot be a matter for one or two astronomers to scrutinize the entire Zodiac down to the telescopic stars.

It took another year for the collaboration – probably without precedent in the history of science – to become a reality. On 21 September 1800 Zach met in Lilienthal with five other astronomers, among them J. H. Schröter, the chief magistrate of Lilienthal, whose world-famous collection of instruments included a Herschel reflector of 27-ft (8.1-m) focal length, C. L. Harding (1765–1834), who was employed by Schröter, and H. W. M. Olbers (1758–1840), a physician from nearby Bremen and longtime collaborator with Schröter. They judged that even six observers were too few for the task ahead, and they decided to coopt practising astronomers from throughout Europe, to make the numbers up to twenty-four. Schröter was to be president of the team and Zach secretary. They divided the entire Zodiac into twenty-four zones each of 15° in longitude and extending some 7° or 8° north and south of the ecliptic in latitude. The zones were allocated to the members by lot. Each member was to draw up a star chart for his zone, extending to the smallest telescopic stars,

and through repeated examination of the sky was to confirm the unchanging state of his district, or the presence of each wandering foreign guest. Through such a strictly organized policing of the heavens, divided into twenty-four sections, we hoped eventually to track down this planet, which had so long escaped our scrutiny – supposing, that is, that it existed and could be seen.

The discovery of Ceres

Zach accordingly set out to recruit the team that was to police the Zodiac. One of those chosen was, naturally, Giuseppe Piazzi (1746–1826) of Palermo in Sicily. As a young man Piazzi had joined the Theatine Order, and had then taught mathematics in a number of Italian cities. In 1780 he had accepted an invitation to take the chair of higher mathematics at the Academy of Palermo.

Although himself inexperienced in astronomy, he had succeeded in convincing his royal patron of the desirability of an astronomical observatory in Palermo, which would give access to stars further south than were visible from any existing European observatory. Piazzi was given leave to travel to northern Europe, to learn the craft of the astronomical observer from such disparate figures as Maskelyne and Herschel, and to purchase instruments of quality. Piazzi succeeded in persuading Jesse Ramsden, greatest of the instrument-makers, to undertake a 5-ft (1.5-m) vertical circle of unique design, and – what was much more difficult – to complete it, a feat that required Piazzi to be present in Ramsden's workshop and literally breathing down his neck (Figure 24.5).

In Palermo, Piazzi found himself privileged to possess this masterpiece of eighteenth-century technology and a site worthy of it. He accordingly set to work to undertake a star catalogue of greater accuracy than any that had gone before (see Volume 3).

The beginning of 1801 found Piazzi patiently at work on the star catalogue. The dramatic events that then unfolded were described by him a few months later in a pamphlet entitled *Risultati delle osservazioni della nuova stella*, which we may quote from the translation made in England for Maskelyne:

... *on the evening of the 1st of January of the current year [1801], together with several other stars, I sought for the 87th of the Catalogue of the Zodiacal stars of Mr La Caille. I then found it was preceded by another, which, according to my custom, I observed likewise, as it did not impede the principal observation. The light was a little faint, and of the colour of Jupiter, but similar to many others which generally are reckoned of the eighth magnitude. Therefore I had no doubt of its being any other than a fixed star. In the evening of the 2d I repeated my observations, and having found that it did not correspond either in time or in distance from the zenith with the former observation, I began to entertain some doubts of its accuracy. I conceived afterwards a great suspicion that it might be a new star. The evening of the third, my suspicion was converted into certainty, being assured it was not a fixed star. Nevertheless before I made it known, I waited 'till the evening of the 4th, when I had the satisfaction to see it had moved at the same rate as on the preceding days. From the fourth to the tenth the sky was cloudy. In the*

evening of the 10th it appeared to me in the Telescope, accompanied by four others, nearly of the same magnitude. In the uncertainty which was the new one, I observed them all, as exactly as possible, and having compared these observations with the others which I made in the evening of the 11th, by its motion I easily distinguished my star from the others. Mean while however I greatly wished to see it out of the Meridian, to examine and to contemplate it more at leisure. But with all my labour, and that of my assistant D. Niccola Cacciatore and [of] D. Niccola Carioti belonging to this Royal Chapel both enjoying a sharp sight, and very expert in the knowledge of the heavens, neither with the night Telescope, nor with another acromatic one of 4 inches aperture, was it possible to distinguish it from many others among which it was moving. I was therefore obliged to content myself with seeing it on the meridian, and for the short time of two minutes, that is to say the time it employed in traversing the field of the Telescope; other observations, which were [being made] at the same time, not permitting the instrument to be moved from its position.

In the mean time, in order to render the observations more certain, while I was observing with the Circle, D. Niccola Carioti observed with the transit instrument. The sky was so hazy, and often cloudy, that the observations were interrupted 'till the 11th of February; when the star having approached so near the Sun, it was not possible to see it any longer at its passage over the meridian. I intended to search for it, out of it [the meridian], by means of the Azimuth; but having fallen ill on the thirteenth of February, I was not able to make any further observations. These, however, which have been made, though they are not at the necessary distance from one another in order to assure us of the true course which the star describes in the heavens, are, notwithstanding, sufficient in my opinion, to make us know the nature of the same, as one may collect from the results, which I have deduced from them.

Piazzi had in fact measured the position of the object on a total of 24 nights between 1 January and 11 February, though some positions were marked as "doubtful" or even "very uncertain". On 24 January, Piazzi had announced his discovery in three letters to fellow astronomers. One was to his friend, Barnaba Oriani of Milan; in it, Piazzi confides that

I have announced this star as a comet, but since it is not accompanied by any nebulosity and, further, since its

24.5. The circle by Ramsden that Piazzi used in compiling his star catalogues and with which he detected the motion of what proved to be the first known asteroid. In this drawing, two of the four supporting arches have been omitted in the interests of clarity.

movement is so slow and rather uniform, it has occurred to me several times that it might be something better than a comet. But I have been careful not to advance this supposition to the public.

To Bode and J.-J. L. de Lalande, however, he limited himself to the cautious claim that the object was merely a comet, though making it clear that the 'comet' had no nebulosity or tail.

When after 11 February he could no longer see the object, he set to work to investigate its orbit, though such mathematical investigations were not his strength. He began with the assumption that it was indeed a comet, and fitted a parabola to three of the observations to see if the orbit would account for the others. It did not. A second attempt with a different group of three observations likewise failed:

From the parabolic hypothesis I passed then to the circular; and having made a few suppositions, I found two radii, 2.7067 and 2.6862; with each of which all the observations were represented a great deal better than any parabola. The planets describing ellipses more or less eccentric, and not circles, it is to be believed that ours will not deviate from this rule. In an ellipsis I should then have continued my calculations; but as the arch observed is very small, the results would be very uncertain, and the labour long and painful. I have therefore preferred the circle. . . .

The agreement of the observed longitudes with the calculated ones in the circular hypothesis, its motion in the Zodiac, from which it only departs a little way in the greatest latitudes, and its position between Mars and Jupiter, leave no doubt that this new star is a true planet. . . .

It is easy to believe that not only Oriani, but more especially Bode, were, in Piazzi's words, "instantly of the opinion that it was a new planet; and settled nearly the same elements of its orbit, as I have done". One can imagine the German's delight that the hoped-for planet had been found, even if the carefully-laid plans of the celestial police had played no part in the discovery.

But now Piazzi was beginning to have doubts. His first estimate of the size of the object was based on the fact that it was almost, but not quite, covered by one of the wires of his telescope, and he concluded that it was larger than the Earth. However, it would seem that in the hazy nights that

followed, the true (and much smaller) size of the object became more evident to the Palermo astronomer, who began to think that the object was diminishing in size and therefore moving rapidly away, so that it must be a comet after all:

As after the 23rd [of January] the star began sensibly to diminish in size and brightness, uncertain whether it was to be attributed to its rapid receding from the Earth, or rather to the state of the atmosphere, which became after that time still more dark and hazy, I began to doubt of its nature, so as even to believe it was a Comet and not a planet.

Eventually, in April, after illness had prevented him from making further progress in the investigation of the object's orbit, Piazzi sent his complete observations to Oriani, Bode, and Lalande in Paris. Like Herschel's, his discovery had been an unexpected – but not therefore accidental – byproduct of a lengthy observational campaign in stellar astronomy; like Herschel, he had been deceived into suspecting the object was a comet; and like Herschel, he had been forced in the end to entrust his discovery to the mathematicians. Nor did the similarities end there; for just as Herschel's attempt to name his planet in honour of his royal patron was rejected by astronomers, so Piazzi's expressed wish that his object – should it ever be recovered – be named Ceres Ferdinandea, Ceres for the patron goddess of Sicily and Ferdinandea for Piazzi's royal patron, has only partially been honoured.

Ceres had been observed for only a fragment of its orbit, a geocentric arc of a mere 3°, and to recover it when it at length emerged from the glare of the Sun seemed beyond the present capacity of mathematicians. Piazzi's own opinion was that recovery depended upon the identification of some earlier sighting when the object had been seen but taken for a star. But fate had provided a brilliant new mathematical talent in the person of C. F. Gauss (1777–1855), and Gauss's analysis of the orbit (see Chapter 25) enabled Zach at Gotha to recover the object on 31 December 1801, a year after its first discovery; the following evening, at Bremen, Olbers also saw it.

Gauss's analysis assigned to Ceres a mean distance from the Sun of 2.767 AU. When this figure was scaled up by a factor of 10 or so, to match the units of the Titius–Bode relation, the fit with the distance predicted for the 'missing' planet –

4 + 3.8 = 28 – was almost perfect: Piazzi's planet, it seemed, matched the description of the missing body sought by the celestial police. The relation – 'law' would by now be not too strong a term – had received its second triumphant vindication within a couple of decades.

It seemed too good to be true, as indeed it was. William Herschel found to his surprise that he could barely perceive a planetary disk in Ceres, and his preliminary conclusion, announced to the Royal Society in mid-February, 1802, was "that the real diameter of the new planet, remarkable as it may appear, is less than five eights of the diameter of our moon", a surprisingly low estimate but one that he was soon to revise sharply downwards.

Worse was to follow. At the end of March, Olbers, who had become very familiar with the region of the sky where the reappearance of Ceres had been predicted, noticed there a small star that formed an equilateral triangle with two stars that were in the catalogues. He was convinced this third star had newly appeared on the scene, perhaps because it was a variable now at maximum; but in hours he could see that it had moved. Within a month Gauss had calculated its mean distance, 2.670 AU, or almost the same as Ceres; its orbit had an eccentricity even greater than that of Mercury, and was inclined to the ecliptic by an astonishing 34°. Meanwhile, Herschel was taking micrometer measurements of both bodies, and concluded that Ceres had a diameter of less than 162 miles (261 km), while the new object, which Olbers named Pallas, was even smaller, with a diameter of less than 111 miles (178 km): underestimates, as we now know, but serving to highlight the contrast between these objects and the known planets. Herschel did not yet know the elements that Gauss had calculated for Pallas, but he was well aware that the orbits of the two bodies were sharply inclined to the ecliptic. "If bodies of this kind", he drily remarked in a paper read to the Royal Society early in May, "were to be admitted into the order of planets, we should be obliged to give up the zodiac; for, by extending it to them, should a few more of these stars be discovered, still farther and farther deviating from the path of the earth, which is not unlikely, we might soon be obliged to convert the whole firmament into zodiac...." Such bodies could not properly be termed planets; since they

were not comets either, they deserved a name of their own, and he proposed the descriptive term 'asteroid' – though this was naturally unwelcome to Piazzi who in consequence would cease to be the discoverer of a planet.

Confronted with not one but two bodies in the gap between Mars and Jupiter, Olbers suggested that they were in fact fragments of a full-size planet that had once occupied the gap – thereby unknowingly reviving the unpublished speculation of Thomas Wright. If this was so, then they might indeed have orbits of various inclinations and eccentricities, but these orbits would all have similar mean distances from the Sun and, more importantly, would all intersect in two positions in opposite parts of the sky, one of which would be the position of the planet at the moment of fragmentation. The two known orbits served roughly to define these two points of intersection, and Harding agreed to compile charts of the faint stars in these regions. In September 1804, while at work on this task, he noticed an object that proved to be a third asteroid, which he named Juno. Its mean distance was again 2.670, while its eccentricity was once more far greater than that of any planet.

Olbers persevered in the search, but not until 1807 was he rewarded by the discovery of a fourth asteroid, for which Gauss, at Olbers's invitation, selected the name Vesta.

Ceres is in fact much the largest of the asteroids, with a diameter now estimated as about 1000km. Pallas and Vesta are the only others with diameter in excess of 500km. The rest are even smaller, and none is as bright as eighth magnitude, so it is not surprising that astronomers wearied of the search. It was revived by a German amateur of extraordinary dedication, an ex-postmaster named Hencke, who began to search around 1830 and was at last rewarded with his first success in 1845, and his second in 1847. By this time the star charts sponsored by the Berlin Academy of Sciences that were to prove decisive in the discovery of Neptune (Chapter 28) were becoming available, and private observers now had the data on the basis of which they could prosecute an effective search. Later that year, J. R. Hind, who was employed as astronomer at the observatory of George Bishop at Regent's Park in London, found two more asteroids, and the next year another was found by Andrew Graham, astronomer at Edward J. Cooper's observatory at

Markree in Ireland. Since then, the number of asteroids known has grown constantly; and as a result of the application of photography, asteroids are today known in their thousands.

The Titius–Bode 'law' had been rescued by Olbers's suggestion that the planet in the Mars–Jupiter gap had fragmented; and it survived to play a key role in the discovery of Neptune. But asteroids were being found whose orbits showed they could not be fragments of a single planet; and the discovery of Neptune, which refused to conform to the 'law' in any simple manner, brought the relationship into disrepute – though it continues to exert a fascination on astronomers of the planetary system.

Further reading

A. F. O'D. Alexander, William Herschel discovers the 'Georgian Planet', 13 March 1781, Chap. 1 in his *The Planet Uranus: a History of Observation, Theory and Discovery* (London, 1965)

Eric G. Forbes, Gauss and the discovery of Ceres, *Journal for the History of Astronomy*, vol. 2 (1971), 195–99

Robert Grant, *History of Physical Astronomy* (London, 1852), 237–43

Stanley L. Jaki, The early history of the Titius–Bode Law, *American Journal of Physics*, vol. 40 (1972), 1014–23

Michael Martin Nieto, *The Titius–Bode Law of Planetary Distances* (Oxford, 1972)

R. Porter *et al.*, History of the discovery of Uranus, in Garry Hunt (ed.), *Uranus and the Outer Planets* (Cambridge, 1982), 21–89

Simon Schaffer, Uranus and the establishment of Herschel's astronomy, *Journal for the History of Astronomy*, vol. 12 (1981), 11–26

25

Eighteenth- and nineteenth-century developments in the theory and practice of orbit determination

BRIAN G. MARSDEN

The eighteenth century

In 1705 Edmond Halley published the results of his investigation of the orbits of twenty-four comets observed up to the end of the seventeenth century (see Chapter 19). As to how he carried out this investigation, we know only that he adapted the graphical method developed by Isaac Newton in his *Principia* for the determination of cometary orbits, and (as Halley puts it) "attempted to bring the same method to an arithmetical calculation", substituting arithmetic for at least part of Newton's graphical construction. Halley gave the first explicit statement of the cubic equation relating the time interval from perihelion passage in a parabolic orbit to the perihelion distance and the tangent of half the true anomaly. He also supplied a table to aid its solution and attempted a generalization to an elliptical orbit.

After the dramatic achievements of Newton and Halley, there was surprisingly little work of either a theoretical or practical nature on the determination of cometary orbits for several decades. James Bradley determined a parabolic orbit for a comet observed in 1723, and Pierre Bouguer (1698–1758), in a piece published in the *Mémoires* of the Paris Academy of Sciences for 1733, computed a hyperbolic orbit for the comet of 1729. While from the point of view of accuracy a step backwards from the results of Newton and Halley, Bouguer's work is of interest because it represented the first attempt at an orbital solution in algebraic terms. Noting that the orbit of the comet is confined to a plane, Bouguer made use of what in vector notation would be written

$$\mathbf{r}_0 = c_1 \mathbf{r}_1 + c_3 \mathbf{r}_3, \tag{1}$$

where $\mathbf{r}_1$, $\mathbf{r}_0$, and $\mathbf{r}_3$ are the position vectors from the Sun to the comet at three observations times t_1, t_0, and t_3, and c_1 and c_3 are scalar constants. He combined Equation (1) with a geometrical condition at each t_i ($i = 1, 0, 3$), namely,

$$\mathbf{r}_i = \rho_i \hat{\boldsymbol{\rho}}_i - \mathbf{R}_i, \tag{2}$$

where ρ_i is the distance from the observer to the comet, $\hat{\boldsymbol{\rho}}_i$ is the unit vector from the observer to the comet, and $\mathbf{R}_i$ is the vector from the observer to the Sun, to obtain

$$c_1 \rho_1 \hat{\boldsymbol{\rho}}_1 - \rho_0 \hat{\boldsymbol{\rho}}_0 + c_3 \rho_3 \hat{\boldsymbol{\rho}}_3 = c_1 \mathbf{R}_1 - \mathbf{R}_0 + c_3 \mathbf{R}_3. \tag{3}$$

The vector quantities in Equation (3) are knowable from observations and solar tables, and if the scalars c_1 and c_3 are also known, one can – at least in principle – solve for the three unknown distances ρ_i. Substitution of the ρ_i into Equation (2) then yields points – only two of the three actually being required – on the heliocentric orbit and eventually the orbital elements. For c_1 and c_3 Bouguer adopted the time ratios

$$c_1^0 = \frac{t_3 - t_0}{t_3 - t_1}, \quad c_3^0 = \frac{t_0 - t_1}{t_3 - t_1}, \tag{4}$$

considering these to be adequate approximations if the time span covered by the observations is small.

For complete accuracy, c_1 and c_3 should be the ratios of the areas of triangles formed by the Sun and respective positions of the comet. The numerator triangle for c_1 involves the comet's positions at t_0 and t_3 and that for c_3 those at t_1 and t_0; in each case the denominator triangle involves the positions at t_1 and t_3. By defining them instead as the ratios of time differences, Bouguer was – as a consequence of Kepler's second law – replacing the *triangles* by the corresponding *sectors* of the comet's orbit. The correct expressions for c_1 and c_3 are therefore

$$c_1 = \frac{t_3 - t_0}{t_3 - t_1} \frac{y_{13}}{y_{03}}, \quad c_3 = \frac{t_0 - t_1}{t_3 - t_1} \frac{y_{13}}{y_{10}}, \tag{5}$$

where each y_{ij} is the ratio of the areas of the sector

and the triangle involving the Sun and the positions of the comet at t_i and t_j.

Nicolaas Stryuck, in a work of 1740 in Dutch, and Giovanni Domenico Maraldi, in the Paris *Mémoires* for 1743, strongly criticized Bouguer's work, largely because of his audacity in claiming a hyperbolic orbit for the comet of 1729. Interest in the subject of orbit determination became widespread during the 1740s, as these and several other astronomers, notably Eustacho Zanotti and Pierre-Charles Le Monnier, began to compute orbits whenever new comets were discovered. Cornelis Douwes and Nicolas-Louis de Lacaille also worked on the orbits of comets observed in previous centuries. Various processes of trial and error were used, involving in particular the adjustment of assumed values for the comet's distance from the Earth at two observations so that the time interval between those observations was consistent with parabolic motion, followed by the selection of the particular parabola that satisfied the comet's observed longitude at a third observation. A method of orbit determination proposed by the Dalmatian Jesuit Rudjer Josip Bošković (in his *De determinanda orbita planetae* of 1749) utilized a geometrical construction based on that used to determine the reflection of light by a spherical mirror.

The first real progress in understanding the orbit-determination process from an analytical point of view was made by Leonhard Euler (1707–83), in his *Theoria motuum planetarum et cometarum* of 1744. Euler appreciated that it was necessary to do something about the y_{ij}, the three sector-triangle ratios in Equations (5). For rather small arcs of an orbit, these ratios are clearly slightly larger than unity, and Euler established for the $t_1 - t_3$ configuration, for example, the approximation

$$y_{13} = 1 + \frac{1}{6} \frac{(\tau_3 - \tau_1)^2}{r_1^{\frac{3}{2}} r_3^{\frac{3}{2}}}, \qquad (6)$$

where τ_1 and τ_3 are the time differences $t_1 - t_0$ and $t_3 - t_0$ specified in 'canonical' units of approximately 58.13 days (that is, such that the Earth's revolution period is approximately 2π).

Euler utilized Equation (6) (and its companion equations) in practice by noting that it means that a small fraction, $-\frac{1}{2}\tau_1\tau_3/r_0^3$, of the comet's radius vector at t_0 is isolated by the chord joining the comet's positions at the first and third obser-

vations. He was thereby able to derive values for ρ_1 and ρ_3 from assumed values for ρ_0. Since eighteenth-century astrometry left a lot to be desired, Euler generally made use of a fourth observation well separated in time from the first three, and the accuracy with which this fourth observation was represented would enable him to select the most satisfactory result. Like Bouguer, Euler settled for any type of conic, and in the case of the comet of 1742 he obtained an ill-defined ellipse with a revolution period of only 42 years.

The fact that Euler made no attempt to force a parabolic solution is particularly surprising in view of his demonstration of the celebrated equation

$$(r_1 + r_3 + s_{13})^{\frac{3}{2}} - (r_1 + r_3 - s_{13})^{\frac{3}{2}} = 6(\tau_3 - \tau_1), \quad (7)$$

where s_{13} is the length of the chord joining the points at t_1 and t_3 in an exactly parabolic orbit. Although Euler established this equation in 1743, he does not seem ever to have put it to practical use. Johann Heinrich Lambert (1728–77), in his *Insigniores orbitae cometarum proprietates* of 1761, rediscovered and generalized it to the case of an elliptical orbit. As was pointed out by Joseph Louis Lagrange (1736–1813), Euler's equation was in fact implicit in Newton's work.

Lambert's other important contribution to orbit literature, published in the Berlin *Mémoires* for 1771, was his demonstration that the apparent trajectory of a comet is convex or concave towards the Sun according as the comet is farther from or closer to the Sun than the Earth. Lambert's theorem, as this is usually called, can be expressed by means of the equation

$$\rho_0 = \rho_0^o \left(1 - \frac{\gamma_0}{r_0^3}\right), \qquad (8)$$

where $\gamma_0 = R_0^3$, the cube of the distance from the observer to the Sun, and ρ_0^o is positive or negative according as r_0 is greater or less than R_0. Bouguer's process effectively supposes that $\gamma_0 = 0$, that is, that $\rho_0 = \rho_0^o$, and it was precisely because the comet of 1729 had such a large r_0 (greater than $4R_0$ even at perihelion) that he obtained a tolerably satisfactory result in this case.

The earliest practical use of Lambert's theorem in orbit determination is usually attributed to Lagrange, who in the Berlin *Mémoires* for 1778 used this *dynamical* relationship between ρ_0 and r_0 in conjunction with the obvious *geometrical* relationship,

$$r_0^2 = \rho_0^2 + 2\rho_0 R_0 \cos \epsilon_0 + R_0^2, \qquad (9)$$

where ϵ_0 is the object's elongation from the opposition point, to obtain an eighth-degree equation for r_0. Since $r_0 = R_0$ (that is, $\rho_0 = 0$, yielding the observer's own motion) is clearly a solution, the equation can be reduced to the seventh degree.

Lagrange also gave the Taylor-series developments of what are nowadays called the 'f and g functions' that allow the heliocentric position vector $\mathbf{r}_i$ at time t_i to be expressed in terms of the heliocentric position and velocity vectors $\mathbf{r}_0$ and $\mathbf{v}_0$, namely,

$$\mathbf{r}_i = f_i \mathbf{r}_0 + g_i \mathbf{v}_0. \qquad (10)$$

Although the series may not converge, the first two terms are

$$f_i = 1 - \tfrac{1}{2}\frac{\tau_i^2}{r_0^3}, \quad g_i = \tau_i \left(1 - \tfrac{1}{6}\frac{\tau_i^2}{r_0^3}\right), \qquad (11)$$

and subsequent terms, which depend also on the radial velocity, rapidly increase in complexity. No practical use seems to have been made of the f and g functions before well into the nineteenth century. We note that the constants c_1 and c_3 of Equation (1) are related to the f and g functions by

$$c_1 = \frac{g_3}{f_1 g_3 - g_1 f_3}, \quad c_3 = -\frac{g_1}{f_1 g_3 - g_1 f_3}. \qquad (12)$$

Lagrange did not present his work in a very practical form, however, and it was Pierre-Simon Laplace (1749–1827) who in the Paris *Mémoires* for 1780 widely publicized the first complete analytical orbit-determination procedure that had a good degree of practicality. In order to set up Lambert's theorem, Laplace utilized Equation (2) and its first and second time derivatives (at time t_0), with then the substitution for the second derivatives of both $\mathbf{r}_0$ and $\mathbf{R}_0$ in terms of $\mathbf{r}_0$ and $\mathbf{R}_0$ themselves from the Newtonian equations for gravitational motion. Although this idea is conceptually very simple, there exists the practical difficulty of the computation of the first and second time derivatives of $\hat{\rho}_0$. In principle, this can be accomplished from only three observations of the comet's position projected on to the sky, but the use of several additional observations is desirable in practice. An added complication is that the observer is also assumed to be travelling on an exact Keplerian orbit about the Sun; for the results to be strictly accurate the observed positions should therefore be reduced to the centre of the

Earth, and this requires prior knowledge of the distances one is attempting to determine. There is one special convenience to Laplace's method, namely, that the computed heliocentric velocity vector can be constrained so that the comet's orbital energy is precisely zero, as is the case in an exactly parabolic orbit.

In spite of its problems, the simplicity and convenience of Laplace's method resulted in its immediate popularity. Alexandre-Gui Pingré, the leading practitioner of orbital computations during the late eighteenth century, in the second volume of his *Cométographie* (1784) gave an extensive critique of methods of orbit determination and concluded that Laplace's was the best. The quick and widespread acceptance of Laplace's method was unfortunate, for it caused a significant concept that had just been presented in the Paris *Mémoires* for 1779 by the distinguished amateur A.-P. Dionis du Séjour (1734–94) to be largely overlooked and certainly misinterpreted.

Dionis du Séjour's method starts with the elimination of the middle terms in both the left-hand and right-hand sides of the vector Equation (3) and then division by the resulting coefficient of ρ_1. This yields an expression of the form

$$\rho_1 + M_{13}\rho_3 = N_{13}. \qquad (13)$$

Eliminating instead the last terms on each side, one obtains

$$\rho_1 + M_{10}\rho_0 = N_{10}. \qquad (14)$$

Here the quantities M_{13}, M_{10}, N_{13} and N_{10} depend, in particular, on c_1 and c_3. Dionis du Séjour followed Bouguer and approximated c_1 and c_3 by c_1^0 and c_3^0. In so doing, he recognized that the same approximations apply equally well to the orbit of the Earth, in which case the right-hand side of the vector Equation (3), and hence also the N_{13} and N_{10} of Equations (13) and (14), vanish. Though not in fact mentioned by Dionis du Séjour, it is this point that shows the fallacy of Bouguer's method, for Equation (3) can therefore be satisfactorily solved only for the *ratios* of the ρ_i, not for the ρ_i individually. Nevertheless, Dionis du Séjour went on to eliminate c_1 and c_3 from Equation (1), and to substitute from Equation (2) and then from Equations (13) and (14) with the zero right-hand sides, obtaining thereby a cubic equation in ρ_1, which he reduced to a quadratic by taking the ecliptic as the reference plane. In reality, this procedure is not significantly different from that of Bouguer, and it can therefore be criticized for the same reasons.

The real significance of Dionis du Séjour's work is that he went on to consider that the square of the length of the chord joining the first and third points on the orbit can be written

$$s_{13}^2 = 4\frac{(\tau_3 - \tau_1)^2}{r_1 + r_3}\left[1 + \tfrac{1}{3}\frac{(\tau_3 - \tau_1)^2}{(r_1 + r_3)^3}\right]. \quad (15)$$

This is an approximation to Equation (7), Euler's equation, recast in a form that is more convenient to use. With the help of Equations (2) and (13) – again with $N_{13} = 0$ – Dionis du Séjour obtained a *second* quadratic equation for ρ_1. By analogy with the *vis viva* integral in the Laplace method, this second quadratic equation provides the additional dynamical constraint necessary to ensure that the orbit is a parabola. This was the missing link in the orbit-determination process, utilized by Newton but not by Euler, even though the latter had written it down in exact algebraic form in Equation (7).

Pingré accorded Dionis du Séjour's work some publicity in his aforementioned critique, but his numerical examples made use of only the first quadratic equation in ρ_1, not the second. He justified his action with the unfortunate comment that "this will almost always suffice", adding that "the observational circumstances almost always make it possible to decide which root one must choose".

The pinnacle of orbit-determination achievement during the eighteenth century was therefore the publication by the Bremen physician H. W. M. Olbers (1758–1840) in 1797 of an essay to which, with a fair amount of justification, he gave the title *Abhandlung über die leichteste und bequemste Methode die Bahn eines Cometen zu berechnen* (Treatise on the easiest and most convenient method of computing the path of a comet). Before introducing the method, Olbers gave a summary of earlier work on the subject. This included a detailed description of Dionis du Séjour's method, much as is given above, although Dionis du Séjour's use of Equation (15) is mentioned only in passing. Very properly dismissing Dionis du Séjour's first quadratic equation as insufficiently accurate, Olbers suggested that the user should simply adopt $\rho_1 = 1$ AU and calculate the corresponding r_1, r_3, and s_{13} from Equation (13) with $N_{13} = 0$ and the obvious geometrical expressions. Substitution into Equation (7) then yields a value for the time interval $\tau_3 - \tau_1$ that is to be compared with the observed value. The correct value of ρ_1 can be established by interpolation (and extrapolation) among the results of further trials.

Olbers formalized the interpolation process into what is frequently known as the "method of the variation of geocentric distances". This involves the examination of two small quantities, for example the differences between the observed and computed values of $\tau_3 - \tau_1$ and between the observed and computed ecliptic longitudes at the second observation, that correspond, not only to some initially assumed values of ρ_1 and ρ_3, but also to the effect of changing ρ_1 and ρ_3 in turn by small increments. By elementary calculus, improved values for ρ_1 and ρ_3 then readily follow from the two increments and the three sets of residuals.

The popularity – even today – of Olbers's method is due to the careful and enthusiastic exposition of it by Olbers, Johann Franz Encke (1791–1865), and numerous later authorities. Olbers acknowledged that the linear two-variable interpolation procedure had previously been employed by Laplace in modifying a comet's perihelion time and perihelion distance to fit the observed longitude and latitude at some time. That all the other basic features of Olbers's method had earlier been espoused by Dionis du Séjour seems to have escaped attention for almost a century, until this fact was pointed out in 1883 by W. Fabritius.

Gauss and the *Theoria motus*

As the nineteenth century dawned there were thus essentially two workable methods for the determination of parabolic orbits for new comets – those broadly attributed to Laplace and Olbers. Although Laplace's method could be utilized for elliptical orbits, the generally poor determination of the necessary derivatives caused difficulties in practice.

Before the nineteenth century there had been little reason for the direct computation of general orbits. The rather small departure of Halley's Comet from parabolic motion could be measured directly from the observed interval between successive passages through perihelion, just as the sizes of the orbits of the known planets were derived by Kepler's third law from observations covering innumerable revolutions. The only significant exception was afforded by the first comet of 1770, for which Anders Johan Lexell, working at St Petersburg under the guidance of Euler, eventually succeeded in obtaining a viable elliptical solution (given in the *Acta* of the St Petersburg Academy for 1777). Even the discovery of Uranus in 1781 caused no difficulty, for when a parabolic solution was found to be untenable a simple circular solution proved to be quite adequate to prevent this slow-moving planet from becoming lost.

Giuseppe Piazzi's discovery of Ceres in 1801 brought to a head the need for a satisfactory method for determining a general elliptical orbit. The discoverer made twenty-four positional observations of Ceres during the course of the first six weeks of the year, and although Johann Karl Burckhardt at Paris made an attempt (published in the *Monatliche Correspondenz* later that year) at an elliptical solution, it was clear that neither it nor the various available parabolic and circular solutions would yield a prediction accurate enough to guarantee the recovery of this object – believed in mid-1801 to be the single, long-sought-for planet between Mars and Jupiter – when it reappeared in a very different part of the sky following its conjunction with the Sun. As is well known, the mathematical prodigy Carl Friedrich Gauss (1777–1855) (Figure 25.1) came dramatically to the rescue, and Ceres was reobserved at the end of the year within half a degree of his prediction.

The procedures utilized by Gauss in 1801 were extensively modified and perfected during the eight years that elapsed before he published the celebrated *Theoria motus*. As to just what Gauss did in 1801, misconceptions have persisted; it has been asserted, for instance, that "we do not know how Gauss actually computed the orbit of Ceres", and also that the methods he used were based on least squares. Clues in the *Theoria motus* – as well as in contemporary publications and documents – show that Gauss was quite reluctant to use least squares in the orbit-determination process.

Much of the third section of the second book of the *Theoria motus* is devoted to the method of least squares and the theory of the distribution of errors. Gauss noted that he had used the method as early as 1795, and the earliest least-squares computation extant seems to be his four-variable application in 1799 to the equation of time. A full six-variable application to the adjustment of an orbit involves more work than even a human supercomputer such as Gauss would undertake lightly, and most of the adjustment he made was simply a two-variable interpolation procedure like the method of

25.1. Carl Friedrich Gauss at the age of 63.

the variation of geocentric distances. The most extensive section of the *Theoria motus* is devoted to a discussion of possible choices of variables P and Q, say, for which initial values are to be adopted, and then of the small quantities, P' and Q', that depend on them. The discussion includes a generalization to the use of three independent sets of P and Q, rather than the variation of P and Q separately for the second and third sets.

The presentation of Gauss's Ceres predictions by Franz Xaver von Zach in the *Monatliche Correspondenz* in 1801 shows that Gauss made separate three-observation solutions from Piazzi's observations on 2 and 22 January and 11 February and on 1 and 21 January and 11 February. From Gauss's notebooks it is clear that Gauss used an interpolation procedure in which P and Q were taken to be Ceres's orbital inclination I and longitude of the ascending node Ω. If these are defined with respect to the origin and reference plane in which x_i, y_i, and z_i are the planet's heliocentric coordinates (at time t_i), it follows that

$$r_i \cos u_i = x_i \cos \Omega + y_i \sin \Omega$$
$$r_i \sin u_i = -x_i \sin \Omega \cos I + y_i \cos \Omega \cos I + z_i \sin I$$
$$0 = x_i \sin \Omega \sin I - y_i \cos \Omega \sin I + z_i \cos I,$$
$$(15)$$

where u_i is the argument of latitude at t_i, or the angle subtended at the Sun by the arc from the ascending node to the planet. On substitution from Equation (2), it follows that the third of Equations (16) can be used to yield ρ_i, and the first two can then be used to give r_i and u_i. The area of the triangle encompassing the Sun and the positions of the planet at the first and third observations then follows from r_1 and r_3 and the angle $u_3 - u_1$. With the help of the sector–triangle ratio y_{13} Gauss converted this to the sector area, which is equal to $2(\tau_3 - \tau_1)\sqrt{p}$, p being the semilatus rectum of the orbit, permitting the computation of p and then the remaining orbital elements. From these elements he could derive dynamical values for r_0 and u_0 and take for the functions P' and Q' required in the adjustment process the differences between these and the geometrical values computed from Equations (16).

Adopting values of approximately 11° and 81° for I and Ω, respectively, and taking the first and third observations to be those of 1 January and 11 February, Gauss's trial computation gave $r_1 = 2.695\,106$ AU, $r_3 = 2.664\,372$ AU, $u_3 - u_1 = 9°.268\,87$, whence Equation (6) indicated $y_{13} = 1.004\,286\,41$; and with $\tau_3 - \tau_1 = 0.703\,478$ this meant that $p = 2.726\,286\,5$ AU. Aware that Equation (6) was an inadequate approximation for the sector–triangle ratio, Euler, in collaboration with Lexell, had refined it in their *Recherches et calculs sur la vraie orbite elliptique de la comète de l'an 1769 et son temps périodique*, published in St Petersburg in 1770. Application of the improved formula to Ceres gave $y_{13} = 1.004\,299\,43$, $p = 2.726\,357\,2$ AU. Recognizing that p would still be good to only six significant figures, Gauss developed and utilized in 1801 an improved procedure that is essentially that given as Formulas I–V of Article 86 of the *Theoria motus*; for Ceres this yielded $y_{13} = 1.004\,299\,82$, $p = 2.726\,359\,3$ AU. The Euler–Lexell formula is given by Gauss as Formula VI.

Around the end of 1805, evidently in the course of the investigation of the hypergeometric function that he published in 1813, Gauss solved the sector–triangle problem completely and exactly. His elegant solution occupies Articles 88–105 of the *Theoria motus* and gives in the Ceres case $y_{13} = 1.004\,299\,85$, $p = 2.726\,359\,5$ AU, which are correct to the last figure stated.

What is normally termed "Gauss's method" of orbit determination is contained in Articles 131–

163 of the *Theoria Motus*, and except for his continued use of the extended variation of distances procedure, it is very different from the method he utilized in his 1801 work on Ceres. As in the Dionis du Séjour – Olbers method, the basic Gauss orbit-determination method starts from the vector Equation (3). This time, however, it is the first and third terms on the left-hand side that are eliminated, yielding an expression of the form

$$\rho_0 = S_0 - c_1 S_1 - c_3 S_3. \tag{17}$$

By adopting as his basic variables

$$P = \frac{c_3}{c_1}, \quad Q = 2r_0^3(c_1 + c_3 - 1), \tag{18}$$

Gauss showed that Equation (17) can be written in the form of Equation (8), with

$$\rho_0^\circ = S_0 - S, \quad \gamma_0 = \tfrac{1}{2}\frac{SQ}{S_0 - S}, \tag{19}$$

where

$$S = \frac{S_1 + PS_3}{1 + P}. \tag{20}$$

Gauss's first approximations for P and Q were

$$P = -\frac{\tau_1}{\tau_3}, \quad Q = -\tau_1 \tau_3, \tag{21}$$

that for P being obvious from Equation (4), while that for Q requires development from Equations (11) and (12) and in the terms of higher order the assumption that $\tau_1 = -\tau_3$. For P' and Q' Gauss simply took the difference between the improved and the initial values of P and Q (or rather their logarithms). The computation could then be repeated with P and Q as the improved values. In a short-arc computation two or three trials might be all that would be required to reduce P' and Q' to zero, but if more than three trials were necessary, use could be made of the generalized interpolation process mentioned earlier.

By utilizing Equation (8) Gauss was effectively acknowledging Lambert's theorem, and indeed, for short-arc orbits $\gamma_0 \sim R_0^3$. Gauss's approach to the problem is much more general, however, and it is applicable over much longer arcs, where γ_0 may be very different from R_0^3. Rather than attempt to solve the equivalent of Lagrange's eighth-degree equation for r_0 (which, contrary to popular belief, does not necessarily include the solution $r_0 = R_0$), Gauss ingeniously transformed it into what is sometimes called Gauss's equation,

$$a \sin^4 \beta_0 = \sin(\beta_0 - \chi), \qquad (22)$$

where the single unknown is the object's phase angle β_0 (that is, the angle at the object in the Sun–Earth–object triangle). Of the possible values of β_0 the one of interest, at least for an object reasonably near opposition, is the one that is less than ϵ_0, the object's elongation from the opposition point.

The *Theoria motus* contains an enormous amount of insight into both the two-body problem and the treatment of observations, and it is still worth serious scrutiny almost two centuries after its publication. Concerned about the indeterminacy that can all too easily arise in a practical orbit solution from three observations, Gauss provided a modified procedure that would utilize four observations. To specify the constant k that converts time intervals in days to those in the convenient canonical units may seem a trivial point, but this introduced a simplification not present in eighteenth-century literature. Gauss's numerical value for k, namely $0.017\,202\,098\,95$, was based on the values, now obsolete, that he used for the various physical quantities from which it is derived. Yet the same value continues to be used today, for the practice, having proven to be immensely useful, has become traditional; thus Gauss's number is now taken as yielding the definition of the astronomical unit. Gauss's clever procedure for handling Kepler's equation (and its hyperbolic equivalent) in nearly parabolic orbits by means of a particularly efficient iterative expansion about the exact parabolic case is another beautiful illustration of what has been described as "the perspicacity that marked his genius", and it is still very useful today.

Gauss addressed at length various possibilities for incorporating into the computations the effects of aberration and parallax. In favouring the idea of removing the annual aberration from the observed positions and predating the observed times, and in hinting at the use of topocentric rather than geocentric solar coordinates, he was more than a century ahead of almost all other workers in the field.

Other nineteenth-century contributions

Since Gauss obtained such a complete and elegant solution to the problem of determining an orbit from three observations, and since he also discussed at length the rationale for the incorporation of additional observations, any subsequent contribution to the orbit-determination process is necessarily an anticlimax. Much of this later work was designed to simplify the practical computation of an orbit by the primitive means then available

In 1806 the Paris mathematician Adrien-Marie Legendre (1752–1833) published his *Nouvelles méthodes pour la détermination des orbites des comètes*. This work is of interest because it includes the earliest published account of the method of least squares, even though there is no consideration of the theory of errors. Like Gauss, Legendre did not advocate a rigorous least-squares solution from a handful of observations, but he did recommend a simplified computation along these lines. Legendre's approach was to calculate the effect on the computed ecliptic longitudes and latitudes of small incremental changes in the initially assumed elements. In modern parlance, he set down the linearized equations of condition and approximated the various partial derivatives by simple differences. To simplify the solution he assumed that the equations corresponding to one of the observations would be satisfied exactly. This meant that the corrections to two of the orbital elements would be expressed in terms of those to the remaining elements. The actual least-squares computation was therefore restricted to the computation of the corrections to the remaining elements, which in the case of an exactly parabolic orbit would imply only three unknowns.

The method Legendre used to obtain his initial set of parabolic elements is also of interest because, although it was not particularly accurate, it seems to have been the earliest method that utilized the f and g expressions in a direct manner. From Equations (10) and (2) it follows that the components of the velocity v_0 can be expressed as a linear combination of ρ_0 and, say, ρ_1. These can be modified by the use of Equations (11) and (14). On calculating the kinetic energy in a parabolic orbit and equating it to the potential energy one can write $1/r_0$ as a quadratic expression in ρ_0, which can then be solved together with Equation (9) to obtain r_0 and ρ_0 and thence the first approximation to the orbit.

A useful modification of Olbers's method was made by James Ivory in 1814, while Encke later improved it, at least for logarithmic computation – by reformulating Equation (7) along the lines of

Equation (15) but in terms of a rapidly convergent power series.

Pursuing Legendre's work, J. W. Lubbock in 1830 noted that $1/r_0$ in a parabolic orbit is also proportional to the second derivative of r_0^2. Equating the expansions for the two, he was able to obtain ρ_0 from a single quadratic equation. However, Lubbock's approximation is valid only when $\tau_1 = -\tau_3$. As already remarked, this assumption of equal time intervals also appears in Gauss's first approximation for P and Q. In the Gauss procedure, however, these initial values are then rigorously corrected. In any case, Encke showed in 1851 that Gauss's initial approximation can be generalized simply by replacing the second of Equations (21) by

$$Q = -\tau_1 \tau_3 \left(1 + \tfrac{1}{3} \frac{P-1}{P+1} \frac{S_1 - S_3}{S} \right). \tag{23}$$

Encke also produced a three-term expansion for the logarithm of the sector–triangle ratio, and he examined at length the practical solution of Equation (22). He noted that there can be no meaningful solutions for β_0 unless a and χ are contained within specific limits and that there may be two meaningful solutions when $\epsilon_0 > 63°.4$.

In 1876 Friedrich Tietjen formulated a differential procedure for solving Equation (22), and he expressed the sector–triangle ratio with the aid of an auxiliary trigonometric relationship. A popular procedure introduced by P. A. Hansen in 1863 developed the sector–triangle ratio as a continued fraction.

The best-known late-nineteenth-century contribution to the subject of orbit determination is that made in 1889 by Josiah Willard Gibbs (1839–1903), professor of mathematical physics at Yale, in his "On the determination of elliptic orbits from three complete observations", which appeared in the *Memoirs* of the National Academy of Sciences. The 'Gibbs ratios' yield c_1 and c_3 in terms of the heliocentric distances and the time differences expanded out to the fourth power. Of more lasting value, however, was Gibbs's introduction of vectors. Gauss was supreme in both his rigorous treatment of the dynamics of the situation and his statement of the relevant equations in a form convenient for logarithmic computation; if there is to be any adverse criticism of Gauss, it must concern his treatment of the geometry, which his brilliant algebraic manipulations could render quite obscure. By recasting much of Gauss's work

in vector form Gibbs paved the way for twentieth-century thought on the subject.

Gibbs is sometimes credited with the introduction of the unit vectors $\mathbf{P}$ and $\mathbf{Q}$ directed from the Sun to the perihelion point of an orbit and to the point in the orbit 90° ahead of perihelion. The use of these particular vectors (which can alternatively be regarded as two rows in a 3×3 matrix) represented a great conceptual advance in the problem of transferring the position and velocity components of an orbiting object from one reference system to another. H. Buchholz incorporated their use by Gibbs in his revision of 1899 of the *Theoretische Astronomie* of Wilhelm Klinkerfues, but otherwise they are not discussed in any of the standard nineteenth-century texts – or even in the authoritative *Die Bahnbestimmung der Himmelskörper* (1906) of Julius Bauschinger. However, J. C. Watson in his *Theoretical Astronomy* of 1868, in recognizing that equatorial values of the position components in an elliptical orbit can be conveniently specified in terms of the eccentric anomaly, came close to using the unit vectors, and Klinkerfues's original edition of 1871 briefly mentions them without explanation. The vectors were in fact originally given in the *Annales de l'Observatoire de Paris* in 1855 by U. J. J. Le Verrier, who expressed them in terms of Ω, I, and the argument of perihelion in the case where the ecliptic is taken as the reference plane. The corresponding expressions with respect to the equator were finally given by C. E. Adams in 1922. Emphasis on logarithmic computation undoubtedly precluded much early use of Equation (10), and the 'closed expressions' for f and g in terms of differences in eccentric anomaly were first published at a surprisingly late date – namely by Franz Kühnert in 1879.

Laplace's method continued to have many adherents throughout the nineteenth century, among them G. B. Airy and James Challis. The still older methods of Lagrange, Newton, and Lambert were revived, respectively, by P. de Pontécoulant in 1829, Emile Plantamour in 1839, and J. Glauser in 1889. In the late 1840s the end of the hiatus in the discovery of new minor planets yielded increased interest in orbit determination, and there were particularly extensive writings during the next decade by A. L. Cauchy, W. C. Goetze, J. A. Grunert, A. de Gasparis, and Y. Villarceau. A. Cayley, in a memoir of 1871, included consideration of the curves described by the pole of an orbit as a function of variations in some of the elements.

Concern about multiple solutions and indeterminacy led to further developments on orbit determination from four observations, notably by Theodor von Oppolzer in his *Lehrbuch zur Bahnbestimmung der Kometen und Planeten* of 1870, and by Adolf Berberich in a paper of 1902. As for fitting orbits to more extensive series of observations, the four discoveries of minor planets early in the century rapidly yielded a wealth of raw material. The study of the motion of Pallas was particularly difficult, owing to high values of both inclination and eccentricity. It was in his *Disquisitio de elementis ellipticis Palladis* of 1810 that Gauss made his principal practical application of least squares to orbital improvement. Here, however, he simplified matters by separating the corrections to I and Ω from those to eccentricity and longitude of perihelion, and he calculated some partial derivatives analytically and others by approximating them by differences. Utilizing single equations of condition for the mean anomaly and for the heliocentric latitude at each opposition, he linked six oppositions during 1803–09 in three groups of four. In a separate memoir, Gauss made a step-by-step integration of the effect of planetary perturbations on Pallas up to 1811 and included a seventh opposition in his least-squares solution for the orbit.

In the mean time Friedrich Wilhelm Bessel (1784–1846) in his *Untersuchungen über die scheinbare und wahre Bahn des in Jahre 1807 erschienenen grossen Kometen* of 1810 had made a six-unknown least-squares determination of the orbit of the comet of 1807, using six normal places representing 70 observations spanning six months and also allowing for perturbations. Since the orbit of this comet was nearly parabolic, Bessel had to modify the equations given by Gauss for the differential effect of adjustments to the elements of an elliptical orbit on geocentric ecliptic longitude and latitude. For this purpose he used his own expansion of near-parabolic motion about the parabolic case. While less inspired than Gauss's near-parabolic procedure, Bessel's form is in many respects more convenient for this purpose; his expansion was modified and extended by F. F. E. Brünnow in 1858. In a memoir of 1816 on the orbit of Olbers's comet of 1815, Bessel first set down and utilized equations for differential correction in terms of right ascension and declination.

By the 1830s several astronomers were regularly making least-squares orbit improvements. Thomas Clausen in 1831 wrote the equations of condition in terms of small rotations about the axes, and Le Verrier in 1846 expressed concern at the need to calculate partial derivatives to as many as five significant figures when the initial uncertainty of an orbit was rather large. He showed that only two figures were needed if one selected for the six orbital 'elements' the geocentric longitudes and latitudes corresponding to three of the normal places, although he did not indicate how he calculated these partial derivatives.

A method of orbit improvement given by M. A. Kowalski in 1859 separated the corrections involving the plane of the orbit from those connected with the object's position in this plane, but his paper did not become widely known. Tietjen in 1877 presented a rather similar method that became particularly popular. The most thorough treatment of the differential-correction process was published by Eduard Schönfeld in the *Astronomische Nachrichten* for 1885. This involved the simultaneous solution for corrections to all six elements, but the partial derivatives of the right ascension and declination with respect to the elements were derived in an eminently convenient and systematic manner, both for an ellipse of low to moderate eccentricity and for the near-parabolic case by the Gauss procedure. Expressions for variations of orbital elements were also set down in the 1880s by J.-C.-R. Radau in 1887 and by L. Schulhof in 1889. H. Kloock, in a paper of 1893, revived Le Verrier's point about the precision of the partial derivatives. P. Harzer in 1896 also indirectly addressed this problem by introducing a differential-correction process into his adaptation of the Laplace method. Harzer used the f and g series to approximate the partial derivatives required when the corrections are made to the position and velocity components at t_0. Later, Henri Poincaré and A. O. Leuschner carried out to even greater lengths procedures for correcting the Laplace first approximation.

Twentieth-century postscript

Following the introduction of mechanical calculating machines, C. Veithen in 1912 and Gerald Merton in 1925 independently modified the Gauss method so that the use of Equations (18)–(22) was bypassed, an expression having the form of Equation (8) being obtained directly from Equation (17) by means of the 'Encke' approximation that led to Equation (23). The Veithen–Merton method, as well as a Lagrange-type method given by Alexander Wilkens in 1919 and a version of Olbers's method presented by Tadeusz Banachiewicz in 1925, were featured in G. Stracke's textbook of 1929, *Bahnbestimmung der Planeten und Kometen*.

F. R. Moulton in 1914 proposed a method of orbit determination in which Equation (2) was substituted into the left-hand side of Equation (10), but he was unable to give a satisfactory procedure for solving the resulting nine equations. This matter was independently rectified in a paper by the Finnish astronomer Y. Väisälä in 1939, and in the unpublished Harvard University Ph.D. dissertation of L. E. Cunningham (1946).

These authors continued to stress the use of approximations; and the Gibbs ratios, the Hansen continued fraction, the Encke power series for Euler's equation, and the derivation of perhaps five or more terms in the f and g series, are still given prominence in recent texts. In fact it is preferable to use Gauss's rigorous hypergeometric-function solution for the sector–triangle ratios, as well as the closed expressions, rather than the series development as in Equations (11), for f and g.

In 1993 R. H. Gooding published a completely new method of orbit determination based on his sophisticated solution of Lambert's elliptical generalization of Equation (7). Gooding's method has proven particularly useful in applications where the orbiting body might make several complete revolutions between observations.

In cases where more than three observations are to be included, the 'standard' least-squares differential-correction development is nowadays considered to be that given by W. Eckert and D. Brouwer in a paper of 1937. A modification by P. Herget in 1939 of the differential-correction procedure in terms of r_0 and v_0 allowed the use of the closed expressions for f and g. With modern computers, however, it is often convenient simply to revert to Legendre's procedure and approximate *all* the partial derivatives used in an orbital differential correction. A full least-squares solution is performed, of course, and the procedure can be particularly effective when perturbations are also included.

Further reading

The most important contributions to the subject of orbit determination during the eighteenth and nineteenth centuries are listed by K. Zelbr in "Bahnbestimmung der Planeten und Kometen", in W. Valentiner's *Handwörterbuch der Astronomie*, vol. 1 (Breslau, 1897), 568–73, and by R. Radau in "Bibliographie relative au calcul des orbites", *Bulletin astronomique*, vol. 16 (1899), 427–45. A bibliography of the literature from 1900 to 1929 is given by G. Stracke in his *Bahnbestimmung der Planeten und Kometen* (Berlin, 1929), 352–61. For the more recent literature, the reader can turn to the references listed in B. G. Marsden, "Initial orbit determination: the pragmatist's point of view", *Astronomical Journal*, vol. 90 (1985), 1541–5, and "The computation of orbits in indeterminate and uncertain cases", *Astronomical Journal*, vol. 102 (1991), 1539–52.

26

The introduction of statistical reasoning into astronomy: *from Newton to Poincaré*

OSCAR SHEYNIN

A feeling for the opposition of randomness and order is presumably older than philosophy if, as Robert Burns put it, "the best laid schemes of mice and men gang aft agley". Chance was a key concept for the pre-Socratic atomists; Aristotle gave an important if strictly limited place to the role of chance. According to some atomists, order arose out of purely chance events, while Aristotle claimed that chance pertained only to the terrestrial region, where it was the result of an intersection of two or more causally determined sequences, each with its own final cause or *telos*. "What happens everywhere and in every case," Aristotle asserted (*De caelo* 289b), "is no matter of chance." Thus the fact that all the stars in their diurnal revolutions, some on greater and others on smaller circles, move with the same angular speed, could not be the effect of chance, but indicates that they are all moved on and by the same sphere.

In medieval and early modern Europe the almost universal acceptance of the doctrine of divine creation meant that any fundamental randomness in the world was denied; and the atomism of the ancients was accordingly anathematized. Johannes Kepler (1571–1630) in a work on the new star of 1604 called randomness "an idol, ... an abuse of God ... and of the world created by Him". Yet when in his teaching on eclipses in the *Rudolphine Tables* (1627) he could not obtain good agreement between observations and his theory, Kepler supposed the discrepancies to be due to small, accidental physical variations *extra ordinem*. Insofar as the interactions of bodies were "physical" or "smelled of matter", he allowed that they might depart slightly from the patterns imposed by the Creator-God.

In the mid-seventeenth century the French priest, astronomer, and philosopher Pierre Gassendi (1592–1655) gained a following for an atomism reconciled with the doctrine of divine providence; and Isaac Newton (1642–1727) was one of those who accepted the new atomism or corpuscularianism divorced from the rule of chance. Yet Newton, projecting a view of the world as ruled by universal laws, was forced to recognize that an initially imposed order could be subject to decay: a pattern of initial conditions could deteriorate owing to the subsequent interactions of bodies.

Order was predominant, but it was mixed with irregularity or randomness. Up to the middle of the eighteenth century, the desire to discover definite rules for discriminating between randomness and divine design determined the very development of the theory of probability. The English mathematician (of French Huguenot extraction) Abraham de Moivre (1667–1754), in dedicating the first edition of his *Doctrine of Chances* (1718) to Newton, declared his intention of giving a

Method of calculating the Effects of Chance ... and thereby fixing certain Rules, for estimating how far some sorts of Events may rather be owing to Design than Chance ... so as to excite in others a desire of ... learning from your [i.e., Newton's] philosophy how to collect, by a just Calculation, the Evidences of exquisite Wisdom and Design, which appear in the Phenomena of Nature throughout the Universe.

That there was a Newtonian 'philosophy' concerning proofs of design is shown also by a letter which the Rev. William Derham wrote to John Conduitt in 1733, in which he speaks of

a peculiar sort of Proof wch Sr Is: mentioned in some discource wch he and I had soon after I published my Astro-Theology [1715]. He said that there were 3 things in the Motions of the Heavenly Bodies, that were plain evidences of Omnipotence and Wise Counsel. I. That the Motion imprest upon those Globes was

Lateral, or in a Direction perpendicular to their Radii, not along them or parallel with them. 2. That the Motions of them tend the same way. 3. That their orbits have all the same inclination.

Thus the near-circularity of the planetary orbits, the uniform direction in the motion of all the planets (counterclockwise as seen from the ecliptic's north pole), and the smallness of all the orbital inclinations to the ecliptic, were, in Newton's opinion, evidences of design.

Earlier, the main fields to which probability theory had been applied were population statistics and games of chance. Now, under Newton's influence, probabilistic reasonings would be applied to the "wonderful uniformity in the planetary system". Even Pierre Simon Laplace (1749–1827), despite his celebrated avoidance of the divine, would continue the trend, proclaiming nature's lawfulness as contrasted with the disorder of randomness. With Daniel Bernoulli and Laplace, the reasonings became explicit calculations.

The inclinations of the planetary orbits

In 1735 Daniel Bernoulli (1700–82), mathematician, physicist, and physician of Basel, argued that there must exist a general cause for the fact that all the planetary orbits lie nearly in the same plane. Here is his reasoning: Let the inclination of the orbit of planet i to the ecliptic be a_i, where $0° < a_i < 90°$, and $i = 1, 2, \ldots, 5$ (only 5 planets were considered since the sixth known planet, the Earth, was by definition in the ecliptic). Further, let A be any number of degrees less than 90. If the values of the a_i were independent and uniformly distributed, the probability that $a_i < A$ for any of the five planets would be $A/90$, and the probability that all were less than A would be $(A/90)^5$. But the latter probability becomes negligible if A is much smaller than 90°, as it is in fact. Consequently, the quantities a_i cannot be independent, and must therefore be the result of a general cause.

Actually, Bernoulli had supposed that the unperturbed situation was one in which all the planetary orbits would be in the same plane, and that the small inclinations of the several planetary orbits to the ecliptic were occasioned by random disturbances of some kind. He also believed that a random event having a sufficiently small probability of occurrence was effectively impossible – a

principle that underlies most applications of probability theory to this day.

The direction of motion of planets and satellites

In 1776 Laplace, like Newton before him, argued that the movement of all the planets and of their satellites (fifteen such bodies in all were known at this date) in the same direction was inconsistent with the assumption that their directions were the effect of chance. Indeed, he claimed that under the latter assumption the probability of the observed uniformity of direction would be only 2^{-15}, and thus negligible.

Laplace's assumption that direct motion has a probability of $\frac{1}{2}$ was in fact arbitrary, although he did not recognize it to be so. If we suppose that the probability of direct motion is 0.9 (say), so that retrograde motion has a probability 0.1, the probability of direct motion of all the bodies would be $(0.9)^{15} \approx 0.2$; and thus a random occurrence of the phenomenon would not be ruled out. The formation and testing of statistical hypotheses is a more delicate operation than Laplace implied. Laplace's reasoning here was to serve as underpinning for his celebrated *nebular hypothesis*, first proposed in 1796, to explain the origin of the solar system (see Chapter 22).

As later discoveries were to show, not all bodies in the solar system revolve in the same direction. In 1787 William Herschel discovered two satellites of Uranus (Titania and Oberon) and, in 1797, he ascertained that they move in the retrograde direction. In 1851 William Lassell discovered two more satellites of Uranus (Ariel and Umbriel), which are also retrograde. However, the orbital planes of all four were in line with the planet's equator, and inclined by 98° to the ecliptic, so that with a slight change in angle the motions would have been described as direct. Triton, a satellite of Neptune discovered by Lassell in 1846, proved to have a retrograde motion, with an orbit inclined at 35° to the ecliptic. Of more recently discovered satellites, four of Jupiter's and one of Saturn's have retrograde motions, on orbits inclined to the ecliptic at various angles less than 35°.

Comets

In 1776 Laplace undertook a mathematical study of the inclinations of cometary orbits to the ecliptic. A senior colleague in the Paris Academy, Achille-

Pierre Dionis du Séjour (1734–94), had posed the question whether the cause that produced the planetary motions – whatever it might be – also produced the phenomena of comets. He had calculated the mean inclination of the 63 known cometary orbits and found it to be 46° 16′. The departure from 45° seemed so small as to be insignificant. Moreover, the ratio of forward to retrograde motions was about 5:4, not very different from 1:1. Dionis du Séjour concluded that the planetary system constituted a distinctive causal ordering, whereas the cometary orbits followed a principle of indifference. Laplace wanted to calculate the degree of certainty of this conclusion.

Assuming the comets to have been originally projected randomly into space, what would be the probabilities, Laplace asked, for the mean inclination of their orbits, and the ratio of direct to retrograde revolutions, to fall within given limits? If the mean inclination were $45° + a$, and on the assumption of a uniform random distribution (there were very large odds that $|a|$ should have a small value), then it could be concluded with *vraisemblance* that some particular cause accounted for the comets moving in one particular plane. Taking up in succession the cases of 2, 3, 4, ..., 12 comets, Laplace constructed the curves representing the probability for the mean inclination to have any value between 0° and 90°. By determining the fraction of the total area under a curve falling between given limits, he could compute the probability that the mean fell within these limits. Thus, the odds for the mean inclination of 12 comets to fall between 37°.5 and 52°.5, he found, were 678 to 322. Was there any evidence, he asked, for a cause tending to make the comets move in the ecliptic? The mean inclination of the 12 most recently discovered comets was 42° 31′; the odds for this mean to exceed 42° 31′ seemed to him too small to reject the independence of the inclinations. He did not go on to apply his reasoning to all 63 of the comets for which orbits had been computed. The calculations failed to provide a conclusive answer as to the dependence or independence of the inclinations. Laplace renewed his attack on the problem in 1810 and again in 1812, but again without achieving a convincing analysis.

In 1792 the German astronomer J. E. Bode claimed that the inclinations of cometary orbits were distributed with near-uniformity. He compared a strictly uniform distribution (the inclinations being ranked as to size and differing by a constant angle) with the actual distribution, and maintained that the differences were negligible. However, he was unable to offer a quantitative estimate of the degree of fit.

Clustering of stars: the work of John Michell

The Rev. John Michell (*c.* 1724–93) was the first astronomer to attempt to calculate the probability of two stars being within a given apparent angular distance of one another. This topic belongs in principle to Volume 3, but we include it here because of the techniques employed.

Michell's first and chief foray into this subject was made in "An Inquiry into the probable Parallax, and Magnitude of the fixed Stars ...", which appeared in the *Philosophical Transactions* of the Royal Society of London in 1767. Earlier Michell had been a fellow of Queens' College, Cambridge, and had written on artificial magnets, the causes of earthquakes, and topics in surveying and navigation. In 1764 he resigned his fellowship for marriage and an ecclesiastical living, and from 1767 onwards held a rectorship in Thornhill, Yorkshire.

The paper of 1767 had as its announced purpose the estimation of an upper limit for the stellar parallax that astronomers might hope to discover. The method Michell used was not new with him, but had been suggested earlier by James Gregory. Michell assumed the size and intrinsic brightness of the fixed stars to be equal to the size and brightness of the Sun; the differences, he granted, might in fact be very great, but an error of 1000 in size (volume) would alter the distance by a factor of only 10. From Saturn's solar distance and apparent diameter, Michell deduced that – supposing Saturn to reflect all the solar light it receives – the Sun would have to be removed to 220 000 times its present distance from us to send back only as much light as does Saturn; at such a distance the Sun's whole parallax on the diameter of the Earth's orbit would be less than 2″. (If Saturn reflects only $\frac{1}{4}$ or $\frac{1}{6}$ of the light it receives from the Sun, as seems likely, the distance of the Sun would have to be increased in the ratio of 2 or 2.5 to 1.) Now Sirius has an apparent magnitude comparable with that of Saturn. Michell therefore concluded that the parallax of Sirius, assuming this star to be of the same size and intrinsic brightness as the Sun, was no greater than 1″. From rough experiments he judged the light received from the faintest naked-eye stars to be between $\frac{1}{400}$th and $\frac{1}{1000}$th of that from Sirius, and so inferred that the parallax of these stars, supposing them to be of the same size and intrinsic brightness as the Sun, was between 2‴ and 3‴.

We come now to Michell's application of statistical reasoning to the distribution of stars in space. If the stars had been scattered by mere chance, the probability that any particular star should be within 1° of any other given star, according to Michell, would be given by a fraction with numerator equal to the area of a circle of 1° radius, and denominator equal to the area of the whole celestial sphere; the result is $p = \frac{1}{13131}$. The probability that these same stars would *not* be within 1° of one another, Michell then reasoned, is $1 - p = \frac{13130}{13131}$; and if there were n stars as bright as the stars in question, the probability that none of them was within 1° of the given star would be $(1 - p)^n$. Finally, Michell inferred (erroneously) that the probability of no two such stars being within 1° of each other was $[(1 - p)^n]^n$.

On this basis Michell concluded that the probability of close optical doubles was very slight. Thus he calculated that the odds against any two stars being within as small a distance of one another as the two stars of β Capricorni, assuming there to be 230 stars equal to these in magnitude, was 80 to 1. The odds against six stars being within as small a distance of one another as are the six brightest stars of the Pleiades he found to be about 500 000 to 1.

From these and similar calculations, Michell drew the conclusion that

the stars are really collected together in clusters in some places, where they form a kind of systems, whilst in others there are either few or none of them, to whatever cause this may be owing ....

From the contrast between deduced probability and observed frequency, Michell inferred that attractive forces – presumably gravitational – are operative among the stars called 'fixed'. This was the first attempt to provide a detailed argument for gravitational action beyond the solar system. When in 1779 William Herschel embarked on a search for double stars to be used in parallax measures, Michell reiterated his point in a second paper, which was published in the *Philosophical Transactions* for 1784:

The very great number of stars that have been discovered to be double, triple, &c. particularly by Mr Herschel, if we apply the doctrine of chances, . . . cannot leave a doubt with any one, who is properly aware of the forces of those arguments, that by far the greatest part, if not all of them, are systems of stars so near to each other, as probably to be liable to be affected sensibly by their mutual gravitation: and it is therefore not unlikely, that the periods of the revolutions of some of these about their principals (the smaller ones ... to be considered as satellites to the other) may some time or other be discovered.

Michell's argument was to be favourably reviewed by Laplace, John Herschel, and Wilhelm Struve, and came to be widely accepted. However, as pointed out by the Edinburgh physicist and geologist J. D. Forbes (1809–68) in two articles of 1849 and 1850, Michell's calculations assume that a random distribution of stars would be a *uniform* distribution, whereas a uniform distribution was quite unlikely as the result of chance. Because of this and other errors in the calculation, Forbes concluded that Michell's mathematical argument had proved nothing at all. The mathematician George Boole (1815–64) took a largely similar stand in an article of 1850.

A correct calculation of the required probabilities can be based on Poisson's distribution. Let the expected number of stars situated not farther than 1° apart be $a = pn$, where p and n have the same significations as before. Then the probability that there is at least one optical double is

$$P = 1 - e^{-a}(1 + a).$$

For $n = 5000$, we find that $a = 0.3808$ and $P = 0.056$. This is a small though not negligible probability, considerably smaller than the several hundreds of optical doubles discovered by William Herschel would imply. Thus Michell's final conclusion was correct in making the probability of star clusters smaller than would be deduced from their actual frequency.

The practical result of Michell's work was to convince people of the essential truth of what he said, and he was vindicated observationally by William Herschel. Re-examining his double stars soon after the turn of the century, Herschel found that some had indeed orbited around each other. By about 1830 several astronomers (including William's son John Herschel) had enough evidence to show that the orbits were Keplerian and the law therefore indeed gravitational.

Sunspots

The first telescopic observation of a sunspot was made by Thomas Harriot in 1610. Galileo exhibited sunspots to observers in Rome in the spring of 1611, and it was by means of sunspots that he showed that the Sun rotates about an axis passing through its centre. In addition, he successfully separated the regular movement of sunspots due to the Sun's rotation from the random component, that is, their proper motions relative to the Sun's disk. His estimate of the period of rotation of the Sun, namely one lunar month, was a good first approximation. Actually, the sidereal rotation period varies with latitude, being about 25 days at the Sun's equator, 27.5 days at latitude $+45°$, and as much as 33 days at latitude $+80°$: the surface of the Sun is not solid.

The first to suspect periodicity in the number of sunspots may have been the Danish astronomer Peder Nielsen Horrebow (1679–1764), who entered a remark about it in his diary. William Herschel in 1801, drawing on data for the period 1650–1717, attempted to determine the relationship between sunspots and the price of wheat, the latter being, presumably, an index of the weather.

In 1844, after studying the occurrence of sunspots for about eighteen years, the German amateur observer Samuel Heinrich Schwabe (1789–1875) announced that their number varies with a period of about ten years; but the future alone, he prudently added, would show whether this period was constant. His yearly data included the number of days of observation, the number of days when no sunspots were seen at all, and the number of groups of sunspots. Schwabe's work appears to have passed unnoticed until Alexander von Humboldt described it in his *Kosmos* (1850).

In 1859 the astronomer Johann Rudolf Wolf (1816–93) of Zurich collected all observations of sunspots beginning with the middle of the eighteenth century, determined the epochs of their extreme numbers, and so derived an estimate of the

periodicity of their occurrence, $T = 11.1$ years. Twenty years later he put on record observations covering 120 years and again calculated the period, testing, in the course of doing so, nineteen hypotheses ($T = 9$ years 6 months; 9 years 8 months; 9 years 10 months; ...; 12 years 6 months). He considered the deviations of the mean yearly data from the general mean number of sunspots and applied as a test the range of the deviations or, alternatively, the root of the sum of their squares divided by the number of deviations. He concluded that there were two periods: $T_1 = 10$ years, and $T_2 = 11.3$ years, their least common multiple (the single common period), as he himself noticed, being 170 years.

Today it is generally accepted that there is an approximate period of about 11 years, but a strict periodicity is denied.

The influence of sunspots on terrestrial magnetism and climate was discovered in the middle years of the nineteenth century. In 1840 Johann von Lamont (1805–79) established a magnetic observatory at Bogenhausen near Munich, and ten years later he announced that the Earth's magnetic field is subject to variation. Wolf later correlated this variation with the cycle of sunspot activity, but without supplying a quantitative measure of the correlation, such measures being as yet unknown.

In 1874 the Scots meteorologist Charles Meldrum announced that

not only the number of cyclones, but their duration, extent, and energy were also much greater in the [years of maximum] than in the [years of minimum sunspot frequency], and ... there is a strong probability that this cyclonic fluctuation has been coincident with a similar fluctuation of the rainfall over the globe generally.

At about the same time the English astrophysicist J. N. Lockyer voiced a similar opinion regarding cyclones, while in 1880 the head of the Meteorological Department of India, the Englishman H. F. Blanford, pointed out a connection between atmospheric pressure and sunspots. But all these authors, lacking as they did a mathematical theory of correlation, were able to give only qualitative comparisons.

The founder of the theory of correlation was the London gentleman-scientist Francis Galton, in a paper on "Co-relations and their measurement, chiefly from anthropometric data", read to the Royal Society in December 1888. The theory was developed mathematically by the Irish statistician Francis Ysidro Edgeworth during the 1890s, and made widely known in its biological applications by Karl Pearson (1857–1936) of University College, London. Between 1908 and 1910 Pearson, partly in collaboration with the statistician, astronomer, and medical researcher Julia Bell, published a number of papers in astronomical journals, applying correlation theory and the Pearsonian test for goodness of fit. They were concerned with the connection between the colours of stars and their spectral classes and magnitudes.

The asteroids

Because of their large number, the asteroids or minor planets constitute the most representative instance of a statistical population in the solar system. The first four were discovered between 1801 and 1807 (see Chapter 24). Since the asteroids were small, astronomers were convinced there must be more. To facilitate their discovery, the Berlin Academy sponsored the construction of star maps of the zodiacal zone containing all stars down to the ninth magnitude. A fifth asteroid was discovered in 1845, and, with the aid of the new star maps, from 1847 onward at least one was discovered each year. By 1852 the number had risen to 20, and by 1870 it reached 110. Since 1891 most of the discoveries have been made photographically (see Chapter 28), and several thousands have been catalogued.

In the 1860s Simon Newcomb (1835–1909), astronomer of the US Naval Observatory in Washington, DC, and a leading celestial mechanician of his day, devoted a number of papers to the study of the asteroids with a view to testing for group properties – characteristics that might apply to asteroids yet to be discovered, or that might indicate a common origin. Thus in 1862, from the Lagrangian formulas for the secular variations of the orbital elements of these bodies, and assuming uniform distribution of certain constants dependent on the totality of the perturbations, he derived probabilities for the perihelia and nodes to fall into each of the four quadrants of the ecliptic. The perihelia, he found, fell within the quadrant containing the perihelion of Jupiter, and the nodes

within the quadrant containing the node of Jupiter, with greater probability than in the remaining quadrants – an expected result, since Jupiter is the greatest perturber of the motions of asteroids. In 1869 he compared the actual distribution of perihelia and nodes with a uniform distribution, and attempted to investigate the possibility that the actual distribution had diverged from a uniform distribution; but he lacked the statistical modes of analysis requisite for such a task.

In 1900 Newcomb returned to the asteroids to test whether the 'Kirkwood gaps' might have arisen by chance. In 1857, and again in 1866, the American mathematician Daniel Kirkwood (1814–95) had remarked that the periods of revolution of the then known asteroids were not uniformly distributed about their mean of 4.7 years; there were gaps at $\frac{1}{3}$, $\frac{2}{5}$, and $\frac{2}{7}$ of Jupiter's period. Later Kirkwood discovered other, somewhat less distinct gaps at $\frac{1}{2}$, $\frac{3}{5}$, $\frac{4}{7}$, $\frac{5}{8}$, $\frac{3}{7}$, $\frac{5}{9}$, $\frac{7}{11}$, and $\frac{4}{9}$ of Jupiter's period. Kirkwood's explanation was that the perturbational forces due to Jupiter would not permit such commensurabilities to exist; an asteroid in one of the gaps would be "pumped" out of it by repeated application of similar forces similarly applied, or its orbit made so eccentric that it would become subject to collisions.

Newcomb selected 354 asteroids with mean motions varying from 600″ to 1000″ per day (the periods thus varied from 3.55 to 5.91 years), and divided them into forty groups according as the mean motions fell into the intervals 600″–610″, 610″–620″, ..., 990″–1000″. Assuming a binomial distribution of the 'population' within the forty groups, Newcomb computed what he called 'the probable number' of groups having x planets each (it is more properly called the mean number). Then, from a qualitative comparison between the observed and probable [mean] numbers of groups with x planets each, he concluded – in agreement with Kirkwood – that the inequalities of distribution could not have arisen in a group of asteroids whose original periods or mean distances had been distributed uniformly.

Not only were the orbital parameters of the asteroids studied statistically, but even the very existence of any as yet undiscovered asteroids was proposed as a subject for statistical inference. The great French mathematician and mathematical physicist Jules Henri Poincaré (1854–1912) in his

Calcul des probabilités (1896, with a revised and expanded version in 1912) undertook to estimate the entire number (N) of asteroids. Suppose that M of these are known and that during a certain year n asteroids were observed, m of which had been known before, so that $N \approx Mn/m$. Let $p = n/N$ be the probability that during the same year a certain (existing) asteroid was observed; ω_i, the prior probability that i asteroids exist (Poincaré assumes that $\omega_i = \text{const}$); and p_i, the probability that n asteroids were observed given that i of them existed ($i \geq n$). In this notation, and writing $q = 1 - p$, Poincaré found

$$p_i = \frac{i!}{n!(i-n)!} p^n q^{i-n},$$

$$\sum_{i=n}^{\infty} p_i = p^n + \frac{n+1}{1!} p^n q + \frac{(n+1)(n+2)}{2!} p^n q^2 + \ldots$$
$$= p^n/(1-q)^{n+1} = 1/p.$$

The probability that there are N asteroids is then pp_N and the expected or mean value of N is

$$E(N) = \frac{n+q}{p}.$$

In his *Calcul des probabilités* and in other writings published at the beginning of this century Poincaré attempted to explain the notion of randomness, and in this connection he once again considered an example concerning the asteroids. He did not provide detailed derivations, and from a methodological point of view his analysis was imperfect; nevertheless, it constituted an appreciable contribution to probability theory and its philosophy.

Why, Poincaré asked himself, are the ecliptic longitudes of the asteroids uniformly distributed? The solar distances of these bodies differ from one another, though often by only small amounts; hence their mean motions do not coincide. If one supposes that they once constituted a single body, over millions of years they would, because of the slight differences in their mean motions, come to be scattered across the entire ecliptic.

Poincaré's proof is too intricate to reproduce here, but it is worth noting that he was dealing with a random process of a kind that later came to be included in the province of probability theory, and that his proof closely approaches that of a celebrated theorem due to Hermann Weyl in 1916 on the distribution of the fractional parts of irrational numbers on the unit interval.

Poincaré offered several explanations of randomness and chance. One was instability of motion, such as may occur in unstable equilibrium, and another was the existence of complicated causes, as in the great number of collisions among gaseous molecules, leading to a relatively uniform distribution. He also mentioned the intersection of chains of events as an explanation of chance.

Conclusion

The examples of statistical reasoning given in the present chapter belong to the prehistory of mathematical statistics. Only in the twentieth century would quantitative tests of correlation and of statistical significance become part of the standard arsenal of investigative techniques. During the period considered here, the steps taken in applying probabilistic reasoning to astronomical phenomena were initial ones, often innovative but sometimes flawed because of the inadequacy of the available analytical tools.

In one area, however, probabilistic reasoning applied to astronomy led to a result of signal and lasting importance: the theory of observational error. It is to the development of this theory that we turn in the following chapter.

Further reading

O. B. Sheynin, Newton and the classical theory of probability, *Archive for History of Exact Sciences*, vol. 7 (1971), 217–43

O. B. Sheynin, On the history of statistical method in astronomy, *Archive for History of Exact Sciences*, vol. 29 (1984), 151–99

O. B. Sheynin, H. Poincaré's work on probability, *Archive for History of Exact Sciences*, vol. 42 (1991), 137–71

27

Astronomy and the theory of errors: from the method of averages to the method of least squares

F. SCHMEIDLER

with additions by Oscar Sheynin

The problem of how best to employ observations in obtaining information about unknown quantities is of fundamental importance for astronomy. Ancient and medieval astronomers, in determining any particular quantity, appear often to have relied upon the smallest number of observations necessary to determine the quantity in question. Thus Ptolemy in the *Almagest* frequently reported only a single observation for the determination of a parameter. In such cases, however, it may be doubted whether he is recounting all the data on which he relies, or merely giving an illustration of how the determination is to be made.

In the centuries after Ptolemy, the accuracy of the observations that he reported remained almost entirely unquestioned. But the problem of observational error was not one that astronomers ignored. Thus al-Bīrūnī (AD 973–p. 1050), in a work on the correction of the distances between cities, says that random errors are inevitable "because celestial observation is a very delicate matter". In one case he chooses a particular value "because it is close to the average between the smaller amount ... and the larger amount, and because the indirect method ... produces an amount which is not far from that amount and [thus] corroborates it". Although referring to the arithmetic mean, he does not adopt it as his general rule; in each case he chooses a common-sense value for the constant sought.

Nicholas Copernicus (1473–1543) maintained toward the observations reported by Ptolemy an attitude of pious acceptance. Tycho Brahe (1546–1601), however, early came to recognize that the predictions of both the Ptolemaic and Copernican theories were often in error by several degrees (see Chapter 1 in Volume 2A), and that the only remedy was an *instauratio*, a renewal of astronomy on the basis of a regular series of accurate observations. But Tycho continued the practice of selecting from among the available observations the one he believed, on whatever grounds, to be the most suitable or trustworthy for the purpose in hand.

Johannes Kepler (1571–1630), when adjusting observations in his *Astronomia nova*, appears to have used the arithmetic mean, and also (in Chapter 10, Volume 2A) a *medium ex aequo et bono*, in which the observations seem to have been weighted according to their presumed likelihood of being in error. But during the seventeenth century such practices did not become general. The Danzig astronomer Johannes Hevelius (1611–87) used the one observation of the position of a star that he considered the best, even when he had made many measurements. And according to Francis Baily, the nineteenth-century biographer of John Flamsteed (1646–1719), England's first Astronomer Royal, Flamsteed

does not appear to have taken the mean of several observations for a more correct result.... [Where] more than one observation of a star has been reduced, he has generally assumed that result which seemed to him most satisfactory at the time, without any regard to the rest.

In the middle years of the eighteenth century, two astronomers stand out as having made a regular practice of using the mean. One was James Bradley, the third Astronomer Royal. According to his own report, "When several observations have been taken of the same star within a few days of each other, I have either set down the mean result, or that observation which best agrees with it." The other, Nicolas-Louis de Lacaille, likewise regularly employed the arithmetic mean in determining observed positions as well as constants such as orbital elements derived from such positions (see Chapter 18).

Early use of equations of condition

Not till the eighteenth century did astronomers and mathematicians undertake systematic studies of how unknown quantities ought to be determined from multiple data. Suppose that a certain number of unknowns, say m, are to be determined, and further, that n observations, where $n > m$, have been made which are algebraically dependent on the unknowns. Then each observation constitutes a condition that must be fulfilled at least approximately by the solution. The equations by which the observed values and the unknowns are connected are in most cases linear; if not, they can usually be linearized. The simplest case is that in which a single unknown quantity can be measured directly; and in this case the arithmetic mean of the several measured values came to be considered the best value – the one on which to rely. During the eighteenth century several attempts were made to refine and generalize the use of the arithmetic mean for more complicated problems. These methods were gradually improved, and finally replaced by the method of least squares which was invented around 1800 and in the following decades came into general use.

The first mathematician to investigate how unknown quantities should be determined when the number of equations exceeds the number of unknowns was the ingenious Roger Cotes (1682–1716), editor of the second edition of Newton's *Principia*, and an assiduous student of Newton's methods. He considered only the case in which a single unknown quantity, x, is to be determined from n equations of condition

$$a_i x = l_i \quad (i = 1, 2, \ldots, n).$$

From each of these equations, the value of the unknown could be calculated as $x = l_i / a_i$. According to Cotes, among these results the more reliable are those for which the coefficients a_i (all taken as positive) have the larger values. Cotes therefore concluded that the best value would be given, not by the mean of the values l_i / a_i, but by the formula

$$x = \frac{l_1 + l_2 + \ldots + l_n}{a_1 + a_2 + \ldots + a_n}.$$

The formula in effect gives a weighted mean, with larger weights for the values l_i / a_i having the larger a_i. Cotes's researches on the subject are contained in his *Harmonia mensurarum*, which was published posthumously in 1722.

How should one proceed if there is more than one unknown? This question was attacked some decades later by the Swiss mathematician Leonhard Euler (1707–83), and by a young cartographer and astronomer in Nuremberg, Johann Tobias Mayer (1723–62), who would later become director of the observatory in Göttingen. Their general idea was to subdivide the equations of condition into as many subsets as there were unknowns to be determined, in such a way that one or two of the unknowns in each subset had relatively large coefficients, while the others had relatively small coefficients; the terms involving the latter unknowns were then neglected, and the resulting equations solved for the unknowns they contained. After the first unknowns were determined, the remaining unknowns could be found consecutively by a similar procedure. This is the so-called 'method of averages'.

Euler's application of the method occurs in his "Recherches sur la question des inégalités du mouvement de Saturne et de Jupiter", which was the winning essay in the Paris Academy's contest of 1748, and was published in 1749. After deriving the perturbational terms in the longitude of Saturn due to the attraction of Jupiter, Euler undertook to compare his theory with some 95 heliocentric longitudes of Saturn reported by Jacques Cassini (Cassini II) in his *Élemens d'astronomie* of 1740; they covered the period from 1582 to Euler's own time. To make the comparison, it was necessary to have values for Saturn's orbital elements – epoch, mean motion, aphelion, and maximum equation of centre; and these can be determined only empirically. Provisionally, Euler adopted Cassini's values for these parameters, but recognized that they stood in need of correction, since "the inequalities caused by the action of Jupiter are there enveloped in the eccentricity and position of the orbit of Saturn". In addition, his theory included (mistakenly) a perturbational term proportional to the time, and Euler, being unable to derive its coefficient from the theory of gravitation, hoped to fix its value from the observations.

All in all, there were eight unknowns, and Euler showed that the differences between the observed longitudes of Saturn and the theoretical values could be represented by linear equations of condition in these eight unknowns. Five of the unknowns were multiplied by sines or cosines – purely periodical coefficients. In order to separate the unknowns, Euler selected pairs of observations 59 years apart, this time interval being equal, very nearly, to two sidereal periods of Saturn and five of Jupiter; in such paired observations the purely periodical coefficients were nearly equal. Subtracting one of the paired equations of condition from the other, Euler obtained an equation in which the terms with the periodic coefficients were negligible; from three such equations he could determine the unknowns having non-periodical coefficients. By other combinations of the equations he then sought to determine the remaining unknowns, and to reduce the maximum error of his theory as much as possible. But the result of these manipulations was only partially successful, because his theory contained incorrect perturbational terms and lacked other terms it should have contained.

Mayer, in undertaking to determine the libration of the Moon, made use of a procedure that was similar in principle, but different in some details. He had at his disposal twenty-seven observations of the angular distance between the crater Manilius and the apparent centre of the lunar disk. These angular distances could be expressed as linear functions of three unknowns: the inclination of the Moon's equator to the ecliptic, the node of the lunar equator on a plane parallel to the ecliptic through the Moon's centre, and the selenographic latitude of the crater Manilius. There were thus twenty-seven equations of condition for three unknowns. To reduce these equations to three, Mayer first added together the nine equations in which one of the unknowns had the largest positive coefficients; then he formed the sum of the nine equations in which the same unknown had the largest negative coefficients; and finally he added together the remaining nine equations, in which the coefficients of this unknown were comparatively small in absolute value. Thus he obtained three equations from which the three unknowns could be directly determined.

Mayer's memoir on the libration of the Moon was published in 1750 in the *Kosmographische Nachrichten und Sammlungen auf das Jahr 1748*. From Mayer's earliest letter to Euler, written in July 1751, we know that Mayer had previously made a thorough study of Euler's prize-winning essay on the inequalities of Jupiter and Saturn (see Chapter 18); but whether he learned his method of combining equations of condition from Euler's essay cannot be settled definitively. In any case his application of the method was more successful than Euler's, since he was not applying a mistaken theory. His procedure was reported in detail and with high praise in the second and third editions of J.-J. L. de Lalande's *Astronomie* (1771 and 1793), the best known textbook of astronomy during this period.

The method of Bošković

Euler's and Mayer's way of combining equations of condition produced reasonable results, but in an important respect it was unsatisfactory. The aim of forming subsets in such a way that particular coefficients were maximal in each subset guided the computation, but there was more than one way in which this aim could be realized. It was therefore possible that, with different combinations of the equations, the solutions would have been different.

It was because of this deficiency in the method that the Croatian polymath Rudjer J. Bošković (1711–87) developed a new method for combining equations of condition. At an early age Bošković had entered the Jesuit order, and studied at the Collegio Romano in Rome, where he became conversant with Newton's *Opticks* and *Principia*, and with a wide range of mathematical disciplines. In 1750 he and a fellow Jesuit, Christopher Maire, were ordered by Pope Benedict XIV to measure the length of an arc of the meridian in the Papal States, and to carry out other operations for the correction of the map of these territories. During the following three years the two of them carried out triangulations, and made determinations of the height of the celestial pole, over an arc of 2° of the meridian in central Italy. In 1755 they published an account of their results in Latin, and here Bošković did not yet make use of his new method for dealing with equations of condition. However, a short summary of this memoir appeared in the *Memoriae de Bononiensi Scientiarum et Artium Instituto atque Accademia* for 1757, and into this Bošković inserted a description of his method. He described it again in his commentary on a Latin poem concerning Newtonian natural philosophy by Benedict Stay, published in 1760. A French translation of the original work of 1755 appeared in 1770 under the title *Voyage astronomique et géographique dans l'État d'Église*, and the description of the method was here given as an appendix. Bošković did not mention the work of earlier authors on this problem, but from his essay on the perturbations of Jupiter and Saturn, which won the *proxime accessit* in the Paris Academy's contest of 1752, we know that he had read Euler's prize-winning essay of 1748.

Bošković wished to compare his result for the length of a degree of the meridian with four other determinations that had been made at different geographical latitudes. His aim was to derive a value for the flattening of the Earth: on the assumption that the Earth is a flattened spheroid, the degrees of the meridian lengthen as one proceeds from the equator towards either geographical pole (see Chapter 15). Bošković sought to represent the five determinations of a degree of the meridian by a function meeting three conditions:

(a) The difference between the length of a degree of the meridian at latitude ϕ_i and the length of a degree of the meridian at the equator should be proportional to the square of the sine of the latitude, or equivalently, to the versed sine of double the latitude (this condition holds for an ellipsoid of revolution). Thus each measurement of a degree of latitude, D_i, should fit, very nearly, a linear equation of the form

$$D_i - D_0 - (1 - x\cos 2\phi_i) = 0, \qquad (1)$$

where D_0 is the (unknown) length of a degree of the meridian at the equator, and x is another unknown dependent on the parameters of the Earth's spheroid. Evidently not all five equations of form (1) could be satisfied exactly, and the best values of D_0 and x would presumably be such that the differences between the left-hand members of Equations (1) and zero (the 'residuals', ΔD_i) were in some fashion minimized.

(b) The sum of the positive residuals should be equal to the sum of the negative residuals, or

$$\sum_{i=1}^{n} \Delta D_i = 0. \qquad (2)$$

(c) The sum of the residuals taken without regard to their algebraic signs should be a minimum, or

$$\sum_{i=1}^{n} |\Delta D_i| = \min. \qquad (3)$$

Bošković's solution of this problem was mainly geometrical. He plotted the lengths of the five meridian arcs against the corresponding values of the versed sines of 2ϕ. His goal was then to fit a straight line to these points in the "best" way. First he determined the 'centre of gravity' of the points, that is, the point with coordinates given by the mean value of the versed sine of the double latitude and the mean value of the measured degrees. Because of Equation (2), the straight line would have to pass through this point. Finally, by trial and error Bošković adjusted the slope of the line in such a way as to satisfy Equation (3).

As we shall see, Laplace was later to give a purely analytic formulation of this method, and Gauss would enunciate a fundamental objection to it.

Early investigations concerning the arithmetic mean and the law of errors

In the long run, progress in minimizing the effect of observational errors would depend on theoretical investigations of a more philosophical character. Galileo in the Third Day of his *Dialogue Concerning the Two Chief World Systems* had formulated some of the basic propositions of a theory of observational errors: their inevitability, the equal probability of positive and negative errors, the accumulation of observations close to the true value, and the advisability of rejecting outliers. The first attempts to develop a theory of errors mathematically date from the middle years of the eighteenth century; they concerned the use of the arithmetic mean and the question of the statistical distribution of errors of observation about the mean.

In the *Philosophical Transactions* for 1756 there appeared a paper by Thomas Simpson (1710-61), a mathematics teacher in a military academy, entitled "An attempt to show the advantage arising by taking the mean of a number of observations in practical astronomy"; a revised version of the paper was published in the following year in Simpson's *Miscellaneous Tracts*. The main merit of the paper was the idea that there is a law of the frequency of errors of different amounts. In the original version Simpson considered two different discrete laws of the distribution of astronomical errors. In the one, the errors

$$-v, \ldots, -3, -2, -1, 0, 1, 2, 3, \ldots, v$$

were supposed equally probable; in the other, the probabilities of these same errors were supposed proportional to

$$1, \ldots, v-2, v-1, v, v+1, v, v-1, v-2, \ldots, 1,$$

hence outlining an isosceles triangle when plotted against the errors. For each of these distributions, Simpson provided generating functions for determining the probability that the sum of the errors of n observations should equal a given integer m. He was here giving the first application of the theory of probability to the investigation of observational errors. In both of the error distributions that Simpson considered, the errors of positive and negative sign occurred with equal frequency, and there was an upper limit to the size of the errors; under these conditions, the arithmetic mean of several observations could be expected to be a good approximation to the true value. Simpson's immediate aim was to refute the opinion that one careful observation was as reliable as the mean of a large number of observations. In the revised version of the paper he considered a continuous distribution along with the discrete distributions.

Johann Heinrich Lambert (1728–77) was another contributor to the theory of errors. The son of a poor tailor, he was entirely self-educated, making intensive studies in many areas of mathematics and philosophy. In his *Photometria* of 1760, which was devoted to laying the foundations for the measurement of the intensity of light, he described the properties of observational errors, remarking that the numbers of errors of opposite sign tend to equality as the number of observations is increased. He proved that an extreme observation should be rejected when it is considerably separated from the rest, because in this way the error of the arithmetic mean is reduced. For a continuous unimodal frequency curve, $\phi(x - x_0)$, he enunciated the principle that x_0, the "most probable" value of x, be determined by the condition that

$$\phi(x_1 - x_0)\phi(x_2 - x_0) \ldots \phi(x_n - x_0) = \max.$$

(This is now referred to as the condition of maxi-

mum likelihood.) In most cases, he stated, x_0 would not differ from the arithmetic mean.

In later works, Lambert carried further his discussion of these matters, introducing the term 'theory of errors', and undertaking to develop this theory as a distinct discipline.

Although the arithmetic mean of several observations had come to be considered the most reasonable approximation to the true value, the first attempt to justify this assumption appeared in 1756–57 (Simpson). Then, in 1776, the mathematician Joseph Louis Lagrange (1736–1813), then head of the mathematical–physical section of the Berlin Academy, published in the *Miscellanea Taurinensia* his memoir "Sur l'utilité de la méthode de prendre le milieu entre les résultats de plusieurs observations". Here he went beyond Simpson in considering many different frequency distributions and applying generating functions to the determination of the probability of errors in each case. For the cases considered he was thus able to compute the most probable value of the error committed by taking the arithmetic mean for the true value. He then turned to the general problem of deriving the most probable correction that must be applied to the arithmetic mean if the law of the frequency of the errors is known, and showed that the arithmetic mean is the most probable value of the unknown if the law of errors is an even function and unimodal. He did not propose a formula for the law of errors.

Further development of the method of averages

The theoretical investigations described in the preceding section had no immediate effect on the practice of astronomers. For them, the method of averages remained the accepted procedure for dealing with equations of condition. Bošković's proposal that the sum of the absolute values of the residuals should be minimized was not generally followed (although Laplace would make use of it in the study of the Earth's figure: see below). But certain refinements in applying the method of averages deserve notice.

In 1787, Pierre-Simon Laplace (1749–1827) presented to the Paris Academy of Sciences his memoir on the inequalities of Jupiter and Saturn – the memoir in which, for the first time, the positions of these planets since antiquity became derivable from the theory of gravitation with only a small margin of error. In fitting the constants of his theory to observations, Laplace used a refinement of Mayer's method. Comparing his theory with twenty-four observations, he compiled twenty-four equations of condition, each having the form

$$\Delta w + a_{i1}\Delta x + a_{i2}\Delta y + \ldots = r_i$$

where Δw, Δx, Δy, ... are differential corrections to the initially assumed values of the constants, the a_{i1}, a_{i2}, etc. are sines or cosines of arguments proportional to the time of the ith observation, and r_i is the 'residual', that is, the difference between the observed position and the position as predicted by means of the initially assumed values of the constants. Unlike Mayer, who had added his equations of condition together in disjoint groups, Laplace combined the same equations together in several different ways, always with a view to maximizing the coefficients. Thus he added all equations in which the coefficient of one particular differential correction was large, either positive or negative, but where the coefficient was negative, he multiplied the equation through by -1 before summation. The aggregation chosen in each case was dependent on *all* the equations. Repeating the same procedure for each term, he obtained as many equations as unknowns, in each of which one term was predominant in comparison with the others. It was then easy to determine the values of the unknowns satisfying the equations.

Evidently the method would fail if, in the equations of condition, some unknowns were to have only positive (or only negative) coefficients; in this case it would be necessary to subtract the arithmetic mean of the equations of condition from each of these equations. In applications to celestial mechanics, however, where the unknowns are normally multiplied by certain values of trigonometrical functions, such a case would almost never arise.

During the late 1780s and 1790s, Jean-Baptiste Joseph Delambre (1749–1822) worked on the construction of tables of the Sun, Moon, Jupiter, Saturn, Uranus, and the satellites of Jupiter; in all cases he followed Laplace's procedure in fitting constants to observations, although he did not mention Laplace. The procedure was adopted by other astronomers, and became the standard practice in the early years of the nineteenth century. Even after the invention of the method of least squares it was still used, mainly by English authors, for instance the Astronomer Royal G. B. Airy (1801–92) and his collaborators. The latest case of its use appears to have been in the correction of B. A. von Lindenau's tables of Venus carried out by James Breen, an assistant at the Greenwich and later at the Cambridge Observatory; the corrected tables were published in 1849.

Laplace's elaboration of Boškovič's method

In a memoir of 1789, and again in the third book of the second volume of the *Mécanique céleste*, published in 1799, Laplace treated the problem of the Earth's figure. First he derived a theoretical shape for the Earth, assuming that its parts had been originally fluid and subject to both centrifugal and gravitational forces. Then he undertook to compare this theoretical shape with observational determinations of lengths of degrees of the meridian, and of degrees of longitude, in different latitudes. For this purpose he had recourse to Boškovič's method; that is, he stipulated that the algebraic sum of the residuals should be zero, and that their sum taken without regard to sign should be a minimum. In the memoir of 1789 Laplace attributed the method to Boškovič; in the *Mécanique céleste*, Boškovič's name went unmentioned.

To determine the values of the unknowns meeting the stipulated conditions, Boškovič had used a geometrical construction. Laplace, in contrast, proceeded analytically, giving a long and tedious description of the algebraic steps of the process, and also providing a proof that his formulas yielded the required result.

The invention of the method of least squares

In a manuscript of the mid-1750s Simpson had posed, but did not further discuss, the following problem: to find a point N as the intersection of four lines drawn from four given points, in such a way that the sum of the squares of the distances from N to the four given points was a minimum. In 1778, in commenting on a memoir of Daniel Bernoulli, Euler suggested the principle of determining a measured value by, practically speaking, minimizing the sum of the squares of the errors, but did not pursue the implications of the principle. In the last decade of the eighteenth century and first decade of the nineteenth, a number of mathematicians independently hit upon the idea of determining the unknowns of a set of equations of condition by minimizing the sum of the squares of the residuals. This *method of least squares*, it was assumed, would yield the most probable values of the unknowns.

It seems the first to hit on the method in this form may have been Daniel Huber, who was born in 1768 in Basle, Switzerland, and became professor of mathematics at the university there in 1791. According to an obituary by Peter Merian in the *Verhandlungen der allgemeinen schweizerischen Gesellschaft für die gesamten Naturwissenschaften* (1830), Huber invented the method of least squares, but published nothing about it:

*Already in his early days (*frühen Zeiten*), through his own reflection, he discovered the method – later made known by Gauss and Legendre – of least squares, for the obtaining of the most probable results of a series of observations.*

Merian gives no date for Huber's invention of the method, but "frühen Zeiten" suggests it was before 1800.

In 1806, Adrien-Marie Legendre (1752–1833) published a book entitled *Nouvelles méthodes pour la détermination des orbites des comètes*; in an appendix bearing the date 6 March 1805, he expounded the method of least squares. This method, he stated, was in his opinion the most general and most accurate way of determining several unknowns from a larger number of equations of condition. He also showed that, in the case of a single unknown, minimization of the squares of the residuals implies the rule of the arithmetic mean. Further, justification he did not supply.

In 1808, Robert Adrain (1775–1843), a teacher of mathematics and principal of an academy in Reading, Pennsylvania, and later professor of mathematics at Queen's College (now Rutgers) in New Jersey, and at Columbia College, New York, published a paper in which he derived the normal law of errors and used it to establish the principle of least squares in the same way as Gauss was to do in 1809. The paper appeared in *The Analyst*, a short-lived journal of which Adrain himself was the publisher. Perhaps he was aware of Laplace's earlier work on error distributions; whether at this point he had read Legendre's memoir is unclear (his library contained the memoir). He included a number of applications of the principle of least squares to problems of navigation and land survey-ing. In 1818 in the *Transactions of the American Philosophical Society* he applied the same principle to the determination of the size and shape of the Earth by the adjusting of meridian arc measurements and pendulum observations. However, his work was little known in Europe before 1872, when J. W. L. Glaisher published a review of it in the *Memoirs of the Royal Astronomical Society*.

In 1809, in the second book of his *Theoria motus corporum coelestium*, the mathematician, physicist, and astronomer Carl Friedrich Gauss (1777–1855) published a probabilistic argument for the method of least squares. He remarked that he had used the method already in 1795. His reasoning, briefly outlined, went as follows. Let V_i $(i = 1, 2, \ldots, \mu)$ be functions of the ν quantities p, q, r, s, etc., to be determined by observation, with $\nu < \mu$. Further, let M_i be the values of the V_i given by observation. Since the observations contain error, we cannot suppose $V_i = M_i$ exactly, but any system of values of p, q, r, s, etc. is possible which gives values of the functions $V_i - M_i$ within the limits of observational error. Gauss's aim was to pick the most probable of these systems.

Suppose there is no prior reason to regard one observation as less reliable than another, and let the probability of an error $v_i = M_i - V_i$ in an observation be $\phi(v_i)$. Gauss assumed that $\phi(v_i)$ is a maximum for $v_i = 0$, is equal for opposite values of v_i, and vanishes for v_i sufficiently large. For ease of calculation, he proposed to treat ϕ as a continuous function. Using the (Bayesian) rule of inverse probability, he argued that the probability for a particular set of values of the V_i on the assumption of given values of the M_i would be equal to the product

$$\prod_{i=1}^{\mu} \phi(M_i - V_i) = \prod_{i=1}^{\mu} \phi(v_i) = \Omega. \qquad (4)$$

He then proposed to maximize (4) by setting the partial derivatives of Ω with respect to the ν unknowns equal to zero:

$$\frac{\partial \Omega}{\partial p} = 0, \quad \frac{\partial \Omega}{\partial q} = 0, \text{ etc.} \qquad (5)$$

Equations (5), divided through by Ω, become

$$\sum_{i=1}^{\mu} \frac{1}{\phi(v_i)} \cdot \frac{\partial \phi(v_i)}{\partial v_i} \cdot \frac{\partial v_i}{\partial p} = 0, \qquad (6)$$

with similar equations for q, r, s, etc. The system of equations (6) will have a determinate solution if the form of the function ϕ is known.

To determine ϕ, Gauss considered a direct measurement of a single quantity, so that all the V_i reduced to a single constant p. Moreover, in this case it was generally assumed that the arithmetic mean of the measurements yielded the most probable value, and on this assumption Gauss showed that (6) implied that

$$\phi(v) = K \exp\left(\frac{kv^2}{2}\right).$$

In order that $\phi(v)$ should have a maximum when v is 0, $k/2$ must be negative, and Gauss accordingly set it equal to $-h^2$. He introduced the normalizing factor $h/\sqrt{\pi}$, and set

$$\phi(v) = \frac{h}{\sqrt{\pi}} \exp(-h^2 v^2), \qquad (7)$$

which brings it about that

$$\int_{-\infty}^{\infty} \phi(v)\, dv = 1,$$

so that the total probability is 1. As Gauss explained, the number h is a measure of the precision of observations. Expression (7) has come to be called the normal (or "Gaussian") density function.

The product (4) now takes the form

$$\Omega = \left(\frac{h}{\sqrt{\pi}}\right)^\mu \exp\left[-h^2(\textstyle\sum v_i^2)\right], \qquad (8)$$

and it follows that for (8) to be a maximum, $\sum v_i^2$ must be a minimum. Hence, Gauss concluded,

that will be the most probable system of values of the unknown quantities p, q, r, s, etc. in which the sum of the squares of the differences between the observed and computed values of the functions V, V', V'', etc. is a minimum, *if the same degree of accuracy is to be presumed in all the observations.*

The principle can be generalized to the case in which the observations are of unequal precision, by making the sum $\sum h_i^2 v_i^2$ a minimum. In a final remark on the logical status of the method, Gauss affirmed that

this principle, which promises to be of most frequent use in all applications of mathematics to natural philosophy, must everywhere be considered as an axiom with the same propriety as the arithmetical

mean of several observed values of the same quantity is adopted as the most probable value.

Gauss further asserted that the alternative principle of Bošković and Laplace – the method of seeking a minimum of the sum of the absolute values of the errors – is unsatisfactory because it implies that v of the μ equations are exactly satisfied, while the remaining equations influence the result only insofar as they help to determine the choice of the equations to be exactly satisfied. This surprising result can in fact be proved, although Gauss did not prove it.

The justification for the method of least squares given by Gauss in his *Theoria motus* of 1809 remained popular over a long period, appearing in textbook presentations throughout the nineteenth and into the twentieth century. Gauss himself, however, soon came to reject it.

Since the method of least squares was discovered by no fewer than four scientists working – it appears – independently of one another, the question of priority arises. The first to publish the method was undoubtedly Legendre. Gauss no doubt knew the method about ten years earlier, and Huber may have found it even some years before Gauss, but there is no documentation for the discoveries before 1800. Among the four scientists who derived the method of least squares, only Gauss published a proof of it.

Later derivations of the method of least squares

In a memoir read to the Paris Academy in April 1810 Laplace undertook to derive a theorem, called since 1920 the *central limit theorem*, for the sum of identically distributed random variables. According to this theorem, such sums will, if the number of terms is large, be approximately normally distributed. A rigorous proof was obtained only in the twentieth century.

Only after writing his memoir did Laplace encounter Gauss's treatment of error theory in the *Theoria motus*. His reaction was swift: he wrote a *Supplement* to his memoir in which he gave a new rationale, different from Gauss's, for the choice of a normal distribution of errors, hence a different argument for the method of least squares. Suppose the number of observations of a quantity is large; let these observations be split into several groups, and take the mean in each group. Then in accord with the central limit theorem, Laplace assumed that these means would be normally distributed. The rest of his derivation, like Gauss's, involved maximization of the likelihood function.

In his *Théorie analytique des probabilités* (first edition, 1812), Laplace gave the following criticism of Gauss's derivation:

M. Gauss ... sought to derive this method [of least squares] from the Theory of Probability, by showing that the same law of the errors of observations which yields the rule of the arithmetic mean among several observations – a rule followed generally by observers – implies equally the rule of least squares of the errors of the observations .... But, as nothing proves that the first of these rules gives the most advantageous result, the same uncertainty exists in relation to the second. The investigation of the most advantageous way of forming final equations is without doubt one of the most useful in the Theory of Probability; its importance in physics and astronomy has led me to concern myself with it.

Laplace did not here note that his own formulas presuppose a large number of observations, whereas Gauss in his derivation made no such requirement.

In his *Essai philosophique sur les probabilités* (1814), Laplace disavowed all methods of adjusting observations used prior to the emergence of the method of least squares, including Bošković's method which he had previously held in high regard. In still later work it appears that he came to regard the normal distribution not merely as a limit but as the actual distribution of errors of directly measured quantities.

Meanwhile, Gauss responded to Laplace's criticism by abandoning the argument for the method of least squares given in his *Theoria motus*, and rethinking the subject statistically. No longer did he consider it appropriate to describe the arithmetic mean as in general "the most probable value"; thus in a letter of 1831 to Encke he urged:

... the problem of finding the most probable value ... intrinsically presumes knowledge of the error law, and it leads to the arithmetic mean only when the error law has the form e^{-kkxx}. To put it abstractly, the probability of the "most probable" value is still only infinitesimal ... and for that reason has little practical interest – much less than the value that on the average makes the

error least harmful. For this reason (as well as for other reasons ...), I have settled upon this latter principle, which should not be confused with the first one.

Gauss's new justification for the method of least squares appeared in his *Theoria combinationis* of 1823. He supposed the density function $\phi(x)$ of the errors to be unimodal and in most cases even. Hence,

$$\int_{-\infty}^{\infty} x\phi(x)\,dx = 0. \tag{9}$$

Then, as a measure of precision, he introduced m, "the mean error to be feared", or (simply) "the mean error", where

$$m^2 = \int_{-\infty}^{\infty} f(x)\phi(x)\,dx = \int_{-\infty}^{\infty} x^2\phi(x)\,dx. \tag{10}$$

In present-day terminology, (9) gives the 'mean value' or 'expectation of x, written $E(x)$, and (10) gives the 'variance', written $E(x^2)$.

The cornerstone of Gauss's new theory was the principle of minimal variance. The choice of an *integral* measure of precision or of error, as in (10), was, according to Gauss, expedient: it avoided the resort to maximized probabilities that were merely infinitesimal. But the selection of $f(x)$ in (10), he allowed, was arbitrary, except for the reasonable restriction that this function should increase more rapidly than the error itself.

Gauss's derivation involves finding multipliers for each of the equations of condition such that the "mean error to be feared" will be a minimum. His somewhat ponderous solution of this problem is tantamount to the solution by the method of least squares. In sum, he shows that the method of least squares leads to minimal variance and is thus justified.

In the *Theoria combinationis* Gauss articulated his objections both to his own earlier deduction of the least squares principle, and to Laplace's deduction of it by way of the central limit theorem, as follows:

The first is totally dependent on the hypothetical form of the probability of the errors; and as soon as that is rejected, the values of the unknown quantities found by the method of least squares really are not the most probable ones any more, not even in the simplest case of the arithmetic mean.... The second ... leaves us completely unenlightened as to what we should do for a moderate number of observations....

He expressed the hope that friends of mathematics would enjoy seeing how

the new justification given here reveals the method of least squares as the most suitable combination of the observations in general, not just approximately but in mathematical rigour, no matter what the function for the probability of the errors is, and no matter whether the number of observations is large or small.

Whereas in the *Theoria motus* he had characterized the least-squares solution as *maxime probabile*, in the *Theoria combinationis* he described it as *maxime plausibile*.

The new substantiation was not widely accepted during the nineteenth century. Even during Gauss's lifetime the so-called 'theory of elementary errors' came to the fore. According to this theory, the error of each observation is the result of a large number of tiny, component ('elementary') errors. Therefore – so the proponents of this theory argued – the central limit theorem (in one version or another) leads to the normality of the errors, and, consequently, to a Laplacian-type substantiation of the method of least squares.

Astronomers tended to assume that if the errors had many causes whose influence was of the same order of magnitude, the law of distribution would be approximately normal. Their opinion on this point was strengthened by findings in other branches of natural science, notably physics: in 1860 James Clerk Maxwell published his celebrated theorem stating that the velocities of gas molecules at equilibrium are normally distributed.

Gauss's substantiation of least squares in the *Theoria combinationis* was revived in 1899 by the Russian mathematician A. A. Markov, and in the twentieth century has come to be generally accepted.

Application of the method

Gauss was probably the first to make decisive and important applications of the method of least squares. He continued in later years to be assiduous in applying the method to astronomic, geodetical, and magnetic measurements – so much so that friends protested at the extent to which he involved himself in the operations for the triangulation of Hanover. He was a master of experimental science, an expert in the analysis of instrumental errors. The inauguration by F. W. Bessel (1784–1846) of a new era of precision in astrometry owed much to Gauss's teaching and inspiration.

Laplace also was a virtuoso in the application of the theory of errors, particularly in the detection of small effects comparable in magnitude with errors of observation. The following quotation from his *Théorie analytique des probabilités* indicates the importance such applications could have in the celestial mechanics of his day:

This [lunar] inequality, although indicated by the observations, had been neglected by most astronomers, because it did not appear to result from the theory of universal gravitation. But, when I had submitted its existence to the calculus of probability, it appeared to me to be indicated with such a strong probability that I felt obliged to investigate its cause.

The great achievements of celestial mechanics in the nineteenth century – the discovery of Neptune, the construction of planetary tables by Le Verrier and Newcomb (see Chapter 28) – were all to involve arduous and painstaking application of the method of least squares.

Further reading

E. Czuber, *Theorie der Beobachtungsfehler* (Leipzig, 1892)

O. B. Sheynin, articles in *Archive for History of Exact Sciences*, vols 7 (1971), 9 (1972), 11 (1973), 12 (1974), 16 (1976), 17 (1977), 20 (1979), 26 (1982), 29 (1984), 31 (1984)

Stephen M. Stigler, *The History of Statistics* (Cambridge, Mass., 1986)

I. Todhunter, *A History of the Mathematical Theory of Probability* (Cambridge, 1865)

PART VIII

The development of theory during the nineteenth century

28

The golden age of celestial mechanics

BRUNO MORANDO

with an appendix on the stability of the solar system by Jacques Laskar

Introduction

The purpose of this chapter is to give an account of the developments in celestial mechanics during the period that extends roughly from the death of Laplace in 1827 to that of Poincaré in 1912. Unquestionably, this period is the golden age of celestial mechanics, and the names that we shall meet with, Hamilton, Le Verrier, Jacobi, Hill, and so forth, and, of course, that of Poincaré himself, bear witness to the fact. However, if we compare this period to the one that preceded it, from Newton to Laplace, we get the impression that it was wholly dedicated to the perserving and skilful exploitation of the splendid achievements of the prior century. Hamilton and Jacobi perfected the analytical mechanics that Euler and Lagrange had founded. Delaunay, Le Verrier, and Newcomb calculated the perturbations of the motions of the Moon and planets by applying methods that derive directly from the works of Euler, d'Alembert, Clairaut, Lagrange, Laplace.... Indeed, Poincaré himself, in the introduction to his *Méthodes nouvelles de la mécanique céleste*, defined with accuracy the goal that this science set for itself in the nineteenth century:

The ultimate goal of celestial mechanics is to resolve the great question whether Newton's law by itself accounts for all the astronomical phenomena; the sole means of doing so is to make observations as precise as possible and then to compare them with the results of calculation. The calculation can only be approximate, and, moreover, it would serve no purpose to calculate more decimal places than the observations can make known. It is thus useless to require more precision from the calculation than from the observations; but one must not, on the other hand, require less.

It would be impossible to improve on this definition of the goals pursued by the celestial mechanicians of the nineteenth century; and it would be ungracious to reproach them for having followed a course approved by such an authority.

Yet the celestial mechanics of that epoch appeared in its own time, and still appears to us today, as a scientific monument of the highest importance. At first its methods consisted in the pure and simple application of methods already known. But they were then improved upon decisively, and from the progress achieved other branches of science were to benefit; this is the case, for instance, with the methods of resolving canonical systems as in the work of Delaunay. Nor should the results obtained be underestimated. The comparisons of observation with calculation of which Poincaré spoke ended by supplying an answer to the question he posed apropos the ultimate goal of celestial mechanics. The discovery by Le Verrier of an advance that he could not explain in the perihelion of Mercury led to the conclusion that the law of gravitation by itself fails to account for all the astronomical phenomena. This is a fact of the first importance, leading to the revolution, at once scientific and philosophical, that is the General Theory of Relativity. The nineteenth-century scientists had an unclouded faith in their science; but they took it as their task to accumulate materials for the theorists of the future, rather than to build a finished structure.

Finally we must not forget that through a large part of the nineteenth century celestial mechanics was the whole of astronomy. Fraunhofer's discovery of the solar spectrum dates from 1814, and the ensuing experiments of Kirchhoff from 1859 but the prodigious development of astrophysics to which they led came only later. For those who lived under Louis-Philippe and Queen Victoria, the discovery of Neptune was comparable in importance to the theory of black holes for us today, a triumphant achievement of the astronomers. And these astronomers, Le Verrier and Adams, were practitioners of celestial mechanics.

Celestial mechanics at this epoch was devoted almost exclusively to the problem of the movements in the solar system – the particular case of the *n*-body problem that Nature had placed most insistently before the astronomers. There were a few exceptions: the work of Edouard Roche (1820–83) on the tidal stability of celestial bodies under the action of the mutual attraction of the particles composing them, leading to his introduction in 1849 of the notion of 'Roche's limit'; the work of G. H. Darwin on the secular acceleration of the Moon; the work of Laplace and of Poincaré on the tides. Here we shall confine ourselves to what the astronomers were primarily concerned with: to formulate and solve the equations of the *n*-body problem for the particular cases where the bodies are planets or satellites of planets.

Equations and solutions

Isaac Newton, in his earliest enunciation of the law of universal gravitation, when considering the multiple attractions influencing the motion of each planet, expressed the opinion that "to define these motions by exact laws admitting of easy calculation exceeds, if I am not mistaken, the force of any human mind". In the second half of the eighteenth century Joseph Louis Lagrange (1736–1813) and Pierre-Simon Laplace (1749–1827) addressed the problem of the complete integration of the equations of motion for *n* bodies attracting one another two by two according to Newton's law, and their studies quickly showed the problem to be insusceptible to a general solution. Not till the very year of Poincaré's death, 1912, was there decisive progress on the theoretical level. This was due to the Finn Karl Sundmann (1878–1949). But Sundmann's method, which consists in developing the coordinates of the bodies, and also the time, in very slowly convergent series in terms of one and the same variable, is hardly applicable in practice. The nineteenth-century theorists, like those of today, struggled to cope with the problem in a variety of ways.

Approximate solutions of the equations of motion can be sought in terms of developments in series. This has been the chief mode of attack for the bodies of the solar system; it requires on the one hand that analytical mechanics be perfected, and on the other that the use of series be justified through the study of their convergence. Another approach is to investigate whether there exist functions of the coordinates and velocities that remain constant over the course of the motion, the 'first integrals' as they are called. Given such integrals it becomes possible to delimit the domains within which the motions can occur. A third mode of attack is to study the configuration that a system of material points had at a time infinitely far in the past, or that it will have at a time infinitely far in the future.

The works of Hamilton and Jacobi

In his *Mécanique analytique* (1788) Lagrange established general equations of mechanics' ('the Lagrangian equations') involving a function of the coordinates and the velocities called the 'Lagrangian'. Also, he established another system of equations especially suited to celestial mechanics. These 'equations of Lagrange' require that the body considered move very nearly in the Kepler-motion that Newton had shown to be the solution of the two-body problem. Such a motion is characterized by six constants determined observationally. In the perturbed motion Lagrange supposed these constants to be augmented by functions of the time having small amplitude or varying with extreme slowness. To determine these functions, he introduced a function of the positions of the mass-points called the *disturbing* or *perturbing function*. This function has played a fundamental role in the celestial mechanics of the nineteenth and twentieth centuries. It is associated with the notion of intermediary orbit (an orbit the departures from which are considered as perturbations) and with the method of variation of constants. For Lagrange the intermediary orbit was the Keplerian ellipse, but as we shall see later on, Hill utilized an intermediary orbit of a different kind to resolve the problem of the motion of the Moon.

Two mathematicians, Jacobi and Hamilton, introduced new concepts and new variables, permitting the Lagrangian equations to be transformed into systems that could be treated more advantageously. Carl Gustav Jacob Jacobi (1804–51), a German mathematician known especially for his studies of the elliptic functions, introduced into mechanics a set of differential equations called canonical equations. The six variables that characterize every problem of mechanics are here associated two by two in three pairs of variables 'canonically conjugated'; the system of differential equations they obey has an especially simple form.

Jacobi investigated the conditions under which a change of variables in such a system would preserve its canonical character. An appropriate change of variables can be a powerful means of resolving a system of differential equations, if it is possible to choose the new variables in such a way as to simplify the system. In his *Vorlesungen über Dynamik* (1866), Jacobi studied the motion of two bodies by a method now called 'Jacobian', which reduces the problem to that of finding a particular solution of a partial differential equation. He identified certain canonical variables, 'Jacobian variables', that are constant in the two-body problem but which by the method of variation of constants can be utilized in the *n*-body problem. Also, inquiring into the conditions under which a canonical system remains canonical under a change of variables, he showed that certain functional determinants, 'Lagrange's brackets' and also 'Poisson's brackets', must have a particular form. Poincaré later returned to this question and found an extremely simple condition, called Poincaré's condition of canonicity, which reduces to the statement that a certain differential form is a total differential.

The works of Jacobi are closely related to those of Sir William Rowan Hamilton, who was born in Dublin in 1805, was elected Andrews professor of astronomy and Astronomer Royal of Ireland at the age of twenty-two, and died at Dunsink (site of Dublin's Trinity College Observatory) in 1865. An eminent mathematician, founder of the algebraic theory of complex numbers and the theory of quaternions, Hamilton also made major contributions to mechanics. He showed that, among all the possible movements of a material point between two instants t_0 and t_1, the real movement is that which minimizes the integral of a certain function between these two instants. This is called 'the principle of least action' or 'Hamilton's principle'; it leads quite naturally to a system of canonical equations, the right-hand members of which are the partial derivatives with respect to the variables of a function called the 'Hamiltonian'.

If, following Hamilton and Jacobi, we write the equations of celestial mechanics in canonical form,

it is then necessary to resolve two difficult problems: to find how to express the Hamiltonian as a function of the canonical variables, permitting us to calculate the partial derivatives appearing in the right-hand members of the equations; and to find a method of integrating the equations.

Development of the disturbing function

To express the Hamiltonian of the motion as a function of the required variables, it is sufficient to express the disturbing function – defined initially in terms of the coordinates of the bodies in the system under consideration – as a function of these same variables. This is possible only if we represent the disturbing function as a sum of a number of terms, in which certain of the variables appear under cosine signs, while the conjugate variables appear in the coefficients of these cosines. To the analytic development of the disturbing function, there were many contributors during the nineteenth century. Laplace had introduced the functions called 'Laplace coefficients' to express as a trigonometric series the reciprocal of the distance between two planets, and the odd powers of this reciprocal. F. W. Bessel (1784–1846) was the first to make use of 'Bessel functions' in a treatise on the development of the disturbing function published in Berlin in 1824. In 1831 Peter A. Hansen – of whom we shall have more to say later – in his *Untersuchung über die gegenseitigen Störungen des Jupiters und Saturns* introduced certain functions, called Hansen's coefficients, which permit the development of the radii vectores of the planets, or any powers of these, as trigonometric series in the mean anomaly, the coefficients being each an entire series in the eccentricity. We note also the important role played by Legendre polynomials in all these questions; they had been introduced in Adrien-Marie Legendre's treatise on the *Figure des planètes* of 1782.

To Laplace is due the first expansion of the disturbing function for the planets to the third order of the eccentricities and inclinations. Pontécoulant carried it out to the sixth order in his *Théorie analytique du système du monde* (1829–46). Benjamin Peirce also published a development to the sixth order in 1849 in the *Astronomical Journal*. Le Verrier pushed the development to the seventh order in the first volume of the *Annales de l'Observatoire de Paris* (1855). In the development of the disturbing function, different angular variables – the variables fixing the position of the orbit in space and the position of the planet on the orbit – may be used. Hansen, in investigating the motion of the minor planet Egeria in 1859, employed the eccentric anomaly in place of the mean anomaly. George W. Hill, for his theory of Jupiter and Saturn in 1890, utilized the eccentric anomaly of one of the planets and the mean anomaly of the other. Simon Newcomb (1835–1909) in 1891 and 1895 gave developments to the sixth order, using first the eccentric anomalies of the four interior planets (Mercury, Venus, Earth, and Mars) then their mean anomalies.

Resolution of the equations of motion
Given the canonical equations and an expansion of the disturbing function to a certain order of precision, how does one obtain a solution of corresponding precision? In the case of the theory of a planet, one can seek an expansion in powers of the masses of the perturbing planets, as these are small relative to the Sun's mass; in the case of the Moon, the small parameter employed is the ratio of the Earth–Moon distance to the Earth–Sun distance, or a related quantity, the ratio of the mean motions. The coefficients of the powers of the small parameters are sums of a large number of periodic terms – sums it is not easy to obtain.

One of the methods devised consists in carrying out a change of variables on the canonical system, in such a way as to simplify it. Thus Delaunay in his theory of the Moon (to be described in a later section) reduces the Hamiltonian to an isolated periodic term, which is a linear combination of three angular variables and the time. This linear combination is taken as the sole angular variable of a new canonical system which is then integrable. This operation is repeated by taking successively the different periodic terms of the Hamiltonian. The method was greatly admired by Hill who in 1907 generalized its application in his memoir "On the extension of Delaunay's method in the lunar theory to the general problem of planetary motions". An important modification of Delaunay's method was made in 1916 by Hugo von Zeipel in his article "Recherches sur le mouvement des petites planètes", which appeared (in French) in a Swedish journal. The modification consists in introducing a generative function in order to effect one of Delaunay's operations, and simultaneously in eliminating several periodic terms even though they are not multiples of the same argument.

All these methods miscarry if the Hamiltonian contains resonant terms, that is, combinations of angular arguments in which the coefficient of the time is zero or nearly zero. This situation was studied by Hill and by von Zeipel in the memoirs referred to above, and also by Poincaré in Vol. 2 of his *Méthodes nouvelles de la mécanique céleste*, published in 1893.

A canonical system can be integrated if the variables are separable, but unhappily this is not the case for systems describing the motion of n bodies. One known case that is integrable – aside, of course, from the two-body problem – is that of a point attracted by two fixed centres; the solution was given by C. V. L. Charlier in Vol. 1 of his *Mechanik des Himmels*, which was published in Leipzig in 1902. The orbits obtained can be employed as intermediary orbits in more general problems of celestial mechanics, but this was not attempted, it appears, until much later, around 1960, when the device was made use of in the study of the trajectories of artificial satellites.

As stated earlier, however, the principal preoccupation of the celestial mechanicians of the nineteenth century – they were mainly astronomers rather than mathematicians – was to construct ephemerides that would give the positions of planets and satellites in as near agreement as possible with the observations. We shall accordingly pass in review all, or nearly all, the theories of the Moon and of the planets and satellites developed during this period – theories that are characteristic of the epoch, and have moreover continued to serve as the basis for the construction of ephemerides into the third quarter of the twentieth century. But first we shall recount the circumstances in which the most spectacular result of the century was obtained: the discovery of Neptune.

The discovery of Neptune

If there is a single outstanding event in the history of celestial mechanics in the nineteenth century, it is surely the discovery of Neptune. The difficulty of the problem to be resolved and the boldness required in addressing it, the excitement produced by the telescopic discovery of the planet very close to the predicted position and, last but not least, the

28.1. Urbain Jean Joseph Le Verrier.

28.2. John Couch Adams (engraving by Samuel Cousins from the painting by Thomas Mogford, 1851)

piquancy added to all this by the controversy between France and England over priority in the discovery – there you have what made the event a sensational affair, widely echoed in the daily press of the time.

The principal protagonists in this drama were two astronomers very different in character, but not in their passion for scientific research.

Urbain Jean Joseph Le Verrier (Figure 28.1) was born at St Lô (département de la Manche) in 1811. As student at the École Polytechnique, then as *élève-ingénieur* of the state tobacco company, he pursued an interest in chemistry. But when a position as *répétiteur d'astronomie* became vacant at the École Polytechnique in 1837, he accepted it, devoting himself henceforth to celestial mechanics and astronomy. In 1839 he published a memoir on the secular variations of the planets. Made famous by his discovery of Neptune, in 1854 he was appointed head of the Paris Observatory, and

remained in that position till 1870, then occupied it again from 1874 until his death in 1877. Despite grave difficulties, owing in large measure to his authoritarianism and difficult disposition, he was able to effect a lasting transformation in the observatory and to turn it into a scientific institution of the first rank (see Volume 4A, p. 116). The principal object of his own scientific writings was the theory of the planets, of which we shall have more to say in a later section. Le Verrier was an ambitious, autocratic man, avid for honours. He played an important role in the inner councils of the Academy of Sciences, and also in politics, for he served as deputy, senator, and president of the General Council of the Département de la Manche.

A quite different man was John Couch Adams (1819–92) (Figure 28.2). In 1839 he was awarded a scholarship to enter St John's College, Cambridge. In 1850 he obtained a teaching post at the University of St Andrews, and in 1859 he was named Professor of Astronomy and Geometry at Cambridge. In 1861 he succeeded James Challis as director of the Cambridge Observatory, and organized there a series of meridian observations. After his work on the perturbations of Uranus he devoted himself to the lunar theory and, in particular, brought to light an error committed by Laplace and, following him, Plana (more will be said of this in a later section). Later he devoted himself largely to mathematics and even calculated Euler's constant (which occurs in the summation of harmonic series) to 236 decimals! Self-effacing and timid, and also concerned about the cost of keeping up appearances consonant with the title, Adams in 1847 refused a knighthood from Queen Victoria.

The planet Uranus had been discovered by William Herschel in 1781, but it was then realized that earlier observations of it had been made by astronomers who took it to be a fixed star. By 1820 there were thus available for analysis some forty years of meridian observations, together with nineteen observations made between 1690 and 1781 by John Flamsteed, P.-C. Le Monnier, James Bradley, and Johann Tobias Mayer. It was Alexis Bouvard (1767–1843), long an assistant of Laplace and the one who had carried out the detailed calculations of his *Mécanique céleste*, who undertook the analysis. In 1808 he had published tables of Jupiter and Saturn, and for years he had supervised the calculation of the ephemerides of the *Annuaire du Bureau des Longitudes*. In 1821 he undertook to recalculate the tables for the motion of Jupiter and Saturn, and to add thereto tables of the motion of Uranus, following the methods developed by Laplace in the *Mécanique céleste* to take account of perturbations. He found it impossible to represent the observations with the expected accuracy, the discrepancy between theory and observation being in some cases as high as 65″. The error, he decided, must lie in the older observations, but on discarding all those made before 1781, he found that errors of about 30″ still remained. He wrote: "... I leave to the future the task of finding out whether the difficulty ... in fact has its source in the inaccuracy of the older observations, or derives from some foreign and hitherto unperceived action to which the planet is subject." But the discrepancies between the theory and the observations continued to increase and reached nearly 2′ in 1845 (see Figure 28.3), despite Bouvard's having discarded the older observations. Beginning in 1835, G. B. Airy, then Eugene Bouvard (nephew to Alexis), François Arago, Bessel, and Sir John Herschel entered into discussions of this subject, and proposed to search for a hypothetical perturbing planet.

Other hypotheses also made their appearance. The discrepancies, it was proposed, could be due to the resistance of the aether, or to a large and unobserved satellite of Uranus, or to a comet that had perturbed the planet's progress. It was even proposed that at such a great distance from the Sun the law of universal gravitation might no longer be exact. Le Verrier in a memoir of October 1846 – the final memoir that he devoted to the discovery of Neptune – disposed of these hypotheses. The resistance of the aether would have manifested itself elsewhere in the solar system and would have produced observable perturbations in bodies of low density. The perturbations due to a satellite of Uranus would have been of short period, contrary to what was observed. Finally, the encounter of a comet with Uranus would have produced its effect during a limited time between the epoch of the older observations and that of the observations used by Bouvard; why then should the discrepancies continue to grow? As for the suggested failure of Newton's law, many, including Le Verrier, simply did not wish to believe in it. There thus remained the hypothesis of the unknown planet, but it was unthinkable to search for it at random. As Le Verrier remarked much later:

How were the astronomical observers to discover, in the immense extent of the sky, the physical cause of the perturbations of Uranus, unless their task could be provided with limits, and their search confined within a determinate region? Who among them would have undertaken to seek a telescopic star successively in the twelve signs of the zodiac? It was thus necessary to begin by showing that the observations should be concentrated within a small number of degrees. Only thus could one expect that the vigils of the observers would not be in vain, and that in the not distant future physical astronomy would be enriched with a star of which theoretical astronomy had previously revealed the existence and fixed the position.

Adams's interest in the divagations of Uranus arose earlier than Le Verrier's. Already in July 1841, while still an undergraduate, he had formed the design of investigating the perturbations of Uranus to see whether they might be caused by a hypothetical perturbing planet. On completing his degree in 1843 he set to work by reviewing Bouvard's tables, correcting errors and pushing further than Bouvard the calculation of the perturbations

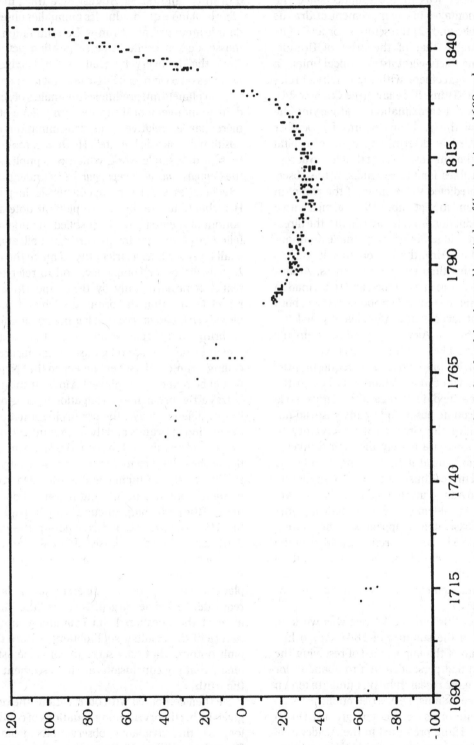

28.3. The discrepancy (in arc-seconds) between the predicted and the observed longitudes of Uranus between 1690 and 1845.

due to known planets. Then he made the following two hypotheses, both of them reasonable: (1) the unknown planet was in the ecliptic, since the latitudinal anomalies in the movement of Uranus were negligible; and (2) the semi-major axis of its orbit was double that of the orbit of Uranus, therefore some thirty-eight astronomical units – a hypothesis in agreement with the empirical relation called 'the Titius–Bode law' (see Chapter 24). Using successive approximations and varying the parameters of the problem, Adams by October 1845 succeeded in determining a set of orbital elements for the unknown planet that led to acceptably small values for the residuals of the observations. He predicted the position of the planet in the heavens, and in September 1845, armed with a letter of introduction from James Challis, the director of the Cambridge Observatory, made an unannounced call on Airy, the Astronomer Royal. As Airy was at this time away on business, Adams returned on a day in late October, and twice made a further attempt to see the Astronomer Royal, both times without success; but before leaving he left a three-page letter for Airy, specifying the elements of his hypothetical planet and its position.

Airy responded on 5 November, requesting the source of the value that Adams ascribed to the radius vector – he did not regard the Titius–Bode law as an adequate basis, and in any case did not want to involve the Greenwich Observatory in what he considered a merely theoretical investigation. Adams judged it futile to reply, and the matter rested there. Thus the honour of the discovery was lost to Adams and to England. The works of Adams on the problem of the hypothetical planet were later published in an appendix to the *Nautical Almanac* for 1851. The theoretical position of the planet at the moment of Galle's discovery of it (of which more will be said below), as calculated from Adams's orbital elements, differs from the observed position by 2° 27′.

Throughout this time Le Verrier was working assiduously. In the summer of 1845 Arago had convinced him of the importance of resolving the problem, and had persuaded him to abandon for the time being the research he had undertaken on comets. Unlike Adams, Le Verrier published successive accounts of his work as it progressed. On 10 November 1845 he presented to the Academy of Sciences a "Premier mémoire sur la théorie d'Ura-

nus"; this was followed on 1 June 1846 by a memoir entitled "Recherches sur les mouvements d'Uranus" and on 31 August by a third memoir "Sur la planète qui produit les anomalies observées dans le mouvement d'Uranus. Détermination de sa masse, de son orbite et de sa position actuelle". After the discovery he read to the Academy of Sciences on 5 October 1846 a final memoir entitled "Sur la planète qui produit les anomalies observées dans le mouvement d'Uranus. Cinquième et dernière partie, relative à la détermination de la position du plan de l'orbite". He then combined all these into a single work, which was published in the *Connaissance des temps pour 1849* under the title "Recherches sur les mouvements de la planète Herschel (dite Uranus)". In a piquant note at the bottom of the first page he justified the title in the following fashion: "In my further publications, I shall consider it as a strict duty to make the name *Uranus* disappear completely, and to refer to the planet henceforth only by the name *Herschel*. I regret keenly that the printing of this memoir is already so far advanced that it is not possible for me to bring it into conformity with this decision, which I shall observe religiously in the future." In calling Uranus "Herschel" he hoped that Neptune would be named after himself. More on this later.

Le Verrier began his investigation by developing, in two different ways, the perturbing function for the motion of Uranus and then, after integration of the second members of the equations, the inequalities of the orbital elements of Uranus as subject to the attractions of Jupiter and Saturn. One of the methods consisted in utilizing the literal development of the perturbing function given by Laplace in his *Mécanique céleste*, and calculating the partial derivatives; the other consisted in calculating the developments numerically by giving particular values to the eccentric anomalies of the interacting planets. The use of two different methods was requisite, Le Verrier urged, because "the importance of the question I am examining, and the nature of the results I shall obtain, require that I omit nothing that may serve to convince astronomers that my conclusions are in agreement with the truth".

He then compared with observations the results given by his theory, forming equations of condition for all the available observations from 22 December 1690 to 27 December 1844. He found residuals as large as 20″.5 – unacceptably high. It was at this point that he introduced the hypothesis of an unknown perturbing planet. Like Adams, he supposed it to be in the ecliptic, with a solar distance double that of Uranus; thus the unknowns remaining were the mass of the planet, the eccentricity and longitude of the perihelion of its orbit, and its longitude at epoch. Actually, in place of the eccentricity and longitude of the perihelion, he used two other variables that Laplace had introduced, namely h', the product of the eccentricity and the sine of the longitude of the perihelion, and l', the product of the eccentricity and the cosine of the longitude of the perihelion. He calculated the perturbations as a function of these unknowns and introduced them into the equations of condition where, except in the case of the mean longitude at epoch, he could give them a simple form. He then expressed m' (the unknown planet's mass), $m'h'$, $m'l'$ as functions of the mean longitude at epoch ϵ'. In order that m' not be negative, it was necessary that ϵ' fall within certain limits. The next step was thus to give a value to ϵ' within these limits such as to reduce the residuals of the observations to a minimum. Finally, he improved the precision of the results by utilizing observations of Uranus at its quadratures between 1781 and 1845 – observations that he had set aside at an earlier stage while retaining only the observations made in the neighbourhood of opposition, for which the equations of condition were easier to form. He also varied the semi-major axis, which he had taken at first as double that of Uranus.

Thus on 31 August 1846 he presented to the Academy of Sciences the following results: mass, $\frac{1}{9322}$ times the mass of the Sun; semi-major axis, 36.1539 astronomical units; eccentricity, 0.10761; longitude of the perihelion on 1 January 1847, 284° 45′ 8″; heliocentric longitude of the planet on the same date, 326° 32′.

It remained to find the planet. This was not as simple as it might seem. Most observers, particularly in France, appear to have been sceptical and unwilling to commit themselves. It has often been claimed that Le Verrier turned to Johan Gottfried Galle (1812–1910), astronomer at the Berlin Observatory, because he knew that the Berlin Observatory had at its disposal the good star charts of Bremiker, but there is no evidence for this; Le Verrier makes no mention of it in his letter to Galle,

and as we shall see, Galle himself did not think of referring to the star chart until the last moment. Galle had sent his doctoral thesis to Le Verrier a short time before, and so was known to him. Le Verrier's letter, dated 18 September, arrived on 23 September and, on the evening of that very day, aided by the young astronomer Heinrich Louis d'Arrest, Galle turned the excellent Fraunhofer telescope of 9-inch (23-cm) aperture toward the indicated region of the heavens; but he did not find the disk of 3 arc-second diameter that, according to Le Verrier, the planet ought to present. There is nothing astonishing in this: as the theoretical resolving power of his instrument was 0″.5, a particularly clear sky would have been required in order to see a disk so small; a magnification of at least 300 to 400 would have been required, and this would have limited the field to several minutes of arc – very little when it is a matter of finding an unknown object!

In was at this point that d'Arrest suggested comparing the field with a chart. Rummaging through the drawers they came on the Berlin Academy's Star Atlas charts, which had been compiled by Carl Bremiker (1804–77), inspector for the minister of commerce of Prussia, and a collaborator in the calculations of the ephemerides of the *Berliner Astronomisches Jahrbuch*. These excellent charts had been printed late in 1845 but had not yet been distributed to foreign observatories. Since the hypothetical planet was within the constellation of Capricorn, it was the chart Hora XXI, corresponding to the zone of 21 hours of right ascension, that they took for the comparison. Galle returned to the telescope and at length found, some 52′ from the position predicted by Le Verrier, a star of magnitude 8 which did not appear on Bremiker's chart. The following day the star had shifted: it was indeed the new planet (Figure 28.4).

On 25 September Galle wrote to Le Verrier a letter that has remained famous: "Monsieur, the planet of which you indicated the position really exists [réellement existe] ...." (Amusingly for us, in citing this letter in an appendix to his final memoir, Le Verrier corrected Galle's French by making him say: "existe réellement".)

The news of the discovery of a new planet resounded throughout Europe like a clap of thunder. The journalists and the cartoonists seized upon it. Airy, meanwhile, had not remained entir-

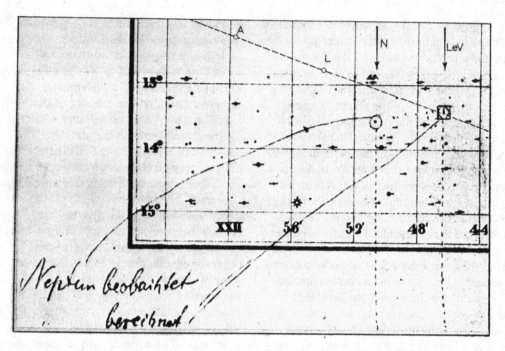

28.4. A section of the newly completed chart Hora XXI used by Galle and d'Arrest to locate the planet predicted by Le Verrier. (The ecliptic and various letters have been added.) The observed position of Neptune is indicated by N, the prediction of Le Verrier by LeV, and that of Adams by A. The handwritten notes are said to be by Galle.

ely inactive since receiving Adams's note. He had read and admired Le Verrier's publications, and on 29 June 1846, at a meeting of the Board of Visitors of Greenwich Observatory, he made known his desire that an observational search be instituted. If we believe what Challis wrote in 1847 in a letter to the editor of the *Astronomische Nachrichten*, it was Le Verrier's publication of 1 June 1846 that persuaded Airy that a search should be begun:

M. Le Verrier by an investigation published in June last, obtained almost precisely the same heliocentric longitude which Mr Adams had arrived at. This coincidence from two independent sources very naturally inspired confidence in the theoretical deductions, and accordingly Mr Airy shortly after suggested to me the employing of the Northumberland Telescope of this Observatory in a systematic search after the planet.

Challis observed at the Cambridge Observatory beginning on 29 July; he did not have very precise charts and was reduced to comparing from one day to another the relative positions of the stars of the field with the aim of discovering a proper motion in some one of them. On 4 and again on 21 August he had Neptune in the field of view and noted carefully its position without realizing that it was the sought-for planet. On 29 September, six days after the discovery that Galle had made on the basis of Le Verrier's prediction, he was again seeking the planet.

The history of the discovery of Neptune must be completed by some account of the controversies to which it gave rise. The name *Neptune* that was in due course adopted did not, of course, please Le Verrier who had hoped to give his own name to the planet. We have seen him playing the saint in attempting to impose the name *Herschel* on the planet Uranus, obviously not without a hidden motive. Arago agreed to support him and promised never to refer to the new planet other than as Le Verrier's. But the names *Janus* and *Oceanus* had been proposed, and Le Verrier himself claimed that the Bureau des Longitudes had chosen the name Neptune – which it does not appear to have done if we are to believe the recorded minutes of its meetings.

Although it does not strictly pertain to the history of celestial mechanics, we must say a word about the controversy that arose between France and England concerning priority of discovery.

Before Le Verrier had even begun his investigation, Adams had completed calculations that would have led to the discovery, but these calculations remained in a drawer, while Le Verrier published the results of his investigations and informed an observer where exactly the planet was to be searched for. A letter from Sir John Herschel to the London journal *The Athaeneum* of 3 October 1846 opened the controversy. The role played by Airy was much criticized: his negligence or rather his prejudice against the work of the young and little known astronomer that Adams then was deprived England of a great discovery. In the issues of the *Astronomische Nachrichten* beginning with no. 585 (1847) one can read Airy's attempts to justify himself.

Today it is generally agreed that equal merit should be assigned to Adams and to Le Verrier. The problem that they were able to resolve, each in his own way, was of great difficulty and importance; its solution constitutes a landmark in the history of science. The two antagonists themselves became good friends after their meeting in Oxford in 1847.

Another controversy, more technical but likewise dictated by envy, arose concerning the circumstances of the discovery. It became possible to determine the true orbit of Neptune on the basis of observations – all the better after it was found that Michel de Lalande (nephew of Jérôme) had twice observed that planet, while taking it for a star, on 8 and 10 May 1795. The discovery of the satellite Triton by William Lassell in 1846 made it possible, in addition, to calculate the mass of the new planet with precision. It was then found that between the reality and the results obtained respectively by Adams and Le Verrier there were notable differences (Figure 28.5). The mass of Neptune is 17 times that of the Earth; Le Verrier had found it 32 times, Adams 45 times. The semi-major axis of the orbit is 30.11 astronomical units, while Le Verrier and Adams, taking as their point of departure the value 38 given by Bode's law, had found it much greater. But this error was compensated by the large eccentricity attributed to the orbit (0.1 in place of 0.01), which brought it about that in the neighbourhood of the conjunction of Uranus and Neptune of 1821, the time at which the perturbations would be especially great, the three orbital arcs (the real orbit, Le Verrier's orbit, Adams's orbit) coincided very nearly, the errors in the distance being compensated by the errors in the mass.

The controversy on this point, which seems futile today, was quite the rage for a while, above all due to the impetus given it by a certain Jacques Babinet, a member of the Academy of Sciences, who went so far as to say that Adams and Le Verrier had discovered nothing at all. Babinet even asserted that there were two perturbing planets, one being a certain Hyperion of which, by means of some ridiculously abridged calculations, he claimed to supply the orbital elements!

The discovery of Neptune by pure calculation was the grand triumph of celestial mechanics as founded on Newton's law of gravitation. The discovery was scientifically important, and the circumstances under which it occurred gave it a spectacular character, such as to make a deep impression on the public imagination. Yet Le Verrier himself, as we shall see later, was the first to put his finger on what would prove a failure of the theory of gravitation, when he studied the motion of the perihelion of Mercury. He did not, however, dare to doubt this law to whose triumph he had contributed so much.

The theories of the movement of the Moon

The Moon is close to the Earth and is of relatively low brightness; hence its displacement among the stars is rapid and easily observable. This explains why a good many of the idiosyncracies of its motion, such as the rotation of the line of nodes the equation of centre, and the evection, became known to Hipparchus, Ptolemy, and the Arabic astronomers. Tycho Brahe observed further inequalities, but it was of course necessary to await Newton and those who pursued celestial mechanics in the eighteenth century before seeing the emergence of mathematical theories of the motion based on the law of universal gravitation.

To what point had the theory advanced at the time of Laplace's death? We recall that Alexis-Claude Clairaut in 1752 published a theory of which the precision with respect to the Moon's position was of the order of $1'.5$. The theories of Jean le Rond d'Alembert and Tobias Mayer and the two theories of Leonhard Euler had a comparable precision. The theory that Laplace published in 1802 had a precision of $0'.5$, and also the merit of partly explaining the secular acceleration (we

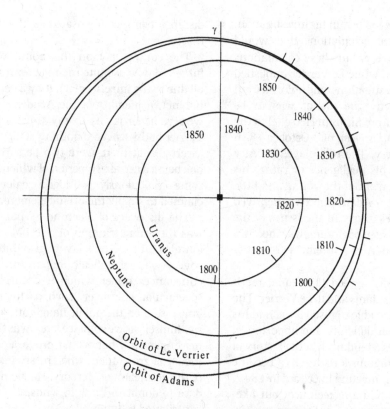

Le Verrier: Semi-major axis 36.2 a.u. Perihelion 285° Eccentricity 0.11
Adams: 299° 0.12 [earlier 0.16] 38.4 a.u. [earlier 316°]
Neptune: 30.1 a.u. (almost perfect circle)
Uranus: between 17.3 and 20.1 a.u. Eccentricity 0.05 Perihelion 170°

28.5. The observed positions of Uranus and Neptune, and the positions of Neptune expected on the basis of the calculations of Adams and Le Verrier.

return to the latter subject in a later paragraph). The tables published by J. T. Bürg in 1806 and those published by J. K. Burckhardt in 1812 were based directly on Laplace's theory. It is starting from this date that we shall study the post-Laplacian lunar theories.

The theories of the Moon between Laplace and Delaunay

In 1820 the Academy of Sciences of Paris set as a prize problem the motion of the Moon. Two theories were entered in the competition, one by Damoiseau and the other by Plana.

Marie Charles, Baron de Damoiseau (1768–1846), an officer of artillery who had emigrated in 1792, returned to France after the Restoration and became director of the observatory of the École

Militaire in Paris. His lunar theory, first published in 1824, attempted no more than to push to a higher order of precision the kind of calculation instituted by Laplace.

The Italian Giovanni Plana (1781–1864) had been a student of Lagrange at the École Polytechnique. From 1811 until his death he was professor of astronomy at the University of Turin. His lunar theory, which was developed in literal notation throughout, was published at Turin in 1832. Like Damoiseau's it was based on the theory of Laplace.

The theory of Siméon-Denis Poisson (1781–1840) was published in 1833. It is not a complete theory of the Moon's motion, for Poisson was unable to carry very far the calculations to which it led him. He supposed in a first approximation that the semi-major axis, eccentricity, and inclination of the orbit were constants and that the mean longitude of the Moon, that of the perigee and that of the node were linear functions of the time. These quantities were substituted in the second members of equations which, on being developed, were integrated term by term. He thus obtained better approximations for the semi-major axis, eccentricity, etc., and these being substituted into the second members, the whole process was then repeated. The calculations, given the limited means available at the time, quickly became too complicated; much later the method was to be usefully employed in certain problems of celestial mechanics, once the drudgery of the calculations could be turned over to electronic computers.

Next to be mentioned is the theory of Pontécoulant. Philippe, Comte de Pontécoulant (1795–1874), a scientist of great merit of whom there will be more to say when we come to the theories of Jupiter and Saturn, published his first memoir on the Moon's motions in 1837. His theory dates from 1846, but revisions were published in the *Monthly Notices of the Royal Astronomical Society* for 1860, and in the *Comptes rendus* of the Academy of Sciences of Paris for 1862. The variables that Pontécoulant used were the longitude, latitude, and radius vector of the Moon; already in the approximate orbit with which he began he introduced the movement of the node and that of the perigee. His theory is entirely literal, that is, he does not assume as known *a priori* the numerical values of any of the variables of the problem.

The most decisive progress was brought about by Peter A. Hansen (1795–1874), one of the great German theorists of nineteenth-century astronomy. At first a watchmaker, he learned French, Latin, and mathematics during his leisure hours. His learning and enterprise brought him to the attention of the Danish astronomer H. C. Schumacher, founder of the *Astronomische Nachrichten*, who had engaged him as a calculator; so Hansen began his career as a researcher. In 1825 he succeeded Encke as director of the observatory of the Duke of Mecklembourg at Seeberg near Gotha, and here he remained until his death. His theory of the Moon was published in 1838 at Gotha under the title *Fundamenta nova investigationis orbitae verae quam Luna perlustrat*. Modifications were published in 1861 and 1864. Hansen himself put the theory in tables which were published in London in

1857 under the French title *Tables de la Lune construites d'après le principe newtonien de la gravitation universelle*. These tables were employed from 1862 onwards for the calculation of ephemerides. As intermediary orbit Hansen chose an ellipse of fixed dimensions, located in the mobile plane of the real orbit, and with uniformly turning perigee. He then calculated the perturbations of the mean anomaly and of the radius vector in this ellipse. The procedure was an advance over the method used and extolled by Laplace in his planetary theories (Hansen employed the same procedure to calculate the perturbations of Saturn and Jupiter).

The theory of Delaunay

Charles Delaunay (1816–72) constructed the most precise and complete of all the literal theories of the Moon's motion, a prodigious piece of work requiring twenty years to finish; it was published in 1860 and 1867 in two huge volumes of the *Mémoires* of the Academy of Sciences. His method was first explained in 1846 in a "Mémoire sur une nouvelle méthode pour la détermination du mouvement de la Lune", then generalized nine years later in "Sur une méthode d'intégration applicable au calcul des perturbations des planètes et de leurs satellites". After Delaunay's death, Airy and later Henri Andoyer (1862–1929) introduced certain corrections into the theory, which was put into tables by R. Radau. It was used for the lunar ephemerides of *La Connaissance des temps* from 1915 to 1925, but was then supplanted by the theory of Hill–Brown (to be dealt with below). The truth is that Delaunay's theory, admirable as it is, cannot, because of the tremendous calculations it entails, lead to sufficiently precise lunar ephemerides.

Delaunay had studied at the École Polytechnique from 1834 to 1836, receiving the Laplace prize which consisted of the complete works of Laplace; the award led to his deciding on mathematics as a career. A mining engineer to begin with, he wrote a dissertation on mathematics and became a professor of the Faculty of Sciences of the University of Paris. Besides his theory of the Moon he published calculations on the second-order terms in the motion of Uranus, and an article on the secular acceleration of the Moon, of which we shall have more to say in the section devoted to this topic.

He was a member of the Royal Society and of the Academy of Sciences; the Academy's *Procès-ver-*

baux re-echo with the strife between him and Le Verrier, whose authoritarianism he opposed. He replaced Le Verrier as director of the Paris Observatory when the latter was removed from the position in 1870, but he died in 1872, drowned at the harbour of Cherbourg while on a promenade with his cousin.

We have already referred to Delaunay's method in the section on the problem of *n* bodies. He employed a canonical system in variables called 'Delaunay variables', and by 57 successive changes of variable eliminated the most important terms of the disturbing function, then eliminated the less important terms by simplified operations. In the end he obtained the development in literal form of the longitude, latitude, and parallax of the Moon to the seventh order in relation to the small parameters (ratio of the mean motions of the Sun and the Moon, eccentricities of the terrestrial and lunar orbits, inclination of the lunar orbit). Because of the slow convergence of certain terms, however, he found it necessary to add complementary terms, extrapolated and calculated in numerical form (some of these terms exceed 1" of arc).

If we accept the comparisons made by F. F. Tisserand and Newcomb, we can judge the precision of the lunar theories reviewed so far by saying that the theories of Damoiseau and Pontécoulant had errors of the order of 4"; those of Hansen and Delaunay, errors of the order of 1".

The theory of Hill–Brown

The works of Hill were the point of departure for a theory of the Moon's motion constructed by E. W. Brown. The tables that Brown derived from the theory were published in 1919 and served as the basis for the British and American lunar ephemerides until 1959, when they were refined by W. J. Eckert.

George W. Hill (1838–1914) (Figure 28.6) was one of the great American astronomers of the latter part of the nineteenth century. He was raised in the countryside at a time when transportation was slow and uncertain, and it was by the good luck of his going to Rutgers College and having Theodore Strong as a mathematics teacher there that he was put in the way of reading Nathaniel Bowditch's English translation of Laplace's *Mécanique céleste*. He went on to read in French the works of Poisson, Pontécoulant, Lagrange, Legendre, and others. Also through Strong's influence he became an

28.6. G. W. Hill.

admirer of Euler. In 1861 he began to work for the *American Ephemeris and Nautical Almanac* in Cambridge (Massachusetts).

When in 1877 Newcomb became director of the Nautical Almanac Office, he transferred the operation to Washington, and, undertaking to remake the theories and tables of the motions of the Moon and planets, he turned to Hill for assistance.

The new method that Hill proposed for the study of the lunar motion was published in 1877 in a memoir entitled "On the part of the motion of the lunar perigee which is a function of the mean motion of the Sun and Moon", and then in a second memoir of 1878 entitled "Researches in the lunar theory". Euler's influence on Hill is seen here in his utilizing, as had Euler in 1772, a system of rotating axes. In setting out the problem Hill made the following simplifications: the Sun describes, round the centre of mass of the Earth–Moon system, a circular orbit with uniform motion; the perturbing function of the Moon's movement is limited to terms independent of the ratio of the semi-major axis of the lunar orbit to the semi-major axis of the terrestrial orbit. The latter condition reduces to saying that the radius of the circular orbit described by the Sun is infinite.

In his system of rotating axes Hill placed the axis

of abscissas in the direction of the Sun. He obtained a differential equation of which he sought a particular solution, dependent solely on the ratio m of the mean motion of the Sun to the mean motion of the Moon. The numerical value of this ratio, being very well known owing to numerous observations from Antiquity onwards, could be introduced at the very start into the equations of the curve that Hill found as a solution for his differential equation. The curve, called the intermediate orbit of Hill, is a closed oval centred at the origin in the plane of the rotating axes. When we later encounter Poincaré's theory of periodic orbits, we shall see that Hill's intermediate orbit is a particular case of Poincaré's periodic orbits of the first kind. Starting from the intermediate orbit of Hill, one can introduce the perturbations that depend on the eccentricity of the Moon's orbit, then those that depend on its inclination to the plane of the Earth's orbit about the Sun.

Ernest W. Brown (1866–1938) was born in England and died in the United States, to which he had emigrated in 1891 and where, as a professor at Yale University, he ended his career. At the instigation of G. H. Darwin, Brown attacked the problem of the Moon's motion, using Hill's method. The theory was completed in 1908; the tables derived from it were published in 1919. Hill had calculated all the terms of which the coefficients were greater than a hundredth of an arc-second, but the precision of the theory was in fact less. In particular, Hill had encountered inequalities in the longitude that would only be explained through recognizing the non-uniformity of a scale of time based on the rotation of the Earth.

The satellites of the Planets

The two satellites of Mars, Phobos and Deimos, were discovered by Asaph Hall in 1877. They are very small and very close to the planet, so that to observe them was difficult. Since these satellites move very nearly in the plane of Mars's equator, which is itself inclined by 27° to the plane of the orbit of Mars about the Sun, the question arises whether this arrangement is stable. The question was investigated by Adams in 1879. He, and Tisserand after him, showed that whatever reasonable value one adopts for the flattening of Mars, the orbits of the satellites remain very close to the planet's equatorial plane.

"The determination of the motions of the Gali-lean satellites of Jupiter", wrote Tisserand in 1896, "constitutes one of the most beautiful problems of celestial mechanics." These four bodies revolve round Jupiter while attracting each other, just as do the planets in revolving about the Sun, but like the Moon in its motion round the Earth they are perturbed by the Sun. Hence their motions present simultaneously a problem of the planetary type, with all its inherent difficulties, and a problem of the lunar type, which also has its own special difficulties. It is also necessary to take account of the perturbations due to the flattening of Jupiter and, above all, to resolve the problems posed by the near-commensurability between the periods of certain of the satellites. For if we designate by n, n', n'' the mean motions of the first three Galilean satellites Io, Europa, and Ganymede, we have almost exactly $n - 3n' + 2n'' = 0$. Laplace showed that this relation is always true. It implies the presence of small divisors which are sources of great difficulty in the theory.

The first true theory of the motion of the Galilean satellites was the one constructed by Laplace in 1788, before the start of the period with which we are concerned (see Chapter 22 above). The second truly complete theory appeared at the very end of this period. It is due to Sampson who published tables in 1910, while his theory itself appeared only in 1921.

Ralph A. Sampson had been, like Adams, a student at St John's College, Cambridge. In 1893 he was named professor of mathematics at the University of Durham; in 1910 he became Astronomer Royal of Scotland and professor of astronomy at the University of Edinburgh. His theory is semi-analytic, that is, the series representing the solution are trigonometric series of which the coefficients are given in numerical form. The method employed derives from that of Hansen. Sampson's theory has recently been corrected and modified, and is still the basis for the ephemerides published today; newer theories now in preparation will eventually replace it.

In the century and more separating the dates of appearance of the theories of Laplace and Sampson, there is little of note in the development of the theory of the Galilean satellites – nothing, indeed, producing a detectable improvement in the precision of ephemerides. J.-B. J. Delambre in 1817, and again Damoiseau in 1836, published tables based on Laplace's theory; but they are scarcely superior to the earlier tables constructed empirically on the basis of analyses of long series of observations. Souillart in 1880 and Tisserand in 1896 introduced important supplements to Laplace's theory; in particular, they modified the values of the principal terms of resonance. Cyrille Souillart (1828–98), after his studies at the École Normal Superieure in Paris, was a lyceum professor, then professor at the University of Nancy from 1867, finally at the University of Lille from 1874. François Félix Tisserand (1845–96), director of the Paris Observatory from 1893 to 1895, is best known for his remarkable *Traité de mécanique céleste*, which appeared in four volumes between 1888 and 1896, and which for astronomers and students became and remained the replacement for Laplace's treatise practically down to the present day. It is in the fourth volume of this work, after having expounded Laplace's and Souillart's contributions to the theory of the Galilean satellites, that he presents his own results.

Of the nine satellites of Saturn known in the nineteenth century, seven had already been discovered by the turn of the century. Hyperion was discovered simultaneously in 1848 by G. P. and W. C. Bond at Harvard and by Lassell in Liverpool in 1848, and Phoebe in 1898 by W. H. Pickering at Harvard. After the works of Laplace on the perturbations of the plane of the orbit of the satellite Japet, we should make special mention of the results published by Tisserand on this same satellite and on the rings of Saturn in 1880, and above all the works of Newcomb on the perturbations of Hyperion by Titan (1891) and those of Hermann Struve on Tethys, Mimas, Dione, and Enceladus. We shall return to Newcomb in the section on planetary theories. As for Hermann Struve, he was a member of the famous line of astronomers that began with Wilhelm Struve (1793–1864), founder of the Pulkovo Observatory, and was in fact the latter's grandson. Himself an astronomer at Pulkovo Observatory for a time, he was subsequently named director of the observatory of the University of Königsberg in 1895, and director of the observatory of Berlin–Babelsberg in 1904.

As the satellites of Saturn are bodies of low brightness, the observations of them are few and of poor quality. The observations do show, however, that the ratio of the mean motion of Tethys to that of Mimas, and the ratio of the mean motion of

Dione to that of Enceladus, are both close to $\frac{1}{2}$, and that the ratio of the mean motion of Hyperion to that of Titan is close to $\frac{3}{4}$. As is well known in celestial mechanics, such small whole-number ratios induce terms of resonance which are a source of difficulty in the theories. Hermann Struve showed that the conjunctions of Enceladus and Dione are always approximately in the direction of the point of the orbit of Enceladus closest to Saturn (the perisaturn) and that they oscillate round this point; further, that the conjunctions of Mimas and Tethys oscillate with a period of 68 years and an amplitude of 45° round the middle of the arc of Saturn's equator comprised between the ascending nodes of the two orbits.

Newcomb, for his part, studied the effects due to the resonance between the periods of Titan and Hyperion, and showed that the point of Hyperion's orbit farthest from Saturn (the aposaturn) has a movement of libration round the middle point of the arc in which the conjunctions of Titan and Hyperion occur.

Two of the satellites of Uranus, Titania and Oberon, were discovered by William Herschel, the discoverer of Uranus itself; two others, Ariel and Umbriel, were discovered by Lassell in 1851. During the period that concerns us there was little progress in the theory of these satellites. Newcomb showed that they move approximately in the same plane, inclined at 98° to the plane of the ecliptic.

In 1846, shortly after the spectacular discovery of Neptune, Lassell discovered its satellite Triton (the second satellite, Nereid, was discovered in 1949). This discovery, as previously remarked, made it possible to calculate the mass of Neptune. Between 1848 and 1892 the longitude of the node of the satellite's orbit advanced by 7°; the advance was attributed to the flattening of Neptune. Newcomb and Tisserand published several works devoted to this question.

The planetary theories

Various works

Before treating of the complete (indeed, magnificent) theories of Le Verrier and Newcomb, we shall review the results obtained in planetary theory during the nineteenth century, except for those having to do with the perturbing function, which were dealt with in an earlier section.

Laplace in 1773, and in a more complete way

Lagrange in 1776, showed that the semi-major axis of the orbit of a planet perturbed by other planets was subject only to periodic variations, and not to secular variations (perturbations, that is, proportional to the time or to the powers of the time), which would have had the long-term effect of dispersing the solar system. This result, however, was demonstrated only for perturbations of the first order with respect to the planetary masses. In 1808 Poisson extended the theorem to the second order of the masses, and Lagrange attempted to give a simpler demonstration; his calculations, however, were mistaken. In 1875–76 Tisserand gave a correct demonstration of Poisson's theorem. S.C. Haretu, however, in the thesis that he defended at the University of Paris in 1878, showed that among the terms of third order with respect to the masses there are some that are secular – a result confirmed in 1889 by D. Egenitis. But this result is somewhat artificial, in that the secular terms appear as such only for reasons of ease of calculation. Newcomb, in fact, showed in 1876 that the orbital elements of a planet can be represented by series in which time is the variable, and which, while satisfying the differential equations of motion, are purely periodic. The question that remained concerned the convergence of these series – a question resolved in the negative, as we shall see later, by Poincaré.

In general, therefore, it appeared appropriate to calculate the secular inequalities of the orbital elements, basing the calculation on the mean value of the perturbing function with respect to all the terms of short period. In 1818 Carl Friedrich Gauss (1777–1855) showed how to resolve this problem by taking the mass of each planet as distributed in a sort of ring along the length of its orbit, the density of the ring at each point being less in the measure in which the planet's speed at this point is greater. He then showed how to determine the deformations of such a ring when subjected to the attraction of other such rings. Applications of this method were published by Hill in 1882 and by P. J. O. Callandreau in 1885.

Before taking up the planetary theories of Le Verrier and Newcomb, we should mention a method of development of such theories proposed by Hansen in 1829 and again in 1859; it is similar to the method he employed for the lunar theory. It permits what no one had known how to do pre-

viously, the application of all the perturbations, whether of long or short period, to the mean longitude, or, what comes to the same thing, to the mean anomaly; the earlier practice, stemming from Laplace, had been to apply the perturbations of long period to the mean longitude and those of short period to the true longitude. The new calculation of the true anomaly, by starting from a mean anomaly containing all the perturbations, thus furnished at one stroke the true longitude as affected by all the perturbations; there remained only a few corrections to be applied to the radius vector and the latitude. To determine these, Hansen considered an ellipse of fixed dimensions situated in the osculating plane at the given instant, that is, in the plane containing the radius vector and velocity vector at this instant; he then obtained by means of a single function the perturbations of this ellipse in the plane. Also original in his procedure is a special mode of calculating certain integrals with respect to the time, whereby he avoids determining the result as the difference of two large numbers – a determination in which precision is inevitably lost. For this purpose he initially uses one symbol for the time in the functions to be integrated, and another symbol for its occurrences elsewhere; after the integration he returns to a single symbol for all occurrences.

The planetary theories of Le Verrier
Le Verrier devoted his scientific efforts almost exclusively to the construction of extremely precise theories of the motion of the planets – theories which served for the calculation of ephemerides of these bodies in the *Connaissance des temps* up to 1984.

In 1839 Le Verrier presented to the Academy of Sciences a memoir "Sur les variations séculaires des orbites des planètes". In 1843 there appeared a second memoir of his entitled "Détermination nouvelle de l'orbite de Mercure et de ses perturbations". From 1844 to 1847 he presented several memoirs on the periodic comets, in particular on Comet Lexell and Comet Faye – studies which he interrupted in order to resolve the problem of the perturbations of Uranus. We have already cited the memoirs that he published on the latter subject.

Having decided to devote himself to the study of the motions of the planets, he established his plan of work in July 1849, and in the same year published a study of the perturbing function in the first volume of the *Annales de l'Observatoire de Paris*. This publication, founded at Le Verrier's initiative, carried in subsequent volumes the results of all his further work on the planets. A little later the *Astronomical Papers of the American Ephemeris* were to play the same role for Newcomb.

The construction of planetary theories took place in stages: first, it was necessary to obtain a solution to the equations of motion, and in Le Verrier's time, to translate this solution into tables, since other means for directly calculating the positions were lacking; second, the positions of the planet under consideration had to be calculated for all the dates of the available observations, and the orbital elements then modified in such a way as to reduce the discrepancies between these positions and those furnished by observation; finally, if it proved impossible to reduce the discrepancies to the level of the observational error by this route, it would become necessary to modify the values of certain of the constants entering into the problem, in particular the planetary masses. It could happen that this expedient of last resort would itself prove insufficient (we shall see an example), and it then became necessary to make new hypotheses as to the causes of such phenomena. If one remembers the limited methods of calculation available at the time, the magnitude of the task will be apparent.

In 1852, Le Verrier published the results of the reduction of the 9000 meridian observations of the planets made by Bradley between 1750 and 1762. In 1858 he published his theory of the Sun's movement. Here he introduced perturbations of the second order with respect to the perturbing masses; the mean longitude of the Sun for 1850 thus received a correction of 5″. In addition he found it necessary to diminish the mass adopted for Mars by 10% and to augment that of the Earth by 10%. We know today, in the light of modern theories based on numerical integration, that the errors in Le Verrier's theory can go as high as 1″.5.

Le Verrier dealt with the theory of Mercury's motion in a volume published in 1859. Here he met with the unexplained advance in the perihelion, to which we shall devote a special section. In 1861 he published his theory of the motion of Venus. One finds here a new inequality discovered by Airy as well as an inequality of the second order due to the Earth and Mars. In developing this theory Le Verrier was led to diminish the mass then adopted for Mercury and to augment the Earth's mass. For determining the constants of integration he employed, besides the meridian observations made from 1751 to 1857, two transits of Venus across the disk of the Sun observed in 1761 and 1769. We know today that the errors in the position of Venus calculated from Le Verrier's theory can go as high as 19″.

In this same year he published his theory of the motion of Mars, based on meridian observations made between 1751 and 1858, and on a conjunction of the planet with the stars ψ_1, ψ_2, ψ_3 of Aquarius, which had been observed in 1672 in France by Jean Picard and at Cayenne by Jean Richer with a view to calculating the parallax of the Sun and thus its distance from the Earth. Le Verrier was led to introduce into the longitude of Mars a hitherto unrecognized term of amplitude 1″.5 and of period 40 years, and two previously unknown inequalities of the second order, one due to the Earth and Venus and the other to the Earth and Jupiter. His memoirs on the theories of the Sun, Mercury, Venus, and Mars won him in 1868 the gold medal of the Royal Astronomical Society.

The problem of the motions of Jupiter, Saturn, Uranus, and Neptune is much more difficult, and Le Verrier was unable to complete his treatment of these planets prior to his death. In November 1872 there appeared a first "Mémoire sur les théories des quatre planètes supérieures" which dealt with the secular terms for these planets. Memoirs dealing with the individual planets appeared in the next two years. But at the time the available means of calculation made it impossible to construct ephemerides by summing trigonometric series, as is done today with the aid of electronic computers. It was necessary to form tables, generally of double entry, and to interpolate in computing the several inequalities at a given instant. The tables for Jupiter appeared in 1874, those for Saturn in 1875, those for Uranus in 1876. The tables for Neptune were completed by A. J.-B. Gaillot and presented by him in October 1877, a month after Le Verrier's death.

The theories of the motions of Jupiter and Saturn are difficult partly because of the magnitude of the perturbing forces, the two planets being massive and far from the Sun, and partly because the ratio of the mean motion of Jupiter to that of Saturn is almost exactly 5:2. As a consequence there is a 'great inequality' in the longitudes of the two planets, of which the period is close to 900 years and the amplitude about 20′ for Jupiter and 50′ for Saturn. C. G. J. Jacobi found a method for isolating the corresponding terms of resonance in the perturbing function, and published this result in 1849. Le Verrier's theory represented the observations of Jupiter made between 1750 and 1869 with an error in longitude less than 1″, and, after certain corrections were introduced by Gaillot, the observations of Saturn made from 1750 to 1890 with an error in longitude less than 3″ in absolute value. In subsequent decades these maxima did not increase; in the 1970s the errors were still found to be always approximately the same.

The planetary theories of Newcomb and Hill

We group under this heading the works of Newcomb on the four planets closest to the Sun, and those of Hill on Jupiter and Saturn tackled by Hansen's method. These planetary theories served as the basis of the *American Ephemeris* until 1984. To such a degree were they the product of the will and direction of Newcomb that they are often referred to by his name alone.

Simon Newcomb (1835–1909) (Figure 28.7), like many Anglo-Saxon astronomers of his time, learned celestial mechanics from Bowditch's English translation of Laplace's *Mécanique céleste*. In 1857 he was appointed to the Nautical Almanac Office in Cambridge, Massachusetts, and starting in 1863 he carried out determinations of the right ascensions of stars with the meridian circle of the Naval Observatory in Washington DC. With a view to improving the constants of the lunar theory, he travelled to the Paris Observatory in 1871, and there located observations of good quality going back as far as 1672. His visit was during the revolutionary insurrection known as the Paris Commune. Delaunay, as director of the observatory, presided over a deserted and menaced institution; on 23 April he wrote to his wife:

All the astronomers have left, and so too the servants, the gardener, and the porter ..., I live almost alone in the building; it has not been invaded and probably will not be. It has been sufficient for me to hoist the red flag in place of the tricolor, following the order given me by the Paris Commune. I have with me here an American

28.7. Simon Newcomb.

astronomer with his wife. He works all day copying in our library, especially such documents from among our older manuscripts as can be of interest in his work.

In 1877 Newcomb was named superintendent of the Nautical Almanac Office, which by then had moved to Washington; he remained in charge until his retirement in 1897, and thereafter continued to work at the Almanac Office as a scientific adviser.

Why undertake to construct new planetary theories at the moment when Le Verrier's theories appeared? First of all, because the empirical basis had been augmented with observations of good quality. But there was a more profound reason in order to make the theory of each planet fit the observations, Le Verrier had been led to attribute to one and the same planet different masses in the different planetary theories in which it figured as a perturbing body, and to employ similarly varying values for other constants entering into the theories. Newcomb, in the preface of his memoir entitled "The elements of the four inner planets and the fundamental constants of astronomy", which appeared in 1895, expressed his reaction to this inconsistency as follows:

The diversity in the adopted values of the elements and constants of astronomy is productive of inconvenience to all who are engaged in investigations based upon these quantities, and injurious to the precision and symmetry of much of our astronomical work .... On taking charge of the work of preparing the American Ephemeris *in 1877 the writer was so strongly impressed with the inconvenience arising from this source that he deemed it advisable to devote all the force which he could spare to the work of deriving improved values of the fundamental elements and embodying them in new tables of the celestial motions.*

Newcomb was thus the first to concern himself with the construction of a coherent system of astronomical constants. After him this idea slowly made progress and ended by being erected into a dogma by the International Astronomical Union, which endeavoured to supply astronomers with such a system. It can seem astonishing that a goal so logical did not come to be adopted more easily. But this goal is much more difficult to realize than initially one might suppose – at the present time perhaps more than ever.

Newcomb's earliest work on planetary theory concerned the development of the disturbing function. Newcomb undertook to generalize the results that Laplace had obtained for circular orbits in introducing the 'Laplace coefficients' – certain functions of the ratios of the radii of these orbits – for elliptical orbits. To do this he created 'Newcomb's operators', which operate on the Laplace coefficients, and he tabulated the functions obtained to the eighth order in the eccentricity. These results appeared in 1891.

He then attacked the problem of the motions of Mercury, Venus, Mars, and Earth, employing 62 000 meridian observations, which he had to reduce, as well as the data obtained from observations of the transits of Mercury and Venus across the solar disk. Like Le Verrier, he too found himself confronted with the unaccounted-for advance in the perihelion of Mercury; the explanations that he proposed will be described later on. The memoirs devoted to these planets, "The elements of the four inner planets, and the fundamental constants of Astronomy", appeared in 1895. The corresponding tables appeared in 1898.

Using the method of Hansen previously described, Hill during this time worked on the theory of the motions of Jupiter and Saturn; the result appeared in 1890. The same volume also contains an exposition of Hansen's method that was clearer

than that supplied by Hansen himself! The corresponding tables appeared in 1898 along with tables for Uranus and Neptune, but Hill is not very explicit as to the theory on the basis of which the latter tables were constructed; he refers merely to an "investigation as yet unpublished". It seems it has never appeared.

What can be said as to the precision of Newcomb's and Hill's theories? In the case of the four inner planets, the maximal error appears to be 3" for Venus; in the case of the outer planets the errors are of the order of a second. Newcomb's and Hill's theories are therefore superior to those of Le Verrier and Gaillot.

Other works

The asteroids

On 1 January 1801, Guiseppe Piazzi of Palermo discovered in the constellation of Taurus a faint star which was not listed in the catalogue of N.-L. de Lacaille: it was the asteroid, or minor planet, Ceres (see Chapter 24). Gauss calculated its orbit and found the semi-major axis to be about 2.8 astronomical units, thus falling in the space left between Mars and Jupiter by the empirical law of Titius–Bode. In March 1802 H. W. M. Olbers discovered Pallas, then in 1804 K. L. Harding discovered Juno and in 1807 Olbers discovered Vesta. By 1860 sixty-four asteroids (a term coined by William Herschel) had been discovered, by 1891 some three hundred.

It was at this point that Max Wolf proposed a method that consists in photographing a region of the sky for several hours, while compensating for the Earth's rotation by rotating the telescope in the opposite sense. The stars appear on the photographic plate as points, but the asteroids, because of their proper motions, leave a rectilinear track. With the aid of this method, the total number of asteroids identified had been increased by 1912 to eight hundred.

In 1772 Lagrange had shown that if an equilateral triangle is formed in the plane of the relative orbits of two bodies, with the bodies occupying vertices, then a body at the third vertex will be in a position of stable equilibrium. Remarkably, starting in 1906, asteroids began to be discovered that occupied precisely such positions in relation to the Sun and Jupiter. These are the 'Trojan' planets, so-called because they have been named after the heroes of the Trojan War: Achilles, Hector, Priam, etc. Their orbits round the Sun are similar to Jupiter's, but they are located in two groups 60° to the east and west of Jupiter.

The problem of the motion of the asteroids, as compared with the corresponding problem for the major planets, presents a number of special aspects. The orbital eccentricities are large and can reach 0.4; the inclinations of the orbital planes to the ecliptic go as high as 35° (the case of Pallas); the principal perturbing planet is Jupiter, the most massive planet in the solar system; the ratios of the semi-major axes of the orbits of the asteroids to that of the orbit of Jupiter are between 0.4 and 0.8 (at least for the asteroids known at the beginning of the twentieth century). Finally, between the mean motions of Jupiter and certain asteroids there are relations of near-commensurability, which introduce resonances, phenomena mentioned earlier in our discussion of 'the great inequality' of Jupiter and Saturn.

Here the classical methods are not very effective, although Damoiseau attempted to apply them to Ceres and Juno (*Addition to La Connaissance des temps pour 1846*), and H. J. A. Perrotin used them with success for Vesta (*Annales de l'Observatoire de Toulouse*, 1880). Gauss, in a letter to Bessel, describes a theory he had constructed for Pallas, which contained 800 inequalities and which he intended to compare with a calculation by quadratures (that is, a numerical integration). However, this theory was not published during Gauss's lifetime.

Le Verrier had found that a resonance term of period 800 years in the motion of Pallas had an amplitude of 895″. A.-L. Cauchy (1789–1857), although a mathematician rather than an astronomer, was charged by the Academy of Sciences with verifying this result, and in six weeks invented an ingenious method which he published in 1845 in the *Comptes rendus*.

Earlier we mentioned Hansen's method and indicated its principal characteristics. This method was employed by F. F. E. Brunnöw for the motion of the asteroids Iris and Flora, by L. Becker for that of Amphitrite, and by O. L. Lesser for the motions of Pomona, Metis, etc. G. Leveau applied the method to the construction of a very complete theory of the motion of Vesta which was published between 1880 and 1892.

Another important method is that due to Hugo Gyldén, director of the observatory of Stockholm,

which appeared in Swedish in 1874. An exposition is given in Vol. 4 of Tisserand's *Traité de mécanique céleste*. Gyldén made it his aim to avoid the appearance of secular terms; also, he took the true anomaly as independent variable instead of the time. He was led to a linear differential equation of the second order which he resolved; in it the second derivative of a function is equal to the product of this function and a periodic function of the variable. Poincaré was greatly interested in Gyldén's method, and devoted to it an important part of Vol. 2 of his *Méthodes nouvelles de la mécanique céleste*, which appeared in 1893. This method was applied by J. O. Backlund in 1875 to the motion of the asteroid Iphigenia, and by Callandreau in 1882 to the motion of the asteroid Hera.

An interesting peculiarity of the ensemble of the asteroids is that the distribution of the semi-major axes of their orbits is not uniform round the mean value of 2.8 astronomical units; there are lacunae, that is, zones empty of asteroids. These zones correspond to values of the semi-major axis for which the asteroid's period is resonant with the period of Jupiter. The American Daniel Kirkwood (1814–95), a professor of mathematics first at Delaware College, later at Washington and Jefferson College in Pennsylvania, and at the University of Indiana, was the first to draw attention to the existence of these lacunae, now called 'Kirkwood gaps' (see Chapter 26). In 1860 he gave examples of commensurability between the asteroids and Jupiter in a note to the *Astronomical Journal*, then proved the existence of the lacunae in an article entitled "On the meteors' which appeared in 1867 in the *Proceedings of the American Association for the Advancement of Science for 1866*. He later made a comparison between these lacunae and the spaces between the rings of Saturn, such as 'the division of Cassini' between ring *A* and ring *B*.

The explanation of this phenomenon is very difficult and has continued to be a subject of investigation down to the present day. Among the numerous memoirs devoted to it in the nineteenth century we mention in particular that of Callandreau, published in 1896, and that of K. Bohlin which appeared in 1888 and gives a method for treating resonances that Poincaré later studied. In 1895 Bohlin applied his method to the asteroids whose mean motion is triple that of Jupiter.

Finally, a word concerning hypotheses as to the origin of the asteroids. It was tempting to think that these small bodies, so well placed according to the law of Titius–Bode, had arisen from the break-up of a large planet. This idea was suggested by Olbers after the discovery of Pallas. But as more and more new objects were discovered the hypothesis ceased to be tenable, and as described in Chapter 26, Newcomb in the late nineteenth century showed that the orbits then known belied it.

The appearances of Halley's Comet in 1835 and 1910

The return of Halley's Comet in 1759 (see Chapter 19) confirmed Newton's hypothesis that comets are part of the solar system like planets, but with orbits that can be very elongated and very steeply inclined to the ecliptic; with this, comets lost something of their mystery, and it seemed possible that the problem of their motions could be resolved by celestial mechanics. However, as we have seen, all the methods elaborated for constructing literal or semi-numerical theories of the motions of the planets, Moon, and satellites of the planets presuppose that the orbital eccentricities and inclinations are small, for they employ series developments in terms of these small parameters. Since the eccentricities and inclinations of cometary orbits are large, the number of terms of the series that would have to be retained becomes exceedingly, indeed prohibitively, large. Worse yet, the convergence of some of the series employed is no longer assured when the values of the eccentricities and inclinations become too large. Thus the developments in series of the functions of the radius vector and of the true anomaly that enter into the disturbing function fail to converge for all values of the time unless the eccentricity is below the critical value 0.662 743. Similarly, if a comet has an orbit of which the semi-major axis is equal to 3 astronomical units, then the inverse of the distance from the comet to Jupiter, which figures in the disturbing function for the perturbations caused by Jupiter, can be developed in a converging series only if the inclination is less than about 30°; and this limit of inclination becomes smaller in the measure in which the orbit of the comet comes closer to that of Jupiter.

For this reason the problem of the motion of comets is almost always treated by the numerical integration of equations. The different methods used and their history are described in Chapter 29

below, and we limit ourselves here to citing the most important of the works on comets to which these methods led.

(i) In June 1770 Messier discovered a comet whose orbit was studied by A. Lexell (1740–84), member of the Academy of Sciences of St Petersburg; it was henceforth called Comet Lexell. The orbit proved to be an ellipse with a semi-major axis equal to 3 astronomical units, so that the cometary period was 5 years and 7 months. With such a short period, the comet should have been discovered much earlier; moreover, it was lost some years after its discovery. It was then shown, in particular by Burckhardt working at Laplace's instigation, that passages of the comet very close to Jupiter had twice profoundly modified the orbit. This posed a peculiar problem which, by the techniques of numerical calculation available at the time, could not be satisfactorily resolved. Laplace, here following an idea of d'Alembert, then introduced the notion of 'sphere of activity'. In a certain part of the trajectory of a comet perturbed by Jupiter, one considers the motion as heliocentric and proceeds to determine the perturbations due to Jupiter. But close to Jupiter, within its 'sphere of activity', it is more advantageous to consider the motion of the comet as jovicentric, and to calculate the perturbations due to the Sun. Le Verrier in 1857 applied this method to Comet Lexell in his memoir "Sur la théorie de la comète périodique de 1770".

(ii) An interesting problem was posed and resolved by Tisserand in 1889, using the rule now known as 'Tisserand's criterion'. The problem is to know whether or not two periodic comets which appeared at two different epochs are the same body, the orbital elements of which were modified by a passage close to Jupiter. Employing a first integral of the restricted three-body problem found by Jacobi, Tisserand showed that the two comets are the same provided that

$$\frac{1}{2a} + \sqrt{\frac{a(1-e^2)}{a'^3}} \cos i = \text{const,}$$

where a is the semi-major axis of the cometary orbit, a' that of the orbit of Jupiter, e the eccentricity of the cometary orbit and i the inclination of the plane of the cometary orbit to the plane of the orbit of Jupiter. The eccentricity of Jupiter's orbit is neglected, but in 1892 Callandreau showed how to take account of the first power of this eccentricity.

The influence of Jupiter on a great many comets besides Lexell's was recognized and studied from the middle of the century onwards. D'Arrest showed in 1857 that in 1842 Comet Brorsen had approached Jupiter with the result that its perihelion distance was diminished by half. Similarly in 1890 R. Lehmann-Filhés studied the effects of the passage close to Jupiter in 1875 of Comet Wolf, discovered in 1884. The problem of the capture of parabolic comets by Jupiter, that is, of the transformation of a parabolic into an elliptical orbit owing to perturbations by Jupiter, was also broached by Tisserand in Vol. 4 of his *Traite de mécanique céleste*. He there cites a memoir of 1891 by H. A. Newton (not to be confused with Isaac Newton!) in which it is shown that the probability of the capture of a parabolic orbit by Jupiter is very small.

(iii) The fact that non-gravitational forces play a significant role in the motions of many comets was first brought into prominence by Encke. Johan Franz Encke (1791–1865) began his career in astronomy in a post that Gauss obtained for him at the observatory of Seeberg near Gotha. In 1825 he became director of the Berlin Observatory, and in 1844 he was named a professor at the University of Berlin. He is known for his work on the first four asteroids to be discovered, and also as founder of the *Berliner Astronomisches Jahrbuch*, providing ephemerides which he published in collaboration with J. P. Wolfers from 1830 until his death.

On 26 November 1818 J. L. Pons discovered a comet that remained observable until mid-January 1819. Encke determined its orbit, and found it to be periodic with a period of 3.5 years. More important, he showed that the same comet had been seen by Pierre-François-André Méchain, Charles Messier, and J.-D. Cassini (Cassini IV) in 1786, by Caroline Herschel in 1795, and by Pons, J. S. G. Huth, and Alexis Bouvard in 1805. He predicted its return in 1822 and again in 1825. The comet, which was named Comet Encke, was the second comet whose periodicity had been established with certainty in the years since the return of Halley's Comet in 1759. It was noticed that its period diminished constantly (and inexplicably) by about two and a half hours from one perihelion passage to the next. Encke suggested that the friction of a resistant medium could account for the phenomenon: a force of friction opposed to the velocity, while diminishing the eccentricity of the orbit,

also diminishes the semi-major axis and hence the period – a problem that was to be studied by A. V. Bäcklund in 1894. On 28 June 1851 at the Leipzig Observatory d'Arrest discovered the comet that bears his name, another comet of short period. It was later shown (in 1939) that it, too, has an acceleration, but that the hypothesis of a resistant medium fails to account for it. Not until 1950, with the appearance of Fred Whipple's model, which takes the nucleus of a comet to be a "dirty snowball" formed of a mixture of ice and dust, was it possible to give a satisfactory account of the nongravitational forces.

(iv) The period with which we are concerned saw two returns of the celebrated Halley's Comet, in 1835 and in 1910.

Among those calculating predictions of the passage of 1835 we meet once more the two competitors in the prize contest on the lunar theory, Damoiseau and Pontécoulant. Damoiseau in 1820 predicted the comet's return for 17 November 1835. He took account of the perturbations by Jupiter, Saturn, and Uranus (a planet unknown at the time of the previous passage in 1759) over the period extending from 1682 to 1835. He then added the perturbations due to the Earth in a second memoir which appeared in 1829; this brought the predicted date of passage to the evening of 4 November. Pontécoulant, with the same hypotheses as Damoiseau, published three memoirs, which appeared between 1830 and 1835. His final prediction for the passage was 12 November at 10 p.m. Others who tackled the problem were J. W. H. Lehman, who predicted perihelion would be on 26 November, and C. A. Rosenberger, who took account of all the planets then known and gave 12 November as the date of the perihelion passage. The comet was rediscovered on 6 August by Father Étienne Dumouchel of the Collegio Romano in Rome, and the perihelion passage occurred on 16 November at 10 a.m.

The prediction of the passage of 1910 was undertaken by Pontécoulant in 1864. Taking account only of perturbations by Jupiter, Saturn, and Uranus, he calculated that the comet would pass its perihelion on 24 May 1910 at 9 p.m. Later, however, P. H. Cowell and A. C. D. Crommelin in England found faults in Pontécoulant's memoir, and re-did all the calculations, which were the subject of numerous articles in the *Monthly Notices* of the *Royal Astronomical Society* in 1907 and 1908. Here they employed the method always used hitherto, which consists in finding by numerical integration the variations of the orbital elements; and they took account of all the planets from Venus to Neptune. Later they undertook a new prediction, employing a method that Cowell had introduced in 1908 for studying the motion of the eighth satellite of Jupiter; it consists in carrying out a numerical integration to determine, not the orbital elements, but the rectangular coordinates. According to their new calculation, the perihelion passage should occur on 17 April. However, by the time this publication appeared, the comet had already been rediscovered by Max Wolf at Heidelberg on a photograph taken on 11 September 1909, and so it was already known that the passage would occur on 20 April. The difference comes from the non-gravitational forces, which once more had been left out of account. The return of 1986 would be predicted, before the rediscovery, to within a few hours.

Two "small difficulties"

At the beginning of this chapter we quoted Poincaré's assertion that the goal of celestial mechanics is to decide whether Newton's law is true. Having completed our survey of the nineteenth-century achievements, let us see how and in what measure the question was answered. Among those who participated in the progress we have surveyed, we generally find that the response was a confident one. Thus, in the preface to the fourth and last volume of his *Traité de mécanique céleste* Tisserand stated: "everything advances with admirable consistency except for one or two small difficulties over which our successors will no doubt triumph".

We now turn to the two most significant of these "small difficulties". We have "triumphed" today over the second of them by taking into account the effects of friction due to the tides, hence due to forces ultimately of gravitational origin, and over the first, not indeed by rejecting the law of universal gravitation, but by taking it as a very good local approximation in the solar system to the behaviour of space-time according to the general theory of relativity.

The advance of the perihelion of Mercury

The Earth passes through the line in which the orbit of the planet Mercury intersects the ecliptic (the line of nodes) in the months of November and May. If at these times Mercury is at inferior conjunction, one can observe a transit of this planet across the disk of the Sun. These observations are easy to carry out and yield, among other results, a very precise determination of the planet's position. Le Verrier employed nine November transits from 1677 to 1848 and five May transits from 1753 to 1845. But he found it impossible to account for these observations theoretically; it was necessary to postulate that the perihelion of the orbit of Mercury rotated round the Sun not 527" per century as the theory of planetary perturbations implied, but 565" per century. There was a discrepancy of 38" that he could not explain.

The first idea that came to mind was to increase the mass of Venus so as to obtain a suitable mean motion for Mercury's perihelion, but for this it would have been necessary to increase the Venusian mass by a tenth, an inadmissible increase: for the mass of Venus contributes very significantly to the secular variation in the obliquity of the ecliptic, and an augmentation by 10% would have led to variations in the obliquity completely incompatible with precise observations carried out since Bradley's time.

There remained the hypothesis of an unknown planet or group of planets, which by disturbing the motion of Mercury would constitute the cause of the anomaly. Such a hypothesis was well suited to seduce the discoverer of Neptune, and he undertook the calculations necessary to confirm it – calculations which he explained and commented on in 1859.

First he showed that the perturbing body must be located between the Sun and Mercury; if exterior to the orbit of Mercury it would have caused empirically inadmissible perturbations in the motions of the Earth and Venus (in the case of Venus, whose theory he had not yet completed, this conclusion was in part based on hypotheses).

Next he established a relation connecting the mass of the unknown planet with the radius vector of its circumsolar orbit (assumed to be circular). This relation, as one would expect, indicated that the closer the planet was to the Sun the greater must be its mass, if its action was to cause the observed effect on the motion of the perihelion of Mercury. An examination of this relation shows that at a distance of 0.17 astronomical units the planet would have a mass equal to Mercury's and be 10° from the Sun at its greatest elongation; it would thus be easily visible, even to the naked eye.

These figures were later revised because Le Verrier was found to have given Mercury a mass too large by a factor of two. If, as Tisserand shows in Vol. 4 of his *Traité de mécanique céleste*, a more accurate mass is taken for Mercury, then the unknown planet, if assigned a mass equal to Mercury's, would be at a distance of 0.21 astronomical units from the Sun, and its greatest elongation would be 13°, so that it would be even more easy to observe. If it were farther from the Sun, its mass, although smaller, would yet be sufficiently great that it could be easily observed, either in its greatest elongation or at the moment of a transit across the solar disk. If it were closer to the Sun than 0.21 astronomical units, its mass would become unreasonably large: for example, at 0.12 astronomical units from the Sun the mass would be six times that of Mercury, while the greatest elongation would still be 7°. Le Verrier, obliged to discard this solution, hypothesized that a ring of small bodies circulating between the Sun and Mercury could have the same effect, the reduced size of the individual bodies being the reason for their not having been observed.

Meanwhile a French amateur astronomer, E. M. Lescarbault, who lived at Orgères in the department of Eure-et-Loir, believed that he had observed on 26 March 1859 the transit of a planet across the disk of the Sun. He had noted the times and places of the contacts and had clocked the transit as lasting 1 hour 17 minutes. Informed of the observation nine months later in a letter from Lescarbault, Le Verrier visited him. From their discussion he concluded that the planet had a period of 19.7 days, and he determined the position of the plane of the orbit. From Lescarbault's estimate of the apparent diameter of the planet, and assuming a density equal to Mercury's, he found that its mass would be $\frac{1}{17}$ that of Mercury. The fact that the greatest elongation was 8° explained why the planet had not hitherto been observed. But the supposed discovery left the problem of the advance of Mercury's perihelion quite unresolved. The planet was given the name Vulcan, the merit of its discovery being attributed to Le Verrier. (Although partially effaced, the planet Vulcan can still be seen in a representation of the solar system engraved on the pedestal of the statue of Le Verrier erected in 1888 in the north court of the Paris Observatory.) No

other observation of Vulcan has ever been made and it is now accepted that it does not exist.

It was clearly a possibility that numerous other observations of spots could be due to small bodies passing in front of the Sun, and in 1856–59 the Swiss astronomer Johann Rudolf Wolf drew up a list of these. Le Verrier sought to isolate among all these reports those that could relate to Vulcan, and even attempted predictions of transits for 1877 and 1882 (he died five years before the second of these dates). Neither of these transits was observed. Accordingly, Tisserand in 1896 concluded that Le Verrier's hypothesis to account for the advance of Mercury's perihelion should be abandoned.

Newcomb, in his reconstruction of the theories of the planetary movements, corrected the mass of Mercury, diminishing it by half. Also, he sought explanations for the anomalous part of the advance of its perihelion, which he put at 41″ per century rather than the 38″ found by Le Verrier. To begin with, he made clear that meridian observations are unsuitable for determining the secular variations of the orbital elements of the inner planets, and that without observations of transits Le Verrier's determination of the advance of Mercury's perihelion would have been impossible. He re-examined the hypothesis of an intro-Mercurial body and, like Le Verrier, concluded that the existence of a single planet in this region and of such mass as to produce the observed motion of Mercury's perihelion is to be excluded. As to the hypothesis of a hitherto unperceived ring, he thought it improbable: given the large mass that would need to be assigned to the ring, it would reflect too much light to remain undetected.

Another attempt to explain the anomalous advance of Mercury's perihelion was made by Hall in 1895. Hall's idea was that the exponent of the inverse of the distance in the law of universal gravitation was perhaps not 2 but a slightly different number. Taking this exponent as unknown, he calculated what value it would need to have in order to account for a 41″ advance of the perihelion of Mercury per century. His result was 2.000 000 151. However, it was then necessary to see what consequences this change, slight as it was, would have for the orbital elements of the other planets. All went well for the perihelia of Venus and the Earth, but an advance of the node of Venus's orbit that Newcomb had found, some 10″ per century, remained unexplained (according to present-day theory, the advance was largely an artefact). In the case of the Moon, Hall's hypothesis proved inadequate to account for certain anomalies which were in fact due to the inadequacies of the assumed measure of time.

It was in late 1915 that Albert Einstein published four articles in Berlin, including "Erklärung der Perihelbewegung des Merkur aus dem allgemeinen Relativistät Theorie". Einstein calculated the extra advance of Mercury's perihelion in the general theory of relativity and found it equal to 43″ per century. He wrote: "I have found an important confirmation of the Relativity Theory which explains qualitatively and quantitatively, without any special hypothesis, the secular precession of the orbit of Mercury discovered by Le Verrier." This result was taken as a brilliant proof of the exactitude of the new theory, just as the discovery of Neptune 70 years earlier had been taken as a brilliant proof of the exactitude of the law of universal gravitation.

The secular acceleration of the Moon

In the 1690s Edmond Halley found discrepancies between the observed and the calculated times for certain ancient eclipses, and suggested that the Moon might be subject to an acceleration. In the mid-eighteenth century, studies by Richard Dunthorne, Tobias Mayer, and J. J. L. de Lalande established that the mean motion of the Moon increases by about 20″ per century.

In 1787 Laplace claimed to have accounted completely for the phenomenon by attributing it to a secular variation in the eccentricity of the terrestrial orbit caused by planetary perturbations. This result was confirmed in 1820 in the theories of both Damoiseau and Plana. The latter, in his literal theory, gave an expression for the secular acceleration as a function of m, the ratio of the mean motion of the Sun to that of the Moon. This expression contained terms in m^2, m^4, and m^5.

In 1854 there appeared in the *Monthly Notices of the Royal Astronomical Society* an article by Adams, who showed that Plana, by failing to take account of the transverse component of the perturbation, had committed an error in calculating the secular acceleration and that the coefficient of the term in m^4, which is negative, must be more than tripled in absolute value. This implied a secular acceleration of 16″ instead of the previously deduced 21″ per

century per century. Thus, a few years after the discovery of Neptune, a new subject of Franco–British quarrel arose in which Adams, the co-discoverer of Neptune, was once more involved!

Pontécoulant bestrode this new war-horse and in 1860 published in the *Monthly Notices of the Royal Astronomical Society* some not very convincing calculations that claimed to prove that Adams was mistaken. In the very same issue Adams replied, analysing lucidly the calculations of both Plana and Pontécoulant and showing that there was indeed an error of principle. The quarrel abated, more especially as Delaunay, for his theory of the Moon, made the same calculations to the eighth order of m and found the same coefficient of the term in m^4 as had Adams. Finally, once everyone was in agreement, the theory was found to imply a secular acceleration of 12″ per century per century, while the analysis of the ancient observations yielded an acceleration about twice as large.

The answer to this dilemma is well known today: the longitude of the Moon was expressed as a function of the time as measured on the scale of Mean Solar Time, and this scale assumes that the rotation of the Earth round its axis is perfectly constant. If, on the contrary, the Earth is slackening in its rotation, the Moon will appear to accelerate. The question of the uniformity of the Earth's rotation had been posed by Immanuel Kant in 1754, and Laplace himself had remarked that a deceleration in the Earth's rotation would give the illusion of an acceleration in the movement of the Moon.

Delaunay in 1865 published an article in which he hypothesized that a slackening in the rotation of the Earth due to the friction of the tides was causing an apparent acceleration of the Moon. Indeed, he took the fact of the Moon's acceleration as proof that the Earth was slowing down, and proposed to evaluate its deceleration by measuring the Moon's acceleration. G. H. Darwin published in 1880 a memoir on the evolution of the Earth–Moon system, and there established the principles of a theory of the effects of the tides on the Earth's rotation. Poincaré cited this memoir in his *Leçons sur les hypothèses cosmogoniques*; but at the time of his death the dogma of the invariability of the Earth's rotation had not yet been definitively rejected by astronomers.

28.8. Henri Poincaré.

The work of Poincaré

Poincaré's name has appeared more than once in the preceding pages; indeed, many of the paths explored by nineteenth-century celestial mechanicians led to works on these same subjects by this mathematician of genius.

Born at Nancy on 29 April 1854, Henri Poincaré (Figure 28.8) belonged to a solid bourgeois family that had long been established in Lorrain. His father was a professor of medicine at the University of Nancy, and one of his cousins, Raymond Poincaré, a man of politics, was to serve as President of the French Republic from 1913 to 1920. A brilliant student, especially in mathematics, and winner in a general competition, Poincaré entered the École Polytechnique in 1873, and then the École des Mines. He occupied a post as engineer of mines while preparing a doctoral thesis which he defended in 1879. At first professor at the University of Caen, in 1881 he was appointed professor in the Faculty of Sciences in Paris, in which position he remained until his death in 1912. He was elected to the Academy of Sciences in 1887 and to the French Academy in 1908.

Poincaré's principal discoveries in celestial mechanics were published in the following works:

"Mémoire sur le problème des trois corps et les équations de la dynamique", for which he received in 1889 the first prize in an international competition on the *n*-body problem that had been proposed by Oscar II, King of Sweden; the three volumes of *Les méthodes nouvelles de la mécanique céleste*, which appeared between 1892 and 1899; and the three volumes of *Leçons de mécanique céleste*, which appeared between 1905 and 1910. It is impossible here to enter into the details, which are necessarily very technical, of so considerable a body of work; we shall rather seek to sketch in outline the principal results he obtained.

We know that the equations of motion of *n* bodies have ten independent first integrals – ten functions of the coordinates and the velocities that remain constant during the motion. Six of these first integrals express the fact that the centre of mass of the system of bodies has a uniform, rectilinear motion in space; three others express the fact that the moment of momentum of the set of bodies is a fixed vector; and the tenth is the theorem of *vis viva*, which states that the sum of the kinetic energy of the ensemble of points of the system (due to the velocities of these points) and the potential energy arising from the gravitational attractions is equal to a constant. These ten first integrals make possible a resolution of the problem of two bodies. Poincaré showed that, whatever the number of bodies, there exist only ten first integrals – or rather, only ten *uniform* first integrals, the only sort that would be utilizable.

In 1878 Hill had found a particular intermediate orbit for the motion of the Moon. It is a closed curve when referred to a plane that rotates uniformly, and is called a periodic orbit in phase-space or simply a periodic orbit. Poincaré generalized this result in magisterial fashion by establishing that there are three families of periodic orbit in the three-body problem provided that two of the bodies are of small mass compared to the third (the case of the solar system). He demonstrated the importance of these orbits by showing how one can construct orbits close to periodic orbits and to asymptotic orbits, that is, orbits which approximate as closely as one pleases to a periodic orbit as the time tends to $+\infty$ or $-\infty$. He showed further that there can be doubly asymptotic orbits, orbits that approximate to periodic orbits as the time tends both towards $+\infty$ and towards $-\infty$. By means of what he called 'characteristic exponents' he showed that, close to any given periodic solution, there are nearby periodic solutions. Using integral invariants, he demonstrated that 'recurrent trajectories', that is, trajectories such that if they pass through a given region of phase space once do so an infinite number of times, are infinitely more numerous than nonrecurrent ones.

Poincaré justified the investigation of periodic orbits as follows:

It seems at first that [this] can be of no interest in practice. In fact there is a zero probability that the initial conditions of motion should be precisely those corresponding to a periodic solution. But it can happen that they differ very little, and this occurs just in the case where the old methods are no longer applicable. One can then advantageously choose the periodic solution as a first approximation, as intermediate orbit, *to use the term introduced by Gyldén.*

Later on in the same memoir we meet the following formulation, which has often been quoted:

Moreover, what makes these periodic solutions so precious to us is that they are, so to speak, the sole breach by which we can attempt to penetrate into a stronghold hitherto believed impregnable.

As we have seen, the celestial mechanicians, in seeking solutions of the differential equations of motion of the celestial bodies, employed developments with respect to the small parameters of the problem, for example the masses; successive powers of these parameters figured as factors in the coefficients of the terms of trigonometric series, the arguments being functions of the time. But the methods of integration employed in practice sometimes made the time appear outside the arguments of sine and cosine, either in the form of 'secular terms', powers of the time, or in the form of 'mixed secular terms' or 'Poisson terms', products of a monomial of the time and a periodic function. Newcomb had shown that theoretically we can obtain the solution as a series of terms periodic in the time, but without any assurance as to the convergence of these series. Poincaré demonstrated that such series are not uniformly convergent, but that they permit us to have solutions approximating as closely as we please to the real solution during a given interval of time.

In addition to this work on the *n*-body problem,

Poincaré carried out investigations of two other aspects of celestial mechanics. In 1885 he published in the *Acta mathematica* a memoir on the figures of equilibrium of fluids in uniform rotation. Colin Maclaurin had found particular solutions of this problem in the form of ellipsoids of revolution, and Jacobi had found another family formed of ellipsoids with three unequal axes. Poincaré found the most general figure answering to the question, to which he gave the name "figure piriforme" (pear-shaped figure); he believed it to be a figure of stable equilibrium. But A. M. Lyapunov and after him James Jeans showed that it is unstable, and ends by breaking into two distinct masses.

In 1892 Poincaré began an investigation into the problem of the tides, and this led him to discover solutions of integro-differential equations of Fredholm's type. In this problem the form of the coasts and the configurations of the sea bottom impose limiting conditions that make the solution of the equations especially complex. Poincaré returned to the problem in his *Leçons de mécanique céleste*; but his investigation of it was never to be completed.

Finally, we should mention the *Leçons sur les hypothèses cosmogoniques*, based on the course given by Poincaré in 1910 at the Sorbonne. It is curious to observe a genius like Poincaré giving himself so much trouble to expound theories, often merely confused, that had been born in the heads of incompetent dilletantes. Yet this work is pleasant to read and full of ingenious insights. We find there at the turn of a page the famous virial theorem which Jeans was later to exploit in founding stellar dynamics.

Such, rapidly surveyed, is the immense work of Poincaré in the domain of celestial mechanics. It is an achievement so concentrated and so profound that it is far from having been assimilated by those present-day investigators who continue to concern themselves – and profitably so – with a more classical celestial mechanics, and who are well aware that the exploitation of Poincaré's ideas, for instance those relative to periodic orbits, can supply the key to many problems.

Conclusion

The death of Poincaré in 1912 marks the end of the triumphal epoch of classical celestial mechanics. An astrophysics whose conquests were becoming ever more numerous and spectacular was thereafter to be the focus of attraction for students. It began to seem impossible to improve on the work of Le Verrier, Hill, Newcomb, and Brown, and for several decades the observers themselves ceased to make observations of the positions of planets and satellites, to the great regret of present-day astronomers. Almost alone J. F. Chazy undertook to make the connection between classical celestial mechanics and general relativity, and William de Sitter re-did the theory of the Galilean satellites.

In 1960, at last, there came a renewal of celestial mechanics – an extremely fruitful one, impelled forward by the prospect of the exploration of space, and aided by the advent of powerful computers. Yet, as we look forward to the moment when present-day investigators will have assimilated the ideas of Poincaré and other modern mathematicians, and vanquished the difficulties involved in bringing celestial mechanics completely under the aegis of general relativity, it is interesting to note that the brilliant results obtained recently are direct prolongations of the achievement of the "golden age"

Further reading

The work of many of the authors referred to appeared in publications that are not widely accessible but may be found in the libraries of nearly all observatories and universities. This is so, at least, for the following: *Annales de l'Observatoire de Paris, Monthly Notices of the Royal Astronomical Society, Astronomical Papers of the American Ephemeris, Astronomical Journal, Comptes rendus de l'Académie des Sciences, Astronomische Nachrichten.*

Tisserand's *Traité de mécanique céleste*, which was published by Gauthier–Villars in Paris between 1888 and 1896, contains expositions of methods and theories that are often clearer than those given by the original authors; this is the case, for instance, for Hansen's theory. The first two volumes of Tisserand's work were reissued by Gauthier–Villars in 1960.

For the discovery of Neptune, we refer the reader to *The Discovery of Neptune* by Morton Grosser, first published in 1962 and republished in paperback in 1979; it contains an extensive bibliography.

There is is also much to be learned from the relevant entries in the *Dictionary of Scientific Biography*, in particular those on Jacobi, Hamilton, Le Verrier, and Poincaré.

Appendix:
The stability of the solar system from Laplace to the present
JACQUES LASKAR

As we have seen in Chapter 22, Pierre-Simon Laplace (1749–1827) established, to his own satisfaction and that of his contemporaries, the stability of the solar system: it was subject only to small, pendulum-like oscillations of short or long period. Before turning to the nineteenth-century developments, we review the nature of the relevant Laplacian proofs.

Motion in accordance with Kepler's three laws can be considered as motion of order zero; this is the motion that the planets would have if their masses were nil, so that they would not perturb each other's motions. Laplace's work consisted in describing the approximation of order 1. He studied the variations of the orbital elements of the ellipses (of order zero) due to perturbations caused by the planets taken as moving on the orbits of order zero. The types of inequality arising in the approximation of order 1, according to Laplace, were two: (1) *short-period inequalities*, which depend on the situations of the bodies either with respect to each other or with respect to their aphelia, and which have periods of only a few years; and (2) *secular inequalities*, which alter the orbital elements almost imperceptibly in each revolution, but end after millions of years by changing entirely the shapes and positions of the orbits. The latter inequalities are those that put in question the stability of the solar system.

An initial question was whether there were secular terms in the semi-major axes or, equivalently, in the periods or mean motions of the planets (according to Kepler's third law, the mean motion n of a planet is related to the semi-major axis a by the equation $n^2 a^3 = \text{const}$). A change in the semi-major axes of the planetary orbits could have disastrous consequences, for a planet might then break loose from the system or two planets collide.

Laplace showed that, to an approximation carried to the third order in the eccentricities, there existed no secular terms in the semi-major axes or in the mean motions of the planets. Lagrange then extended this result to all powers of the eccentricities. According to Laplace, "M. de la Grange has extended [this result] to an unlimited time, in showing by an ingenious and simple analysis that the mean solar distances of the planets are immutable." But this statement exaggerates: the approximation carried out by Laplace and J. L. Lagrange takes into account only terms of the first order with respect to the planetary masses – a point to which we shall return.

These proofs were apparently contradicted by an observational result: Edmond Halley – here rediscovering an anomaly already detected by Kepler in 1625 – in the 1690s had found Jupiter to be accelerating, and Saturn decelerating, at rates such that the displacements in 2000 years amounted to $+3°\ 49'$ for Jupiter and $-9°\ 16'$ for Saturn.

Laplace resolved the discrepancy. He showed that the apparent variations in the mean motions of Jupiter and Saturn arose from an inequality involving the positions of both planets, but having a period sufficiently long to resemble a secular inequality. The inequality depended on the combination of longitudes $2\lambda_J - 5\lambda_S$; its period was about 900 years. His complete theory of the couple Jupiter–Saturn proved to be in excellent agreement with both ancient and modern observations. He drew the important corollary that the masses of comets must be very small, for otherwise they would have perturbed the motion of Saturn. Thus the new theory confirmed the constancy of the mean semi-major axes, and established Newton's law as the ultimate formula for the precise description of the motion of the celestial bodies in the solar system.

But constancy of the semi-major axes was not

sufficient for stability. For if the eccentricities of two orbits could vary considerably, these orbits could intersect, even though their semi-major axes remained constant. Laplace therefore turned to the problem of the eccentricities and inclinations.

For greater simplicity, we use complex notation, similar to that introduced by Simon Newcomb; it differs in nothing essential from the notation used by Laplace. We set

$$z = e \exp(\sqrt{-1}\,\tilde{\omega}),$$
$$\zeta = \sin(i/2)\exp(\sqrt{-1}\,\Omega),$$

where e, i, $\tilde{\omega}$, and Ω are the eccentricity, inclination, longitude of the perihelion and longitude of the node. Laplace showed that, if one considers terms of the first order in relation to the masses, and of the first degree in the eccentricities and inclinations, the solutions for the variables z_i and ζ_i of the different planets are given by a system of linear differential equations with constant coefficients:

$$\frac{d}{dt}\begin{bmatrix} z_1 \\ \cdot \\ \cdot \\ z_k \\ \zeta_1 \\ \cdot \\ \cdot \\ \zeta_k \end{bmatrix} = \sqrt{-1}\begin{bmatrix} A & 0 \\ 0 & B \end{bmatrix}\begin{bmatrix} z_1 \\ \cdot \\ \cdot \\ z_k \\ \zeta_1 \\ \cdot \\ \cdot \\ \zeta_k \end{bmatrix}$$

where A and B are real matrices with constant coefficients. The two systems in eccentricity and inclination are therefore decoupled, and their solutions are combinations of complex exponentials of the form

$$z = \sum_{j=1}^{k} a_j \exp(\sqrt{-1}\,g_j t),$$

$$\zeta = \sum_{j=1}^{k} \beta_j \exp(\sqrt{-1}\,s_j t),$$

where a_j, β_j are constant values, the g_j are the eigenvalues of the matrix A, and the s_j the eigenvalues of the matrix B.

The whole problem of the stability of the solar system thus reduced to the calculation of the eigenvalues of the two real matrices A and B; these were 7×7 matrices in Laplace's time, since Neptune had not yet been discovered. If the eigenvalues were all real and distinct, then the solutions were quasi-periodic, and the eccentricities and inclinations were subject only to periodic variations about their mean values. But if one of the eigenvalues had a non-zero imaginary part, then the solutions contained an exponential instability, and an eccentricity could become very great so that the possibility arose of two planets colliding. The possibility of collision also arose if two of the eigenvalues were identical.

In Laplace's time, the numerical calculation of the eigenvalues of a 7×7 matrix was not an easy matter. Laplace avoided this problem by finding a very simple demonstration showing all the eigenvalues of these matrices to be real and distinct – a demonstration valid, moreover, for any number of planets revolving about the Sun in the same direction. It is based on the constancy of the angular momentum. The total angular momentum of the system, if we neglect terms of order 2 with respect to the masses, is

$$C = \sum_{i=1}^{k} m_i \sqrt{a_i(1-e_i^2)}\,\cos i_i.$$

Noting that the semi-major axes are constant, and neglecting terms of degree 4 in the eccentricities and inclinations, Laplace derived the result that

$$\sum_{i=1}^{k} m_i \sqrt{a_i}\left(\frac{1}{2}e_i^2 + 2\sin^2\frac{i_i}{2}\right) = \text{const.}$$

But to this approximation, the eccentricities are independent of the inclinations. Laplace could thus write

$$\sum_{i=1}^{k} m_i \sqrt{a_i}\,e_i^2 = \text{const,}$$

$$\sum_{i=1}^{k} m_i \sqrt{a_i}\,\sin^2\frac{i_i}{2} = \text{const.}$$

These equations must hold for all time; hence there could be neither a term proportional to the time nor a real exponential in the solutions, and Laplace therefore concluded that the eigenvalues of the matrices A and B were all real and distinct.

In sum, Laplace had shown that (1) the semi-major axes of the planets were constant, except for periodic variations of small amplitude and relatively short period; (2) the eccentricities and inclinations were subject only to periodic variations about their mean values. The latter variations included both short-period oscillations dependent on the positions of the planets, and secular variations of the orbits with periods ranging from about 50 000 to some millions of years.

Laplace's calculations, however, involved two fundamental approximations. He carried out the developments only to the first order in the masses, neglecting terms that depended on the squares and products of the planetary masses. And while the invariance of the semi-major axes was proved for all powers of the eccentricities and inclinations, the result concerning the smallness and quasi-periodicity of the variations in the eccentricities and inclinations was obtained only by neglecting terms of degree greater than the second. Because of these approximations, the duration of the validity of Laplace's proofs was limited. Although conscious of the dependence of his proofs on approximations, Laplace appears to have underestimated their importance.

The calculations of Le Verrier: a new formulation of the problem of the stability of the solar system

Urbain Jean Joseph Le Verrier (1811–77), besides playing a major role in the discovery of Neptune and carrying out the immense task of constructing new and more precise theories of all the planets, also pursued, in Vol. 2 of the *Annales de l'Observatoire de Paris* (1856), the study of the stability of the solar system. Having calculated the perturbing function between two planets to degree 7 in the eccentricities and degree 6 in the mutual inclination of the planets, he returned to Laplace's calculation of the secular variations of the eccentricities and inclinations, and pushed it to a higher order of approximation. Laplace had carried out the calculation only for the quadratic part of the perturbing function, and only for the planets Jupiter and Saturn. Le Verrier pushed the calculations to degree 4, and took all the planets into account. He found that in this case Laplace's beautiful demonstration of the stability of the solar system no longer held, or rather, no longer had the same implications.

Laplace had employed the conservation of angular momentum, truncated at degree 2 in the eccentricities and inclination, to obtain

$$\sum_{i=1}^{k} m_i \sqrt{a_i} e_i^2 = \text{const.} \qquad (1)$$

This relation puts a limit on possible values of the eccentricities, and hence leads to the conclusion

that the eigenvalues of the secular system are all pure imaginaries; but the limit obtained is of practical interest only if the masses are of approximately the same value, as in the case of Jupiter and Saturn. As Le Verrier put the matter,

The eccentricity e will always remain very small if the planet m constitutes a considerable part of the sum of the masses of the system of planets. But no analogous conclusion can be drawn with regard to the planets whose masses are a small fraction of the total mass of the system; only the complete integration of the equations can show, with respect to these planets, whether their eccentricities will remain confined within narrow limits.

Thus the existence of a quasi-periodic solution for the secular system no longer sufficed to establish the stability of the solar system; in addition it was necessary that the amplitudes of the coefficients not be too large – a condition not guaranteed by the conservation of angular momentum. Le Verrier thus reformulated the conditions for stability:

The conditions necessary for the stability of our planetary system, relative to the eccentricities, are of two kinds: the first have to do with the nature of the roots of the equation in [g_i and s_i as above]; the second with the absolute magnitude of the coefficients, N, $N_1, \ldots, N'$, N'_1[α_j and β_j as above], etc.

Le Verrier also raised the question of the convergence of the series approximations he employed. The approximations carried out *a priori*, he knew, had a meaning only if the eccentricities and inclinations remained small:

Only by calculating the coefficients N, N', ... can we know whether any of them will be large, and whether, as a consequence, any of the eccentricities can increase in such a way as to make the series – on which this entire analysis is based – but slowly convergent or even divergent. For the equation [(1) above], in which the value of the constant in the right-hand member is given by actual observations, shows us that the eccentricities of planets of large mass cannot increase beyond rather narrow limits; but it tells us nothing about the limits of the orbital eccentricities of planets of small mass.

Le Verrier therefore launched into a calculation of the eigenvalues; it consisted in resolving an equation of degree 7 (he omitted Neptune, initially, from the account). The task was extremely complicated, and involved successive approximations, with a decoupling of the systems of the planets. He obtained a complete solution for the linear secular system of the seven planets, and so was able to derive maximal values for the eccentricities and inclinations, and thus to show that these when calculated by the secular equations of the first order remained "eternally small'. What was new in his study was the consideration of the bodies of small mass, in particular the interior planets. For them, this result was obtained, Le Verrier emphasized, "only for the relations of the major axes that we have considered: we do not know what the result would be for other mean distances of the planets."

Le Verrier also laid the basis for a study of the secular resonances of the minor planets (asteroids).

There exists, for example, a position between Jupiter and the Sun such that if one placed there a small mass, in an orbit initially but little inclined to the orbit of Jupiter, this small mass could depart from its original orbit, and attain a large inclination relative to the plane of the orbit of Jupiter, owing to the action of Jupiter and Saturn. It is remarkable that this position is located approximately at a distance double that of the Earth from the Sun, that is, at the lower limit of the zone where all the minor planets have so far been found.

For the solar distance of the position of resonance Le Verrier found 1.977 AU, a value very close to that accepted today. Later, Le Verrier also suggested that the large inclination of Mercury's orbit could be due to secular resonances.

Le Verrier thus pushed the study of the linear system much further than Laplace. In a yet more important development, he raised the question of the influence of non-linear terms, that is, terms of degree 3 in the secular equations, or of degree 4 in the perturbing function.

Laplace's analysis had in effect yielded only the first term of an infinite series. This term was no doubt the most important part of the solution; but, did the later terms have only a negligible effect? Le Verrier posed the question of convergence in the manner of an astronomer who examines only the early terms of the series to see if they decrease with sufficient rapidity: he did not raise the problem of *mathematical convergence*. He then recognized that, for some initial conditions, the series failed to converge; he considered also that Laplace's approximation was insufficiently precise to yield a

solution valid for a very long time; thus it was necessary to have regard to terms of higher degree, namely of degree 3 in the eccentricities and inclinations.

Here is the programme Le Verrier set for himself:

We propose to determine whether, by the method of successive approximations, the integrals do in fact develop in series sufficiently convergent so that one can answer the question about the stability of the solar system; and, in the second place, to give to these integrals all the accuracy possible in the present state of our knowledge of the masses of the planets.

We shall give an indication of Le Verrier's calculations in a simple example illustrating the problem of 'small divisors' in celestial mechanics. Suppose that we have already carried out Laplace's first reduction, and let us consider only a single non-linear term in one of the equations:

$$\frac{dz_j}{dt} = i(c_j z_j + z_1 z_2 z_3)$$

with $z_j(0) = A_j$, and where $z_1 z_2 z_3$ is a term of degree 3. If one considers only the linear part, one obtains by integration

$$z_j(t) = A_j e^{ic_j t}.$$

We seek a solution for small eccentricities e; in this case $z_1 z_2 z_3$ is very small, and we can assume that this term leads to only a small correction in the solution of the linear part. We introduce a small formal parameter, say ϵ, in carrying out a change of variables $z_j = \sqrt{\epsilon} z_j$ (the square root makes things prettier). By changing also the initial conditions $A_j = \sqrt{\epsilon} A_j$, we then obtain

$$\frac{dz_j}{dt} = i(c_j z_j + \epsilon z_1 z_2 z_3)$$

with the initial condition $z_j(0) = A_j$. We seek a solution in the form of an integral series in ϵ:

$$z_j = z_j^{(0)} + \epsilon z_j^{(1)} + \epsilon^2 z_j^{(2)} + \dots.$$

There are many possibilities for the choice of the initial conditions of the $z_j^{(0)}$, and to simplify, we choose $z_j^{(0)}(0) = A_j$, $z_j^{(i)}(0) = 0$ for $i \neq 0$. By carrying out the development to the order 1 in ϵ and identifying terms in the same powers of ϵ, we find

$$\frac{dz_j^{(0)}}{dt} = ic_j z_j^{(0)}; \quad z_j^{(0)}(0) = A_j$$

$$\frac{dz_j^{(1)}}{dt} = i(c_j z_j^{(1)} + z_1^{(0)} z_2^{(0)} z_3^{(0)}); \quad z_j^{(1)}(0) = 0.$$

The whole problem reduces to the resolution of the last equation with

$$z_j^{(0)}(t) = A_j e^{ic_j t}.$$

This gives, if we denote $A_1 A_2 A_3$ by B,

$$\frac{dz_j^{(1)}}{dt} = i(c_j z_j^{(1)} + B e^{i(c_1 + c_2 + c_3)t}).$$

This last equation is resolved easily by the method of variation of constants introduced by Lagrange, yielding

$$z_j^{(1)}(t) = \{iB \int e^{i(c_1 + c_2 + c_3 - c_j)t} dt + C\} e^{ic_j t},$$

where C is a constant of integration. If $c_1 + c_2 + c_3 - c_j$ is not zero, we obtain

$$z_j^{(1)}(t) = \frac{1}{c_1 + c_2 + c_3 - c_j}(B e^{i(c_1 + c_2 + c_3)t} - B e^{ic_j t}),$$

for we have $z_j^{(1)}(0) = 0$. We see that the terms of degree 3 add a term of frequency $c_1 + c_2 + c_3$ and correct the amplitudes of the linear part. No difficulties arise unless the combination of frequencies $c_1 + c_2 + c_3 - c_j$ is very small, in which case the effect of the degree 3 can become greater than that of the linear part, strongly contradicting a hypothetical convergence. This is the problem of the small divisors in celestial mechanics. Le Verrier did not face this problem for the moment because he had a more serious problem to deal with. For in the part of degree 3 there also exist terms for which the combination of frequencies is of the form $c_j + c_k - c_k - c_j = 0$. Such a term is said to be resonant, and one then has

$$z_j^{(1)}(t) = iBt e^{ic_j t}.$$

Thus a mixed term appears in the solution, with an amplitude Bt which increases with time; if this term corresponds to a physical reality, the stability of the solar system is compromised.

Le Verrier proceeded to eliminate the mixed term:

We have taken as point of departure the rigorous integrals obtained by keeping only the terms of the first order, and it is by varying the arbitrary constants introduced into these equations by integration that we have been able to take into account the terms of the third order. This procedure produced arcs de cercle *outside the sine and cosine signs; but we can eliminate them by changing appropriately the values of the arguments introduced in the first approximation.*

He introduced here the premisses of the method

that Poincaré would call "the method of Lindstedt" for the resolution of such systems. The solutions causing secular terms to enter were (if we take account only of these terms) of the form

$$z = A e^{ict} + iBt e^{ict}.$$

Le Verrier eliminated the mixed terms in $Bt e^{ict}$ by changing slightly the frequency of precession c, which amounts to considering $1 + itB/A$ as the beginning of a power series in $e^{i\delta ct}$ ($e^{i\delta ct} = 1 + i\delta ct \ldots$), where $\delta c = B/A$ is the correction of the frequency. He found corrections amounting to $0''.3231$ per year for the largest frequency g_6 related to the Jupiter–Saturn couple ($g_s = 22''.427$). A modification this large surprised him, for he knew that Laplace and Lagrange had thought their formulas to be valid for an infinite time: "[these] illustrious mathematicians ... were far from thinking that the terms of the third order could introduce into the arguments such large corrections".

To give an order of magnitude, one can say that an uncertainty of $0''.3$ per year in one of the frequencies of precession limits the duration of the validity of the solution to about three million years. But Le Verrier's judgment of his predecessors should perhaps be tempered, since the age of the Earth admitted in Laplace's time attained this value only in the boldest speculations of Georges-Louis Leclerc, Comte de Buffon (in print the latter, concerned perhaps to avoid shocking his readers, went only so far as to propose a duration of 75 000 years).

These corrections, however, were small in relation to the values already determined, and only the value of g_6 was notably changed. Le Verrier therefore believed that at least for the exterior planets, whose masses were well known, these values were very close to the true values of the frequencies:

We limit ourselves therefore to concluding from the preceding calculations that the corrections of the arguments are very small in relation to the arguments themselves, so that the series in which the integrals are developed are regarded as convergent.

Le Verrier appears to have been persuaded that further corrections would be much smaller, and that his solution would be utilizable for a time comparable to the age attributed to the Earth – some tens of millions of years in Le Verrier's time.

By means of this change, one can answer, at any epoch whatever, for the accuracy of the results, to within the limits fixed by means of the formulas we have given for taking account of the uncertainty of the values assigned for the masses. However, we make abstraction of the influence of the constant terms in the perturbing function, that depend on the higher powers of the masses, with which we are not concerned here.

Le Verrier was well advised to enter this *caveat*, for George W. Hill in 1897 showed that the contribution of the second order in the masses increased the value of g_6 by about 5″ per year because of the quasi-resonant term $2\lambda_J - 5\lambda_S$ between Jupiter and Saturn.

Finally, Le Verrier posed the question of the existence of small divisors in the system of interior planets – an especially important question because some of the masses were poorly known, and an admissible change of mass could make a divisor very small. The lack of determination of these masses, however, prevented him from settling the matter, and all he could do was to appeal to the mathematicians:

It thus appears impossible, by the method of successive approximations, to decide whether, in virtue of the terms of the second approximation, the system composed of Mercury, Venus, Earth, and Mars will enjoy stability indefinitely; and one must hope that the mathematicians, by the integration of the differential equations, will give a means of removing this difficulty, which could very well be only a matter of form.

Poincaré: the response of mathematics

Laplace had established the stability of the solar system in taking into account the linear approximation of the equations of its motion. Le Verrier, we have seen, was not so optimistic. He showed that the problems of small divisors met with in the system of interior planets prevented him from drawing a conclusion as to the stability of the solar system, and he asked the mathematicians to find new methods for integrating the differential equations of the motion, so as to be able to resolve this important problem.

Le Verrier had examined the influence of the non-linear terms in the development of the equations of the secular motion of the planets; his method was to investigate solutions in the form of series depending on a small parameter. He did not pose the problem of the convergence of these series as a mathematician would do it today. For him, a series converged, or rather converged sufficiently, if the successive approximations introduced successively smaller corrections. He thought that he would thus obtain a formula which would approach the true motion of the planets. For Le Verrier, this convergence did not hold universally, but depended on initial conditions. He knew that for certain of these initial conditions there emerged very small divisors that compromised the convergence of the series. The implied resonances could correspond to real instabilities, such as that of the orbits of minor planets with semi-major axes less than about 2 AU. But he was unable to reach systematic results, and his methods, despite improvements later introduced by Newcomb and Lindstedt, could not be extended easily to higher orders. Astronomy had arrived at an impasse; hence Le Verrier's appeal to the mathematicians.

The response to the question posed by Le Verrier was given some years later by Henri Poincaré (1854–1912). In the three volumes of his *Méthodes nouvelles de la mécanique céleste* (1892–99), Poincaré took an altogether new approach, and transformed celestial mechanics totally. He undertook to simplify the equations in order to study them in their greatest generality. For this purpose, he took advantage of the canonical formalism introduced by C. G. J. Jacobi (1804–51) and W. R. Hamilton (1805–65). He proposed to the astronomers a set of variables, very close to the variables that they had used since Laplace, but permitting a considerable simplification in the writing of the equations of motion, which thus became

$$\begin{cases} \dfrac{dI_i}{dt} = \dfrac{\partial H(I,\theta)}{\partial \theta_i} \\[2mm] \dfrac{d\theta_i}{dt} = -\dfrac{\partial H(I,\theta)}{\partial I_i} \end{cases} \tag{2}$$

where (I_i, θ_i) are called canonical variables. The variables I_i are related to the semi-major axes, eccentricities, or inclinations of the planets, while the angular variables θ_i represent the longitudes of the planets, perihelia and nodes. The function $H(I,\theta)$, or Hamiltonian, represents the energy of the system. In the case of the solar system, this function takes the special form

$$H(I,\theta) = H_0(I) + \epsilon H_1(I,\theta), \tag{3}$$

where $H_0(I)$ represents the energy of the planets orbiting round the Sun when one neglects their interactions, while $\epsilon H_1(I,\theta)$ represents the mutual interaction of the planets which is much smaller (the small parameter ϵ is of the order of the planetary masses). If one neglects the interaction of the planets, the second term is suppressed, and the equations can be integrated very simply because the variables of action I_i are then constant and the angles θ_i are linear functions of the time ($\theta_i = \nu_i t + \theta_{i0}$). In this case the motion can be considered as compounded of uniform circular motions with radii I_i and angles θ_i; the number of couples (I_i, θ_i) is called the number of degrees of freedom of the system. In the case of such a system with two degrees of freedom, the motions will be represented by curves on a torus. One sees here the power of this formalisation, for to seek to integrate these equations of motion now reduces to finding transformations which reduce the motion (which can appear complex) to the products of uniform circular motions. Poincaré showed that all the perturbational methods available to Laplace and Le Verrier reduced to seeking changes of variables in the form of series in the small parameter ϵ

$$\begin{cases} I' = I + \epsilon S_1(I,\theta) + \epsilon^2 S_2(I,\theta) + \epsilon^3 S_3(I,\theta) + \ldots \\ \theta' = \theta + \epsilon T_1(I,\theta) + \epsilon^2 T_2(I,\theta) + \epsilon^3 T_3(I,\theta) + \ldots \end{cases} \quad (4)$$

so as to keep the simple form of Equations (2), while seeking to transform the Hamiltonian $H(I,\theta)$ into a new Hamiltonian which depends only on the variables of action (I_i). The solutions would then be combinations of uniform circular motions, corresponding in the original variables to quasi-periodic motions, combinations of periodic motions.

The simplification introduced by Poincaré permitted him to respond to Le Verrier's plea, but his response was negative. For in his memoir "Sur le problème des trois corps et les équations de la dynamique", he showed the impossibility of integrating the equations of motion of three bodies (the Sun, Jupiter, and Saturn, for instance) as one could integrate the equations for two bodies. The only thing that could be done was to seek approximate solutions with the help of perturbational methods.

Poincaré showed that the "methods of successive approximation" used by astronomers could be extended to all orders. But he also showed that the series of Lindstedt, which in general defined these changes of variables and which made it possible to

obtain quasi-periodic solutions, were divergent, because of the arbitrarily small divisors that always appeared in the expression of them. The result was that, contrary to Le Verrier's hope, it was impossible to find series that converged towards quasi-periodic solutions for the whole domain of initial conditions.

These results profoundly disturbed the astronomers of the time, for they had been persuaded that their series would converge provided that they took certain precautions. But Poincaré left them some hope, for he knew that their methods permitted them to calculate the motions of the planets with precision for a finite time; he proceeded to clarify the implications of his discoveries. He made precise the difference in meaning of the expression 'convergent series' for the astronomers and for the mathematicians, and laid the basis of the formal calculation that could be made with divergent series:

To take a simple example, let us consider the two series that have for general term

$$\frac{1000^n}{1 \cdot 2 \cdot 3 \cdots \cdot n}, \qquad \frac{1 \cdot 2 \cdot 3 \cdots \cdot n}{1000^n}.$$

The mathematicians will say that the first series converges, and even that it converges rapidly, because the millionth term is much smaller than the 999 999th; but they regard the second as divergent, because the general term can increase beyond all limits.

The astronomers, on the contrary, regard the first series as divergent, because the first 1000 terms show an increase; and the second as convergent, because the first 1000 terms show a decrease, and this decrease is at first very rapid.

The two rules are both legitimate: the first, in theoretical investigations; the second, in numerical applications.

The series used by the astronomers are thus in general divergent, but they can serve to approximate the motion of the planets for a certain time, which can be long, but not infinite. Poincaré did not seem to think that his results could have much practical importance, except in the study of the stability of the solar system:

The terms of these series, in fact, decrease very rapidly at first, and then begin to increase; but since the astronomers stop with the first terms of the series, long before these terms have ceased to decrease, the approximation is sufficient in practice. The divergence of these expansions would be inconvenient only if one wished to use them to establish rigorously certain results, for instance the stability of the solar system.

By "the stability of the solar system" Poincaré here meant its stability over an infinite time, which is very different from its stability in a practical sense, that is, for an interval of time comparable to its expected duration.

Le Verrier had reformulated the question of stability by pointing out that it was necessary to take account of terms of higher degree than those considered by Laplace and Lagrange; Poincaré was still more demanding, in requiring the convergence of series. Possibly, under certain conditions, the series used by the astronomers might be convergent, but he regarded this as improbable.

Postscript: the modern results

Around 1960, the mathematicians A. N. Kolmogorov, V. I. Arnold, and J. Moser demonstrated that, contrary to Poincaré's intuition, certain initial conditions can lead to convergent series and thus to quasi-periodic solutions. But these conditions required very small values for the planetary masses, eccentricities, and inclinations, and could not be applied directly to our solar system.

In 1988 the Americans G. Sussman and J. Wisdom at MIT, using a computerized numerical integration of the equations of motion for the exterior planets (Jupiter, Saturn, Uranus, Neptune, and Pluto) over a duration of 875 million years, found Pluto's motion to be chaotic. That is, this motion proved to be extremely sensitive to initial conditions: the uncertainty in the initial conditions increases by a factor of 3 every 20 million years, so that prediction beyond about 400 million years becomes impossible. Because Pluto's mass is relatively small ($\frac{1}{130\,000\,000}$ of the solar mass), no macroscopic instability results for the rest of the solar system; the motion of the large planets remains very regular.

A little afterwards, using a very different method, J. Laskar obtained similar results for the interior planets (Mercury, Venus, Earth, and Mars). Because of the very rapid orbital motion of these planets, direct numerical integration over comparable intervals of time was not within the reach of computers. To obtain results for the whole of the solar system, it was first necessary — using methods similar to those of Laplace and Le Verrier — to transform the Newtonian equations into a new and much larger system (it contained 150 000 terms), which no longer represented the motions of the planets but rather the mean motions of their orbits. It was then possible, in only a few hours of calculation, to integrate numerically the motion of the solar system over 200 million years. The outcome was surprising. For the large planets, the result was a regular motion, as given also by direct numerical integration, but for the interior planets the behaviour of the trajectories was chaotic. Initial uncertainties were found to increase by a factor of 3 every 5 million years, preventing all prediction beyond about 100 million years. An error of 15 metres in the initial conditions produced an error of only 150 metres at the end of 10 million years, but an error of 150 000 000 kilometres after 100 million years.

More recently, the origin of this chaotic motion has been identified: it comes from resonances between the periods of precession of the orbits of Mars and the Earth on the one hand, and between those of Mercury, Venus, and Jupiter on the other. In the neighbourhood of such resonances, there exist regions in which the dynamics is complex and highly sensitive to initial conditions.

Thus celestial mechanics, which Laplace erected as the model *par excellence* of predictable science, has shown its limits. A new formulation of the problem of stability is imposed on us today. The solar system, we have shown, is unstable; it is a matter now of knowing with precision the effects of the instabilities for a time comparable to the system's age. To do this, it is necessary to study globally all the neighbouring trajectories, and thus follow the way opened a hundred years ago by Henri Poincaré. A better knowledge of the ensemble of these motions will not permit us to predict whether a catastrophic event, such as a sudden increase in the Earth's orbital eccentricity, will actually happen in the next billion years; it will authorise us only to say whether such an event, within such a period of time, is possible or not.

Further reading

P. S. Laplace, *Exposition du système du monde* (five different editions) (Paris, 1796–1835)

J. Laskar, A numerical experiment on the chaotic behaviour of the solar system, *Nature*, vol. 338 (1989), 237–8

J. Laskar, The chaotic behaviour of the solar system: A numerical estimate of the size of the chaotic zones, *Icarus*, vol. 88 (1990), 266–91

J. Laskar, La stabilité du système solaire, in *Chaos et déterminisme*, ed. by A. Dahan Dalmedico *et al.* (Paris, 1992), 170–211.

U. J. J. Le Verrier, *Annales de l'Observatoire de Paris*, vol. 2 (Paris, 1856)

Henri Poincaré, *Méthodes nouvelles de la mécanique céleste* (Paris, 1892–99)

G. J. Sussman & J. Wisdom, Numerical evidence that the motion of Pluto is chaotic, *Science*, vol. 241 (1988), 433–7

PART IX

The application of celestial mechanics to the solar system to the end of the nineteenth century

29

Three centuries of lunar and planetary ephemerides and tables
BRUNO MORANDO

Almanacs giving predictions of notable celestial events for the successive months of the year have an origin lost in Antiquity. Ephemerides – tables of day-to-day or month-to-month positions of the Moon and planets – were already being constructed in the fourth century BC in Seleucid Babylon. But a new era in lunar and planetary prediction began in the late sixteenth century with Tycho Brahe's observational work; on the basis of the large body of accurate observations that Tycho compiled, Johannes Kepler was able to discover his laws of planetary motion, and so to develop unprecedentedly accurate planetary tables. Later in the century, the introduction of new and more precise instruments of observation – the filar micrometer, the pendulum clock, and telescopic sights applied to graduated arcs – led to further increments in predictive accuracy. Finally, in the 1750s, the law of universal gravitation that Isaac Newton had announced in 1687 at last began to have an effect upon prediction, as the approximate solution of differential equations was brought to bear upon the problem (see Chapter 18). Successive refinements in theory, as well as in instruments and techniques of observation, would characterize the further history of lunar and planetary prediction up to the present day.

These increments in predictive accuracy had as their consequence the appearance of ephemerides of an increasing accuracy and precision. Simple traditional calendars and almanacs – often filled with absurd information – continued to appear, but where accuracy of prediction mattered, as it crucially did in navigation and astronomy, the more accurate ephemerides were increasingly in demand. This trend culminated in the emergence of the great national nautical almanacs.

The new ephemerides fell into two classes. On the one hand, there were *annual* ephemerides, their tables giving the day-by-day coordinates of the bodies of the solar system over an entire year. Their form, admitting of easy interpolation, varied little from year to year. They were constructed at first by isolated astronomers, then by national bureaux such as the Bureau des Longitudes and the British and American Nautical Almanac Offices.

On the other hand, there were tables of various astronomical data, collections of planetary ephemerides or of star positions, which covered periods of time other than the year. These were intended for particular clienteles, often mariners. They became very numerous during the eighteenth century, so that it would be difficult to make an exhaustive list, but we shall be citing some of them.

But the word 'tables' in astronomy has yet another important meaning. The theories of celestial mechanics give the positions of the planets and the Moon in the form of series composed of a large number of terms, usually trigonometrical; in accordance with an astronomical tradition that predated celestial mechanics but persisted into the nineteenth century, the individual terms were called 'equations'. The summation of these series for each body, and for each date of a projected ephemeris, entailed calculations that were – given the means of computation available – truly horrendous. To simplify the calculations, astronomers derived, on the basis of the theory, tables of single or double entry, each one yielding either a single term or several terms of a series by linear interpolation; the several 'equations' had then only to be added together. Until the advent of electronic computers in the mid-twentieth century, these tables played an enormous role. We shall devote a section to the most important of them, and to the astronomers who compiled them.

Astronomical ephemerides and tables

Ephemerides before the national nautical almanacs

As we have remarked, the extraordinary progress in astronomy that began with the work of Tycho Brahe and Kepler led to ephemerides of a new order of accuracy. But, as also mentioned, the new ephemerides were successors in a tradition that stretched back to Antiquity. Collections of diverse predictions arranged according to the calendar had long been in common use. They were called *almanacs* (a term of Arabic origin that appeared in Medieval Latin by the thirteenth century) or *ephemerides* (a name of Greek origin that was already used for the *Ephemerides* of Johannes Regiomontanus published in 1474). The almanacs tended to contain instruction of all sorts, mixed with astronomical predictions such as those of eclipses and phases of the Moon. The positions of the planets and Moon, whether in almanacs or ephemerides, were calculated from tables based in part on empirical data and in part on purely kinematic theories employing epicycles and deferents to describe the motions. The most widely used of these tables up to the seventeenth century were the *Alfonsine Tables*, traditionally associated with the thirteenth-century King of Castile, Alfonso X, and the *Prutenic Tables*, prepared by Erasmus Reinhold on the basis of Nicholas Copernicus's *De revolutionibus* and published in 1551.

The advent of printing with movable type gave rise to numerous printed ephemerides. As early as 1471, a wealthy inhabitant of Nuremberg, Bernhard Walther, founded a printing-house for the express purpose of publishing ephemerides, and engaged Regiomontanus who computed ephemerides for the years 1475–1506. Subsequently, a succession of astronomers took up the task: Johann Stöffler, Cyprian Leovitius, Joannes Stadius, and others.

A forward leap in predictive accuracy came with the introduction of Kepler's elliptical orbit and areal rule, incorporated in his *Tabulae Rudolphinae (Rudolphine Tables)*, which were published at Ulm in 1627. Kepler himself calculated and published ephemerides for the years 1617–37. These were continued in Danzig by Lorenz Eichstadt and then by Johann Hecker, who computed positions for the period from 1666 to 1680.

When no Danzig successor to Hecker appeared,

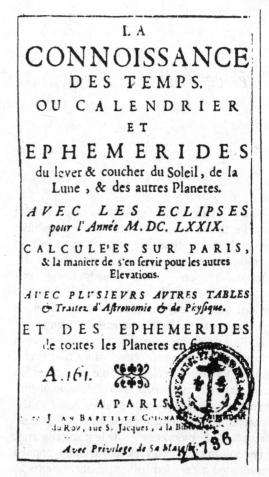

LA
CONNOISSANCE
DES TEMPS.
OU CALENDRIER
ET
EPHEMERIDES
du lever & coucher du Soleil, de la Lune, & des autres Planetes.

AVEC LES ECLIPSES pour l'Année M.DC. LXXIX.

CALCULE'ES SUR PARIS, & la maniere de s'en servir pour les autres Elevations.

AVEC PLVSIEVRS AVTRES TABLES & Traitez d'Astronomie & de Physique.

ET DES EPHEMERIDES de toutes les Planetes en

A.161.

A PARIS,

J AN BAPTISTE COIGNARD
du Roy, rue S. Jacques, a la Bib...

Avec Privilege de Sa Majesté

29.1. Title-page of the first volume of the *Connaissance des temps*.

the astronomers of the newly established Paris Observatory decided that Hecker's ephemerides, which had been reprinted in Paris, should be continued. So appeared the first ephemerides worthy of the title *Connaissance des temps*: a regular annual publication, exact, simple to employ, and conceived specially for the use of those most urgently concerned, astronomers and navigators.

The "Connaissance des temps" and "The Nautical Almanac"

The title of the new publication (see Figure 29.1) may be translated as follows:

The Knowledge of Times or calendar and ephemerides of the rising and setting of the Sun, Moon, and the other planets. With the eclipses for the year 1679 calculated for Paris and the way to use them for other elevations [of the celestial pole]. With several other tables, and treatises on astronomy and physics, and ephemerides of all the planets in figures.

In the preface addressed to the King we read that the ephemeris has been "purged of all the ridiculous things with which works of this sort have been filled till now". The preface is signed only with three stars (***), and the 'privilege' (authorization for publication necessary at the time) states: "It is permitted to Sieur *** to cause to be printed by whatever printer he shall choose ... etc." It was long believed that the person so designated was Jean Picard (1620–82), but it appears that it was rather Joachim Dalencé (*c.* 1640–1707). Of Dalencé we know very little: he purchased a telescope in England in 1668; became the intermediary between Henry Oldenburg, Secretary of the Royal Society, and Christiaan Huygens; and in 1685 settled in the Low Countries where he wrote a treatise on the magnet in 1687 and proposed a thermometer whose points of reference were to be the melting points of ice and butter!

Although the *Connaissance des temps* was directed primarily to astronomers, the general public was not altogether forgotten: according to the preface preceding the tables the first part contained "everything that has been considered useful and necessary to the public, and so easy to apply that the least intelligent can make use of it".

Dalencé retained the publication privilege till 1685, but Picard seems to have carried out many of the calculations, leading A.-G. Pingré to believe that he had been the founder of the ephemeris. The privilege next passed to Jean Le Fèvre (1652?–1706) and after him to Jacques Lieutaud (*c.* 1660–1733), then in 1702 was vested in the Academy of Sciences. The volume for 1702 appeared under the title: *Connaissance des temps pour l'année 1702 au méridien de Paris publiée par l'ordre de l'Académie Royale des Sciences et calculée par M. Lieutaud de la même académie.* Lieutaud continued the calculations for the years through 1726. Among the astronomers who were subsequently charged with the publication were Joseph-Jérôme Lefrançais de Lalande (1732–1807), for the years from 1760 to 1776, and Pierre-François-André Méchain (1744–1804), for the years from 1788 to 1795. The Bureau des Longitudes at its creation in 1795 was assigned the responsibility of publishing the

Connaissance des temps; two centuries later it continues to carry out this task.

Other ephemerides appeared: the Bolognese in 1715, the Viennese in 1757, the Milanese in 1775, those of Berlin in 1776. But the most renowned publication of this kind was the *Nautical Almanac* (Figure 29.2) founded in England in 1767 by Nevil Maskelyne (1732–1811).

Maskelyne, appointed Astronomer Royal in 1765, was a firm defender of the method of lunar distances for the determination of longitude at sea. For this purpose he proposed using the lunar tables of Johann Tobias Mayer (1723–62). Mayer first published his tables in 1753 in Göttingen, but he subsequently refined them. Neither their theoretical basis nor the mathematical methods employed in them were especially novel (see Chapter 18). But in fitting the theory to observations, Mayer was eminently skilful, and consequently the tables were very accurate. After Mayer's death, his widow submitted the amended tables to the British Board of Longitude, which was charged with improving the determination of longitude at sea. In October 1765 Maskelyne persuaded the Board that Mayer's tables should be officially endorsed, and proposed the creation of the *Nautical Almanac*. In the preface to its first volume the series is announced as follows:

The Commissioners of Longitude, in pursuance of the powers vested in them by a late Act of Parliament, present the Publick with the NAUTICAL ALMANAC and ASTRONOMICAL EPHEMERIS for the year 1767, to be continued annually; a Work which must greatly contribute to the Improvement of Astronomy, Geography, and Navigation.

The preface (of which Maskelyne was the author) also claims that Mayer had so perfected the tables of the Moon that one could hope to determine the longitude at sea with an accuracy of the order of a degree.

For a long time the Americans made use of this same almanac, but in 1849 a Nautical Almanac Office was created by act of the Congress of the United States. The first volume of *The American Ephemeris and Nautical Almanac*, for the year 1855, was published in 1852. Since 1981 the two publications have been united under the title *The Astronomical Almanac.*

THE

NAUTICAL ALMANAC

AND

ASTRONOMICAL EPHEMERIS,

FOR THE YEAR 1767.

Published by ORDER of the

COMMISSIONERS OF LONGITUDE.

LONDON:

Printed by W. RICHARDSON and S. CLARK,
PRINTERS;

AND SOLD BY

J. NOURSE, in the Strand, and Meff. MOUNT and PAGE,
on Tower-Hill,
Bookfellers to the faid COMMISSIONERS.
M DCC LXVI.

29.2. Title-page of the first volume of the *Nautical Almanac*.

Other tables and ephemerides

As indicated earlier, besides the regular annual ephemerides there appeared numerous tables for periods longer than the year, intended for special clienteles. Eustachio Manfredi, an astronomer at Bologna, published ephemerides for successive decades from 1715 to 1750. His work was carried on till 1786 by Eustachio Zanotti, Petronio Matteucci, and others. Apropos of these ephemerides the Baron Franz Xaver von Zach (1754–1832) remarked in 1819 (on p. 474 of Vol. 2 of his *Correspondance astronomique*):

It is especially to be remarked that it was these ephemerides that constituted the science of the Jesuits in China. There they used them, we know, as a vehicle for introducing and propagating the Christian religion among that idolatrous and ignorant people, whose great knowledge in astronomy they have always so ridiculously and deceitfully praised.... Manfredi's ephemerides were then distributed to the four quarters of the world and a prodigious number of them were sold in all the missions.

The Abbé Nicolas-Louis de Lacaille published ephemerides in three volumes for the years from 1745 to 1774. Then Lalande published three volumes for the years from 1775 to 1804. Another famous producer of ephemerides was Father Giovanni Inghirami (1779–1851), professor of astronomy at the Ximenes Institute of Florence and director of its observatory. He calculated ephemerides of the two brightest planets, Venus and Jupiter, for the years 1820 to 1824; they were intended for the use of navigators, and were published in Zach's *Correspondance astronomique*.

In 1786 Zach had entered the service of the Duke of Saxe-Coburg, who had erected for him an observatory on the Seeberg near Gotha; Zach was director of this observatory until 1806, and there trained a number of astronomers who would later produce planetary tables. In 1809 he himself published in Florence his *Tables abrégées et portatives du Soleil calculée pour le méridien de Paris sur les observations les plus récentes d'après la théorie de M. Laplace*. These tables are followed by tables of the Moon ... *d'après la théorie de M. Laplace et d'après les constantes de M. Bürg*. The idea of publishing ephemerides in a small volume, which has become a preoccupation of the almanac offices in the late twentieth century, is thus not new; Zach, in his preface, pointed out its advantages and disadvantages:

In the older tables of the Sun it was believed sufficient to give four or five equations of perturbation; the new theories require twenty-two. Astronomers, navigators, topographical engineers, amateur astronomers who, either because of their responsibilities or their interests, travel a great deal, find themselves in the necessity and embarrassment of having to go abroad with whole libraries.... One must concede, however, that the advantage of abbreviated tables can be purchased only at the price of lengthening the calculations. For there are only two possibilities: either the volume of the tables must be increased, in order to abbreviate the calculations, or the calculations must be increased, in order to diminish the size and number of the tables.

Tables for the construction of ephemerides

Tables employed to the middle of the nineteenth century

In France a law of 1795 assigned to the Bureau des Longitudes the perfecting of astronomical tables. In 1806 the Bureau published *Tables du soleil* by Delambre and *Tables de la lune* by Bürg. These tables are described in the *Connaissance des temps* for 1808.

Jean-Baptiste-Joseph Delambre (1749–1822) had established the basis of his solar tables in the *Mémoires* of the Academy of Berlin for 1784 and 1785; he then calculated them in 1790 using 300 observations of Maskelyne, and they were published in the third edition of Lalande's *Astronomie* in 1792. In the revised version published in 1806 under the aegis of the Bureau des Longitudes, Delambre had recourse to a much larger number of observations: 718 observations made by James Bradley from 1750 to 1755, 700 new observations made by Maskelyne, and some observations of Alexis Bouvard. According to Delambre, "All these calculations have resulted in only 1″.7 to be subtracted from the mean longitude in 1800 and 3″ to be subtracted from the place of the apogee; the equation of centre remains the same to within a few tenths of an arc-second, but the mean motion has undergone a greater diminution." The constant of precession previously accepted, he found, needed to be reduced by 15″ per century. He was also able to correct the masses of Venus, Mars, and the Moon.

Johan Tobias Bürg (1766–1834), after serving as a calculator at the observatory of Seeberg under the direction of Zach, became a professor at the University of Vienna, then an astronomer at the university's observatory. The lunar tables of Bürg derived from Tobias Mayer's tables, which had been improved by Mason. Charles Mason (1728–86) had been Bradley's assistant at the Royal Greenwich Observatory; after observing the transits of Venus across the Sun in 1761 and 1769, he worked on the *Nautical Almanac*. In 1778, making use of 1200 unpublished observations of Bradley, he published his *Lunar Tables in Longitude and Latitude according to the Newtonian Laws of Gravity*; a second, improved edition appeared in 1780. Bürg found some new 'equations', but despite the fact that he had at his disposal 2000 new observations of Maskelyne, he modified but little the values of Mayer's and Mason's coefficients. In 1806 the Bureau des Longitudes instituted a prize of 6000 francs for the best lunar tables, and this prize was awarded to Bürg after his tables had been compared with P.-S. Laplace's lunar theory as published in the third volume of the *Mécanique céleste*.

In 1812 the Bureau des Longitudes published Burckhardt's *Tables de la lune*. Johan Karl Burckhardt (1773–1825), born in Leipzig, had learned astronomy under the tutelage of Zach. In 1797 Lalande induced him to come to Paris; he became an adjunct member of the Bureau des Longitudes in 1799, then a full member in 1817. His tables were based on Laplace's theory, but many of the coefficients of perturbational terms were still (as in Mayer's and Bürg's tables) adjusted to fit the observations. The Bureau des Longitudes had appointed a special commission consisting of Laplace, Delambre, Bouvard, S.-D. Poisson, and D. F. J. Arago to study Burckhardt's tables; it concluded that they were much more accurate than Bürg's, and the Bureau des Longitudes thereupon adopted them.

In 1824 and 1828 there appeared the *Tables de la lune formées par la seule théorie de l'attraction* by Marie Charles, Baron de Damoiseau. They were derived from the theory entered by Damoiseau in the Paris Academy's contest of 1820, a contest in which Giovanni Plana also had presented a theory of the Moon's motions.

In 1808 Bouvard, who had been Laplace's assistant and had carried out the calculations for the *Mécanique céleste*, published tables of Jupiter and Saturn based on Laplace's theory. They were laid out in the same form as Delambre's solar tables, and were based on observations made at opposition between 1747 and 1804 by Bradley, Maskelyne, Lacaille, and the astronomers of the Paris Observatory. In 1821 Bouvard published new tables of Jupiter and Saturn, utilizing 126 observations of oppositions and quadratures of Jupiter, and 129 of Saturn, made between 1747 and 1814. He justified this new publication not only by the greater number of observations brought to bear but more importantly by the fact that since the publication of his first tables in 1808 Laplace had corrected an error in the sign of the terms of the fifth order in the "great inequality".

To the tables of Jupiter and Saturn of 1821 Bouvard added tables of Uranus. He of course employed the observations made since William Herschel's discovery of the planet in 1781, but he also used seventeen earlier observations, made by astronomers who had taken Uranus for a fixed star: Flamsteed in 1690, 1712 and 1715; Bradley in 1753; Mayer in 1756; and P.-C. Le Monnier who sighted the planet twelve times, including eight times in 1769 alone. Laplace had applied Laplacian theory to this planet to reach a result he considered satisfactory. Barnaba Oriani (1752–1832), director from 1802 of an observatory established by the Jesuits at Milan, and dubbed a count and senator of the realm of Italy by Napoleon, had published (among numerous other astronomical tables) tables for Uranus that appeared to satisfy all the observations of this planet, including the older ones. But Bouvard's tables of 1821 showed an irreducible discrepancy between the older and the more recent observations. This discrepancy, we know, would only increase, and was to result in the discovery of Neptune (see Chapter 28).

Bernhard von Lindenau (1779–1854), after working at the observatory of Seeberg near Gotha under the direction of Zach, became its director in 1808, and remained in this position till 1818 when he was succeeded by Encke. For a time he served as editor-in-chief of the *Monatliche Correspondenz zur Beförderung der Erd- und Himmelskunde*, a journal founded by Zach in 1800; then in 1816 he founded with J. G. F. von Bohnenberger the *Zeitschrift für Astronomie und verwandte Wissenschaften*. His activities were many and various. His astronomical work dealt principally with the problem of improving the constants of aberration and nutation, and with the construction of planetary tables. His tables were based on Laplace's theory: *Tabulae Veneris novae et correctae ex theoria cl. de Laplace et ex observationibus recentissimis in specula astronomica Seebergensi habitatis erutae* (Gotha, 1810), tables of Mars with an analogous title (Eisenberg, 1811), and *Investigatio novae orbitae a Mercurio circa Solem descriptae cum tabulis planetae* (Gotha, 1813). Lindenau's tables were employed in the *Nautical Almanac* from 1834 to 1863 for Mercury, to 1864 for Venus, and to 1865 for Mars.

The solar and planetary tables of Le Verrier, Newcomb, and Hill

The solar and planetary tables employed in ephemerides during the first half of the nineteenth century were all more or less directly the outcome of Laplace's work as set forth in the *Mécanique céleste*. Urbain Jean Joseph Le Verrier (1811–77) was the first to make a new beginning and to undertake to treat the motions of all the planets (see Chapter 28). Le Verrier's theory and tables of the Sun appeared in Vol. 4 of the *Annales de l'Observatoire de Paris* (1858), his theory and tables of Mercury in Vol. 5 (1859), and his theory and tables of Venus and Mars in Vol. 6 (1861). In Vol. 12 (1876) his tables of the motions of Jupiter and Saturn "based on comparison of the theory with the observations" appeared. The theories of the motions of Uranus and Neptune appeared in Vol. 13 (1876), and the corresponding tables in Vol. 14 (1877), some months after Le Verrier's death.

A. J.-B. Gaillot (1834–1921), Le Verrier's sole collaborator from 1861, played an important role in the improvement of these tables. In 1873 he became director of the Bureau of Computation of the Paris Observatory; in 1897 he became assistant to the director of the observatory; and in 1908 he was made a *Correspondant* of the Academy of Sciences. The discrepancy between the observed positions of Saturn and the positions calculated by Le Verrier had reached 4″. Gaillot attributed this discrepancy to that fact that Le Verrier had neglected the terms of the third order in the masses, and undertook to recompute the tables, not by the method of Le Verrier which had led to inextricable calculations, but by means of a 'method of interpolation'. Thus Gaillot established his *Tables rectifiées*

du mouvement de Saturne, published in 1904 in Vol. 24 of the *Annales de l'Observatoire de Paris*. In Vol. 28 (1910) he published his *Tables nouvelles des mouvements d'Uranus et de Neptune*, in which he followed Le Verrier's method but re-did the entire calculation in order to take account of modified values of the perturbing masses. Finally, Gaillot rounded off this very considerable body of work with his *Tables rectifiées du mouvement de Jupiter*, which were published in 1913 in Vol. 31 of the *Annales de l'Observatoire de Paris*. He justified these new tables, like those for Saturn, on the grounds that, Le Verrier having neglected the terms of the third order in the masses, the departures of the calculated from the observed positions varied periodically in an unacceptable way.

The tables of Le Verrier and Gaillot were employed in the computation of ephemerides for the *Connaissance des temps* until 1984, and up to the same date, tables derived from the theories of Le Verrier were the basis of the ephemerides published in the *Nautical Almanac*; the errors in position being of the order of 1″ to 3″ for the large planets but reaching 19″ for Venus. *The American Ephemeris and Nautical Almanac* did not employ Le Verrier's tables, although Joseph Winlock (1826–75), who worked in the Nautical Almanac Office of the US Naval Observatory from 1852 and was superintendant of this office from 1858 to 1859 and from 1861 to his death in 1875, established tables of the motions of Mercury founded on Le Verrier's theory, and these were the basis of the ephemeris of this planet in *The American Ephemeris* from 1865 to 1899.

The new theories constructed in the United States under the direction of Simon Newcomb (1835–1909) led to the establishment of tables which appeared in the *Astronomical Papers of the American Ephemeris* (hereinafter abbreviated as *Astronomical Papers*). The tables of Jupiter and Saturn were constructed by George William Hill on the basis of his theory which employed the method of Peter Andreas Hansen (1795–1874). They were published in Vol. 7, parts 1 and 2, of the *Astronomical Papers* (1895). Provisional tables constructed on the basis of observations made before 1830 had been published in Vol. 4 of the *Astronomical Papers*; the tables of 1895 employed observations extending to 1888.

Parts 3 and 4 of Vol. 7 contain the tables of

Uranus and Neptune, respectively. They were the work of Newcomb. After having established the tables of Neptune, then the last known planet of the solar system, Newcomb investigated whether the residues in the observed longitudes might not betray the action of an unknown planet. He found these residues very high, but he could not discern in them any systematic character enabling him to exploit his hypothesis. In 1855 Marian A. Kowalski, a professor at the University of Kazan, had published a theory of Neptune based on his doctoral dissertation, which he had defended in 1852. Tables deduced from this theory were employed in *The American Ephemeris* till 1870. Newcomb used Kowalski's theory of Neptune for his global theory of the planets.

Newcomb's tables of the Sun, Mercury, Venus, and Mars were introduced into *The American Ephemeris* beginning in 1900. His tables of Uranus had been utilized for the ephemerides beginning in 1877, and his tables of Neptune beginning in 1870 (therefore before their publication in *Astronomical Papers*).

The tables of the Moon after 1850

In 1853 the lunar tables of Benjamin Peirce (1809–80) appeared. While still a student at Harvard, Peirce had found his vocation in astronomy when Nathaniel Bowditch asked him to re-read the proofs of the famous Bowditch translation of Laplace's *Mécanique céleste*. His tables of the Moon were based on Plana's theory, with corrections by G. B. Airy and M. F. Longstreth, and two inequalities of long period due to Venus that Hansen had calculated, and also the secular variations of the mean motion and perigee found by Hansen.

Through his works on the Moon published in 1838 and in 1862–64, Hansen made possible a considerable improvement in the precision of the lunar theory. In 1857 he published tables based on his theory, and these were employed from 1862 in the *Nautical Almanac*. Corrections to the right ascension and declination, which Newcomb had calculated, were introduced in 1883, and corrections in the parallax and semi-diameter were introduced in 1897. These tables were employed for the ephemerides until 1922, when they were replaced by Brown's tables based on the theory of Hill–Brown.

Charles-Eugène Delaunay's purely analytic

theory of the motion of the Moon was published in the *Mémoires* of the Academy of Sciences, Vols. 28 (1860) and 29 (1867). In presenting Vol. 29 to the Academy, Delaunay stated that he proposed to publish in a third volume the solution of various accessory questions. There followed his "Expressions numériques des trois coordonnées de la lune" in the *Connaissance des temps* for 1869; and various additional notes appeared up to his accidental death in 1872.

But Delaunay had not awaited completion of his theoretical researches to initiate preparation of tables based on his theory. The first sheets, calculated with the support of the Bureau des Longitudes, were presented by Victor Puiseux in 1873 after Delaunay's death. No further progress was then made till 1878, when F. F. Tisserand supervised the resumption of calculations, responsibility for which he assigned to Leopold Schulhof. Progress, however, was slow. Hansen's tables of 1857 were believed to be very exact, and were of a convenient form; an incentive was thus lacking for the construction of other tables. In 1880 Newcomb and J. Meier published in *Astronomical Papers*, Vol. 1, part 2, a work entitled "Transformation of Hansen's lunar theory, compared with the theory of Delaunay", in which they showed that the two theories were in excellent agreement, at least for the century 1750–1850, which was the period of the observations used by Hansen. A note due to Tisserand, "Sur l'état actuel de la théorie de la Lune", published in *Bulletin astronomique* in 1891, and numerous corrections effected by Henri Andoyer, gave impetus to the carrying through of the calculations for the lunar tables derived from Delaunay's theory.

These tables, completed under the direction of E. Radau, appeared in 1911 in the *Annales du Bureau des Longitudes*. Brown's theory, a completion of Hill's lunar theory, had been published in 1908, although not yet put into tables; and Radau, in a long introduction to Delaunay's tables, gave grounds for concluding that tables derived from Brown's theory would prove more accurate. This judgement was later confirmed; the tables of Delaunay–Radau were employed in the *Connaissance des temps* from 1915 to 1925, and were then supplanted by those of Brown.

The theory of Hill–Brown, which took for its starting-point an intermediate orbit calculated in a system of rotating coordinates (see Chapter 28), thus showed itself superior to Delaunay's theory, in particular because the introduction from the beginning of the well-established numerical value of the ratio of the mean motions of the Sun and Moon avoided severe problems of convergence. This theory was put into tables by Brown himself, assisted by H. B. Hedrick. The tables were published in six volumes in 1919 under the title *Tables of the Motion of the Moon*, and were employed for the construction of ephemerides from 1923 to 1959. The decision was then made to renounce tables and, starting in 1960, to employ directly Brown's series, as corrected and completed in the *Improved Lunar Ephemeris*.

Conclusion

The year 1960 can be considered as marking the end of a period stretching over nearly three centuries, which saw the appearance of ever more carefully worked-out ephemerides, derived from tables which in the eighteenth century had come to be deduced from analytic or semi-analytic theories. The advent of electronic computers made it possible, first to program the existing theories, then directly to construct much more accurate ephemerides, using either numerical integrations or new theories that depended on algebraic manipulations previously unthinkable. In 1984 the ephemerides based on the works of Le Verrier, Gaillot, Hill, Newcomb, and Brown at last disappeared altogether from the *Connaissance des temps* and *The Nautical Almanac*.

To be sure, these publications, the form of which has greatly changed, will continue to exist; the number of useful ephemerides they contain has increased. But henceforth the venerable tables that were used to construct them and which cost so much effort to so many illustrious astronomers will repose on the highest and dustiest shelves of our libraries for eternity.

30

Satellite ephemerides to 1900

YOSHIHIDE KOZAI

By the end of the nineteenth century, the telescope had added to the Moon some twenty-one satellites of primary planets of the solar system. The first four had been the 'Galilean' satellites of Jupiter, discovered by Galileo in 1610 (see Chapter 9 in Volume 2A); the last was the Saturnian satellite Phoebe, the first satellite to be discovered photographically (by W. H. Pickering in 1898). These twenty-one satellites, each with its currently accepted name, period and orbital radius, along with the discoverer and year of discovery, are listed in Table 30.1.

Before the introduction of photographic methods astronomers made position observations of these satellites by measuring the satellite's distance from the planet's centre (or its nearest limb), and determining the angle between the meridian and the line joining the two components (called the 'position angle' of the line), by means of a heliometer or micrometer; or by similarly measuring the mutual distance between two satellites and the position angle. During the nineteenth century the mean error in the distance measurement was reduced to about $0''.5$ when measurements were made with respect to the centre of the planet, while for the mutual distances of satellites the r.m.s. error could reach as low as $0''.1$ in the best cases. For the Galilean satellites of Jupiter, observations of the eclipses, transits and occultations by Jupiter could provide even more accurate position information.

An early motive for careful observation of the Galilean satellites was to develop ephemerides for use in determining geographical longitudes. With the publication in 1687 of Newton's *Principia*, another motive emerged, namely, determination of the masses of the primaries as fractions of the Sun's mass by way of Kepler's third law – values for these masses being crucial to the determination of mutual planetary perturbations. Yet another motive was potent from the first days of telescopic

observation, an interest in satellite systems for their intrinsic beauty and for the analogy they supplied to the solar system as a whole.

The orbital period of any satellite is shorter – usually much shorter – than that of its primary; thus the motions of the satellites can be traced over many orbital periods. Therefore, without the need for any analytical theory, it was usually possible to derive fairly accurate expressions for the motions from observations. Ephemerides of the satellites could then be computed by extrapolation.

Perturbations

In some cases, observation when extended over a few years, besides confirming mean motions, epochs, orbital inclinations and the longitudes of pericentres and nodes, yielded with considerable accuracy the secular motions and the amplitudes and periods of several long-period perturbations. Thus in 1719 James Bradley completed a study of the Galilean satellites of Jupiter, in which he detected a small inequality in the eclipses of the first two satellites, with a period of 437 days, and also discovered the eccentricity of the orbit of the fourth satellite. In the 1740s Pehr Wilhelm Wargentin detected, in addition to the aforementioned inequalities of the first two satellites, a similar inequality in the motion of the third satellite.

The theoretical question was then posed: were these inequalities derivable from Newton's law of gravitation? In a prize essay of 1766 Joseph Louis Lagrange succeeded in deriving the perturbational inequalities that Bradley and Wargentin had detected, but failed to account for a peculiar relation among these inequalities (see Chapter 21). Another foray into theoretical explanation was made by Jean-Sylvain Bailly, using the equations of Alexis-Claude Clairaut's lunar theory. But the first to put the analytical theory of the motion of

Table 30.1

Satellite	Period (days)	Distance from planet (kms)	Discoverer	Year of discovery
*Mars (a = 1.52 AU, P = 1.88 years)**				
Phobos	0.319	9 378	A. Hall	1877
Deimos	1.262	23 459	A. Hall	1877
Jupiter (a = 5.20 AU, P = 11.86 years)				
Io	1.769	422 000	Galileo	1610
Europa	3.551	671 000	Galileo	1610
Ganymede	7.155	1 070 000	Galileo	1610
Callisto	16.589	1 883 000	Galileo	1610
Amalthea	0.498	181 000	E. E. Barnard	1892
Saturn (a = 9.55 AU, P = 29.46 years)				
Mimas	0.942	185 400	W. Herschel	1789
Enceladus	1.370	238 200	W. Herschel	1789
Tethys	1.888	294 600	Cassini I	1684
Dione	2.737	377 400	Cassini I	1684
Rhea	4.518	526 800	Cassini I	1672
Titan	15.945	1 200 000	C. Huygens	1655
Hyperion	21.277	1 482 000	W. C. Bond, W. Lassell	1848
Iapetus	79.330	3 558 000	Cassini I	1671
Phoebe	550.480	12 960 000	W. H. Pickering	1898
Uranus (a = 19.22 AU, P = 84.02 years)				
Ariel	2.520	191 800	W. Lassell	1851
Umbriel	4.144	267 300	W. Lassell	1851
Titania	8.706	438 700	W. Herschel	1787
Oberon	13.463	568 600	W. Herschel	1787
Neptune (a = 30.11 AU, P = 164.77 years)				
Triton	5.877	353 600	W. Lassell	1846

Note:

*a is the mean solar distance of the planet in the astronomical unit, which is 1.496×108 km; P is the planet's period.

satellites on an adequate mathematical basis was Pierre-Simon Laplace; his final account of the theory appears in Vol. 4 of his *Traité de mécanique céleste* (1805). A refinement of the theory is given in Vol. 4 of François Félix Tisserand's work of the same title (1896).

Celestial mechanics, here as elsewhere, undertook to compute positions with an accuracy equal to that of the observations. Among satellites, different cases may be distinguished according to whether the solar perturbation or the non-sphericity of the planet has a preponderant effect. The solar perturbing force is proportional to $(n'/n)^2$, where n' is the mean motion of the planet about the Sun, and n is that of the satellite about the planet.

The dynamical oblateness factor is measured by J_2/a^2, where J_2 is the second-order zonal harmonic for the planet and a is the mean distance from the satellite to the planet's centre. The latter factor is evidently greater for satellites close to their primaries, while the solar factor is larger for satellites with smaller mean motions, hence farther from the primary.

Computation of periodic perturbations as required for planets, is usually unnecessary for satellites. However, in the case of satellites for which the solar perturbations are appreciable, long-period perturbations must be included. In the case of a satellite for which the dynamical oblateness factor is dominant over the solar factor, the inclination of the satellite's orbital plane to the planet's equator is nearly constant; in the contrary case, the inclination of the two orbital planes is nearly constant. For intermediate cases, the pole of the orbital plane moves around a point between the pole of the planet's equator and that of its orbital plane.

The pericentre and the ascending node move slowly with a rate proportional to the oblateness factor or the solar action. If the eccentricity and the inclination are very small, they move with the same rate but in opposite directions. From the secular motions of the longitudes of the pericentres and the nodes, the dynamical oblateness can be determined.

There are several pairs of satellites whose motions are especially interesting from a dynamical point of view. These are pairs for which simple relations hold between their mean motions, namely, the inner three of the Galilean satellites, and the Saturnian satellites Mimas–Tethys, Enceladus–Dione, and Hyperion–Titan. Since Rhea is very close to Titan, a giant satellite, its motion suffers a large perturbation due to Titan. By determining amplitudes and/or periods of certain of the perturbation terms, the masses of several of the disturbing satellites can be derived. The interactions in many cases involve pendulum-like librations in one parameter or another, providing case-studies in mechanical resonance and stability. The relative shortness of satellite periods as compared with planetary periods provides a magnification of the time scale, making satellite systems of special interest for the study of secular and long-period perturbations.

Ephemerides

In their satellite ephemerides, the earlier editions of the *Nautical Almanac* and *Connaissance des temps* gave only predictions of eclipses and diagrams for configurations of the Galilean satellites of Jupiter. From 1855, the first year for which it was issued, to 1881, the *American Ephemeris* gave ephemerides only for these four satellites; in 1882 it added diagrams for the configurations of these satellites, and at the same time introduced ephemerides of the epochs of elongations of the satellites of Mars, Saturn, Uranus and Neptune as well as diagrams of their apparent orbits. In 1896 the *Connaissance des temps* introduced ephemerides of the epochs of

elongations for the satellites of Saturn, Uranus and Neptune, and in 1899 it added several tables useful for computing ephemerides of these satellites. In 1899 the *Nautical Almanac* also introduced diagrams of the apparent orbits at the time of opposition for the satellites of Mars, Saturn, Uranus and Neptune. Ephemerides for the satellites were computed annually by Albert Marth and published in *Monthly Notices of the Royal Astronomical Society* and in *Astronomische Nachrichten*: for Uranus from 1870, for Saturn from 1873, for Neptune from 1878, and for Mars from 1883.

Mars

The two satellites of Mars were discovered in 1877 by Asaph Hall (1829–1907) at the US Naval Observatory, using the 26-inch (66-cm) refractor built shortly before by Alvan Clark & Sons. Hall first observed the outer satellite on 11 August, while he was measuring positions of white spots on Mars; but cloudy weather prevented its certain recognition as a satellite that night. On 16 August he observed it again, and established its motion over a period of two hours, during which it moved over 30 arc-seconds. The following night he discovered the inner satellite. The discoveries were confirmed by the Clarks using the 26-inch refractor they then had under construction for the Leander McCormick Observatory, and by E. C. Pickering and his assistants at Harvard. On 19 August the Smithsonian Institution was informed, and from there the news was transmitted worldwide.

The orbital elements of the satellites (and the mass of Mars itself) were computed by Simon Newcomb using the data of ten days of observations, and published in *Astronomische Nachrichten*. Early the following year Hall named the satellites Phobos and Deimos, thereby adopting a suggestion emanating from Britain.

The satellites were again observed at the US Naval Observatory in 1879, 1892, 1894 and 1896. In 1894 W. W. Campbell made numerous micrometric measurements with the 36-inch (91-cm) refractor of Lick Observatory. Meanwhile, the satellites came under the scrutiny of Karl Hermann Struve (1854–1920) and others at Pulkovo in 1894 and 1896.

For the two Martian satellites, the dynamical oblateness effects are dominant over the solar perturbations, and therefore the inclinations to the equator (0°.96 and 1°.73) are almost constant. The eccentricities, 0.017 and 0.003, are also small. Hence their motions are relatively simple and without large perturbations. However, the longitudes of the pericentres and the ascending nodes move secularly with rates of 158° per year for Phobos and 6°.37 for Deimos. In the twentieth century the mean motions of the satellites have been found to be subject to secular accelerations.

Jupiter

Amalthea

The fifth satellite of Jupiter, Amalthea, was discovered on 9 September 1892 by the noted observer, E. E. Barnard, using the Lick Observatory 36-inch refractor. Its existence was confirmed the following month by Ormond Stone of the Leander McCormick Observatory, and by Taylor Reed of Princeton. Barnard himself made extensive observations from first discovery through to the end of 1894, and meanwhile Struve observed it at Pulkovo from October 1893 to January 1895. The observations of Barnard and Struve were analysed by Fritz Cohn of Königsberg to determine the orbital elements of the satellite and the mass of Jupiter itself. In addition, he was able to determine the dynamical oblateness of Jupiter from the annual secular motion of the ascending node, which was derivable from observations.

The satellite's eccentricity (0.003) and inclination with respect to the equator of Jupiter (0°.4) are small. The secular motions of the longitudes of the pericentre and the node are 914° per year, and are mainly due to the oblateness of Jupiter. Thus the motion can be expressed in very simple form, except for a secular acceleration term in the mean longitude, which was discovered in this century.

The Galilean satellites

In the eighteenth and nineteenth centuries many observations were made of the Galilean satellites of Jupiter. Most of these observations were of eclipses of the satellites by the parent planet, and they were usually published in *Monthly Notices of the Royal Astronomical Society* or in *Astronomische Nachrichten*. Among the mean motions of the inner three satellites there is a remarkable relation. The mean motion of Io, n, is a little more than twice as large as that of Europa, n', which in turn is a little more

than twice as large as that of Ganymede, n''. Moreover, the following relation:

$$n - 2n' = n' - 2n''$$

almost exactly holds, so that

$$n - 3n' + 2n'' = 0. \qquad (1)$$

In fact, if one forms the expression $\Theta = \lambda - 3\lambda' + 2\lambda''$, where λ, λ' and λ'' express the mean longitudes of the satellites, then Θ is found to librate around 180° with a very small amplitude and a period of six years. The near constancy of this relation, which had been recognized by Lagrange, was first explained by Laplace, and so it is termed 'the Laplace relation'.

Laplace rounded out the theory of motion of the four satellites by determining from observations the mass and the dynamical oblateness of Jupiter and the masses of the satellites. J.-B. J. Delambre published in 1817 a table for computing positions of the Galilean satellites based on Laplace's theory and on observations made between 1662 and 1802. W. S. B. Woolhouse published new tables in the *Nautical Almanac* for 1835, and other tables were published by C. T. de Damoiseau the following year. In the *Nautical Almanac* for 1881 John Couch Adams extended Damoiseau's tables to later dates; these extended tables were then used in the preparation of the ephemerides of eclipses and the drawings of configurations of the satellites that appeared in the *Connaissance des temps* and the *Nautical Almanac*. A more complete theory was developed by Cyrille Souillart of Lille and published in *Memoirs of the Royal Astronomical Society* in 1880; Souillart's theory is the chief basis for the account of the motions of the Galilean satellites in Tisserand's *Traité de mécanique céleste*.

The eccentricities of the orbits of the Galilean satellites and their inclinations to the equator of Jupiter are very small: for Io, Europa, Ganymede and Callisto, they are 0.0000, 0.0003, 0.0015 and 0.0075, and 0°.03, 0°.47, 0°.18 and 0°.27 respectively. Therefore, in spite of the commensurabilities, the inequalities in the mean longitudes with the arguments $2\lambda' - \lambda$ and $2\lambda'' - \lambda'$ are small, and were barely detectable in observations. However, the terms with such arguments produce appreciable perturbations in the eccentricities and the longitudes of the pericentres for the three inner of the four satellites. In consequence, the following terms appear in the expressions for the true longitudes of these satellites:

$$25'.9 \sin 2(\lambda - \lambda'),$$
$$-61'.5 \sin (\lambda - \lambda'), \qquad (2)$$
$$-3'.8 \sin (\lambda' - \lambda'').$$

Even in the eighteenth century, the effects represented by these terms could be detected in the timing of eclipses. From the period of the inequality of each satellite, together with the synodic period of that satellite with respect to the Sun, one can compute the period in which the inequality returns to the same phase in eclipses; for all three satellites this proves to be 437.6 days (see Chapter 21). The amplitudes of the inequalities, expressed in Equations (2) as maximum angular deviations in the Keplerian motion about Jupiter, imply differences in the times of onset of eclipses; the greatest maximum time-difference is that for the second satellite and comes to about 14 minutes. These time-differences were the inequalities that Bradley and Wargentin had detected.

Souillart corrected the expressions in Equations (2) by adding terms in the disturbing function with the arguments $4\lambda' - 2\lambda - 2\varpi$, $4\lambda' - 2\lambda - \varpi - \varpi'$, $4\lambda' - 2\lambda - 2\varpi'$, $4\lambda'' - 2\lambda' - 2\varpi'$, $4\lambda'' - 2\lambda' - \varpi'$ $- \varpi''$, and $4\lambda'' - 2\lambda' - 2\varpi''$, where ϖ, ϖ' and ϖ'' are the longitudes of the pericentres of the three satellites. The resulting corrections for the amplitudes in Equations (2) were $-91''$, $+186''$ and $-36''$.

When these perturbations are taken into account, it can be shown that the critical argument, $\Theta = \lambda - 3\lambda' + 2\lambda''$, satisfies the following equation:

$$\frac{d^2\Theta}{dt^2} = -K\left(\frac{m'm''}{a^2} + \frac{9m''m}{a'^2} + \frac{4mm'}{a''^2}\right)\sin \Theta, \quad (3)$$

where K is a positive constant depending on the ratios of the semi-major axes, and m, m' and m'' are the masses of the three satellites, their products two by two indicating that mutual perturbations of the satellites are involved. Since Equation (3) is the equation for a pendulum, it follows that Θ librates; the period of this libration is computed to be about six years, which is of the order of the orbital period divided by the mass of one satellite.

The secular motions of the pericentres and nodes for the four satellites were determined from observations, and from these data the mass of the fourth

satellite and the dynamical oblateness of Jupiter were derived. Finally, the equations for the secular perturbations were solved simultaneously as in the case of planetary theory, the mutual perturbations being taken into account, along with the actions due to the Sun and the oblateness of Jupiter, and even the precession torque exerted on Jupiter by the Sun and the satellites, as expressed in the node-inclination equations. In this way, as new observations became available, the masses of the satellites and the mass and oblateness of Jupiter, as well as the orbital elements and their time variations, were improved step by step.

Saturn

Phoebe

Phoebe was one of nine satellites of Saturn known by the end of the nineteenth century (see Table 30.1), and the first to be discovered on photographic plates, the plates in question having been taken on 16 and 18 August 1898 by W. H. Pickering using the 13-inch (33-cm) Bruce photographic telescope at the Boyden Station of Harvard Observatory at Arequipa, Peru. In 1904 A. C. D. Crommelin was able to infer, from the 1898 position and from four positions determined in 1904, that Phoebe's motion is retrograde; the orbital inclination was later found to be 175° with respect to the ecliptic, and the orbital eccentricity to be 0.166. The solar perturbation is rather large and the period of long-period perturbations is long, of the order of ten years; as a result the perturbations are expressed by polynomials with time as the variable, as in planetary theory. The secular motions for the pericentre and node are slow: 0°.27 and 0°.435 per year, respectively. The expressions for the orbital elements are therefore very simple, being in fact linear functions of time.

Iapetus

Iapetus was discovered by G. D. Cassini (Cassini I) at Paris on 25 October 1671, with a small glass of 17-ft (5.2-m) focal length. Soon after, Cassini noticed a difference in brightness of the satellite when in different parts of its orbit – a phenomenon now well known, not only for this satellite but also for several others. He hypothesized that the reflectivity of the satellite was different on opposite sides, and that the period of the satellite's axial rotation was equal to its orbital period, as is the case for the Earth's Moon. These hypotheses have been corroborated by later investigators. Continuing the observations, Cassini was able by 1673 to obtain an approximate value of 80 days for the orbital period of Iapetus.

Reports by Jacques Cassini (Cassini II) of observations of the satellite consist of drawings of the planet and its ring with the relative position of the satellite indicated by a cross or mark; as to time of observation, only the date is supplied. Nevertheless, from these observations it is possible to infer the approximate times of the superior and inferior conjunctions of Iapetus in March and May 1685 and in May 1714, and these are still useful for the determination of the orbital period when employed in combination with observations of later years.

Few observations of Saturnian satellites were made between 1714 and 1787, when attention was again directed to them by J.-J. L. de Lalande in a paper on the motion of Iapetus. Lalande's call for data resulted in observations by P. J. Bernard at the Marseilles Observatory during the opposition of 1787 and by William Herschel during the opposition of 1789. Lalande analysed these observations, and on the basis of the knowledge thus acquired published tables of the motion of Iapetus in the *Connaissance des temps* for 1791 and 1792.

There then followed a forty-year gap, before observations by F. W. Bessel with a heliometer were published in volumes of the *Astronomische Beobachtungen . . . zu Königsberg* in the 1830s. John Herschel, during his stay at the Cape of Good Hope, observed the position angles of the satellite between 1835 and 1837, and another English amateur, the wealthy brewer William Lassell (1799–1880), made a series of observations in 1850. An extended series of observations was made at Madras by W. S. Jacob in 1856–58 and published in the *Memoirs of the Royal Astronomical Society*. Then, in 1874, the US Naval Observatory commenced a series of observations of Iapetus with the 26-inch refractor. In 1885 Hall published an analysis based on observations up to and including those of 1884 in an Appendix to the *Washington Observations*.

The peculiar characteristics of the motion of Iapetus's orbital plane are due to the fact that the oblateness and the solar factors are of the same order of magnitude. As a consequence, the inclinations with respect to the equator and the orbital plane of Saturn – the latter two planes being inclined to one another by 28° – are not constant, but the pole of the orbital plane moves slowly along an ellipse defined by

$$K \cos^2 i + K' \cos^2 i' = \text{const}, \qquad (4)$$

where K and K' represent the oblateness and the solar factors, respectively, and i and i' represent the inclinations with respect to the equator and the orbital plane of Saturn. As previously mentioned, if K is dominant, i is nearly constant, while if K' is dominant, i' is nearly constant; but if, as here, K and K' are of the same order, then the motion of the orbital plane is complicated.

This strange motion was noticed by Laplace, who undertook to explain it by a theory of the secular perturbations in his *Traité de mécanique céleste* (Vol. 4, Bk VIII, Chap. 17). However, as the orbital period of Iapetus and, therefore, the secular period of the ascending node are rather long, the inclination to the ecliptic, i, and the longitude of the ascending node, Ω, as well as the longitude of the pericentre, ϖ, can be expressed by linear functions of the time, the constants in the expressions being derived from observations at several oppositions:

$$i = 18° \ 33' \ 39''.5 - 10''.80t,$$
$$\Omega = 142° \ 26' \ 41''.4 - 126''.00t, \qquad (5)$$
$$\varpi = 353° \ 14' \ 56''.5 + 86''.28t,$$

where t is measured in the unit of a year from the epoch (1880 March 17.0 GMT) and i, Ω and ϖ are referred to the mean equinox of 1880.0.

Hyperion

The faint satellite Hyperion was discovered independently in September 1848 by W. Lassell and by W. C. Bond of the Harvard Observatory. In 1884 Hall pointed out in the Royal Astronomical Society's *Monthly Notices* a peculiar phenomenon in its motion: the pericentre moved in a retrograde direction at the rate of 18° per year, an effect that cannot be explained as due either to the oblateness of Saturn or to the solar action. Newcomb noticed that the critical argument, $\Theta = 4\lambda' - 3\lambda - \varpi'$, where λ' and λ are, respectively, the mean longitudes of Hyperion and Titan and ϖ' is the longitude of the pericentre of Hyperion, librates around 180°

with an amplitude of 36° and a period of 640 days; the following equation holds:

$$\frac{d^2\Theta}{dt^2} = -K\,m\,\sin\Theta, \qquad (6)$$

where m is the mass of Titan and K is a positive constant.

Because Hyperion's mean motion (n') is to Titan's (n) very nearly as three to four, Θ librates around 180° with a period of the order of the orbital period divided by the square root of m. As a consequence, Hyperion can avoid very close approach to Titan, because any conjunction of the two satellites takes place only near the apocentre of Hyperion (the eccentricity of Hyperion is large, namely 0.109). The secular motion of the pericentre of Hyperion's orbit is not due to the oblateness of Saturn but is equal to $4n' - 3n$ because of the libration of the critical argument.

The motion of Hyperion was studied in the 1880s by Stone, Tisserand, and G. W. Hill, all of whom sought to solve the problem by computing perturbations. Then in 1892 W. S. Eichelberger published in the *Astronomical Journal* a complete analysis of observations at the US Naval Observatory for the period 1875–90, and derived the orbital elements and their time variations as well as the masses of Saturn and Titan.

Titan and Rhea
Titan, the second largest satellite in the entire solar system, was discovered on 25 March 1655 by the Dutch physicist Christiaan Huygens with his 12-ft (3.6-m) refractor, while Rhea was found by Cassini I on 23 December 1672.

As Titan is a bright object, position measurements of it are relatively easy. However, Huygens, and likewise Bernard in 1787, gave only estimates of conjunction times. In 1829 Bessel began observations, later published in *Astronomische Nachrichten*, with the heliometer of the Königsberg Observatory; he was able to determine with accuracy the orbital elements of the satellite and the mass of the planet. Bessel also derived expressions for the solar perturbations of Titan, which are appreciable, and he published tables for computing positions of this satellite.

In 1887 Hall published in the *Washington Observations* an analysis of observations of Titan made at

the US Naval Observatory between 1874 and 1884, and determined its orbital elements. The following year H. Struve published his analysis of the orbits of Saturn's satellites including that of Titan. For this his main sources were recent observations by Bessel, Jacob, Newcomb, W. Meyer and himself, but he took into account eighteenth-century observations of conjunctions by Cassini I, Bradley, William Herschel and J. G. Köhler, and even two each by Edmond Halley and Cassini I from the 1680s. He derived the orbital elements and their time variations including the solar and secular perturbations with the difference of the longitudes of the pericentres of Titan and Iapetus as argument, and determined the mass of Iapetus as well as the position of the equator of Saturn from the motion of the orbital plane.

As the orbital elements of Titan had now been well established and are rather stable, and as Titan is a very bright object, positions of other satellites besides Iapetus could thereafter be measured with respect to Titan. Such observations led to repeated improvements in the orbital elements of both Titan and Rhea. (Another satellite that has been adopted as a reference is the inner satellite Tethys, the orbit of which has an extremely small eccentricity, in most computations equatable to zero.)

For Rhea, H. Struve made an analysis similar to the one he had made for Titan. He used observations of conjunctions made by Cassini I in 1689 and 1704 and by Bradley in 1719, as well as more recent ones, and derived the orbital elements and their time variations. But since the orbital eccentricity of Rhea is only 0.0010, he could not determine the longitude of the pericentre precisely, and so for the secular motion of the pericentre he gave the theoretical value due to the oblateness of Saturn and the actions of the Sun and Titan. Later, however, it was found that the eccentricity of Rhea is the forced one due to the perturbation by Titan, and as a consequence the longitude of the pericentre of Rhea moves secularly with the same rate as that of Titan. Although the mean motion of Rhea does not have any special relation with that of Titan, the orbits of the two satellites are very close to each other and Titan is massive, which is why the orbit of Rhea suffers this large perturbation.

The four inner satellites

Despite the enormous length of the aerial tele-scopes he was using, Cassini I succeeded in disco-vering Tethys and Dione on 21 March 1684, just before the ring disappeared. Enceladus was in fact seen by William Herschel in 1787, but went un-recognized at the time. On 18 July 1789, when the ring had almost disappeared (the condition most favourable for observing faint inner satellites) he saw it again, but took it to be Tethys. He did not identify it as a new satellite until 28 August, when he was inaugurating his 40-ft (12.4-m) reflector. A few weeks later he discovered Mimas. Since Mimas is very near the ring, and faint, it was very difficult to observe, and no routine observations were poss-ible until the middle of the nineteenth century with even the best instruments. Herschel's 1789 obser-vations of Mimas and Enceladus were analysed in 1836 by Wilhelm Beer and Johann Heinrich Mädler and the orbital elements derived.

Until the 1789 discoveries, the four satellites previously known were referred to by number in order of the increasing size of their orbits. The new discoveries made this scheme obsolete, and the names we now use were later introduced by John Herschel.

Beginning in the middle of the nineteenth century, the positions of the four inner satellites were the subject of micrometer measurements by numerous observers, among them Lassell. In 1887 Hall determined the orbital elements and the mean motions of Rhea and the four inner satellites from observations made at the US Naval Observatory. He also published tables for the computation of the positions of the six inner satellites.

The ratio of the mean motions of Enceladus and Dione is nearly 2 to 1, and the critical argument, $2\lambda' - \lambda - \varpi$, librates around $0°$ with a small ampli-tude. The observed eccentricity of Enceladus, 0.0045, is the forced one due to Dione, the free one being much smaller (0.0001); the longitude of the pericentre of Enceladus moves at the rate $2n' - n$. From this resonance arise detectable inequalities in the mean longitudes of the two satellites, observed to be $15'$ at maximum for Enceladus and $1'$ for Dione; from these empirical constants were derived the masses of the two satellites.

The ratio of the mean motions of Mimas and Tethys is also nearly 2 to 1, and for this pair the critical argument, which librates about a mean value, is $4\lambda' - 2\lambda - \Omega' - \Omega$, where Ω and Ω' are the longitudes of the ascending nodes, and the ampli-tude and period of the libration are $98°$ and 72 years respectively. As a consequence of this lib-ration, large inequalities appear in the mean longi-tudes, the amplitudes being $43°.4$ for Mimas and $2°.1$ for Tethys. From the observed amplitudes and period of the libration the masses of the two satellites were derived.

H. Struve carried out a complete analysis and theoretical investigation for the two pairs in 1890; for Mimas–Tethys it involved the study of obser-vations going back to William Herschel's of 1789. He also derived the dynamical oblateness of Saturn by using the observed secular motions for the six inner satellites, the masses of most of them being known from other sources. In addition, he deter-mined from satellite observations the position of the equator of Saturn, which is believed to coincide with the ring plane.

Uranus

The outer two satellites of Uranus, now called Titania and Oberon, were discovered by William Herschel in January 1787, nearly six years after his discovery of the planet itself. On 7 February he observed the satellites for nine hours to confirm their nature, and from subsequent observations he determined the mean radii of the orbits, the periods of revolution, and the mass of Uranus. He found also the remarkable fact that the orbital planes of the satellites are almost perpendicular to the eclip-tic. Later he announced, but not very confidently, the discovery of four other satellites, which he termed *supplementary*; because of difficulties in measuring their distances from Uranus, he was unable to assign sizes to their orbits. At least some of these 'supplementary satellites' must have been phantoms or incorrectly identified objects.

Herschel continued to make micrometric mea-surements of the satellites of Uranus until 1798, but thereafter no one attempted to observe them until his son John took up the challenge in 1828. Uranus, having then acquired southern declina-tion, was in an unfavourable position for obser-vation of the satellites, and he was able to observe only the brightest two, Titania and Oberon. He was unable to measure their distances from the planet even after he repolished his telescope mirror in 1830, but on the basis of his father's observations and his own, he was able to determine accurately their orbital periods.

In 1837 Johann von Lamont of the Royal Observatory at Munich observed the satellites for a month, using a 10.5-inch (22-cm) refractor with a micrometer. He reported that he observed three satellites, one of them only once. Using the orbital sizes and periods of Titania and Oberon, he determined a value for the mass of Uranus; this is smaller, however, than the value accepted today.

In September 1847 Lassell began observing and measuring the positions of the satellites. Besides the two bright satellites discovered by William Herschel, he believed he was seeing one or two inner satellites – satellites that Herschel had probably observed – but he was unable to find any satellite with orbit larger than that of Oberon. He succeeded in obtaining data on one of the faint inner satellites. That same autumn, Otto Struve at Pulkovo observed the satellites with a 15-inch (38-cm) refractor, and succeeded in measuring the size and period of the orbit of one of the inner satellites.

Towards the end of the same year, W. R. Dawes, who had observed with Lassell, undertook an analysis of Lassell's and Struve's observations, showing that they were not consistent if they were of the same satellite, and concluding that the two observers had been observing two different satellites, one with a period of 2.1108 days and the other with a period of 3.9236 days. Because of different reflectivities of the satellites on different sides, and the difference in the times at which Lassell and Struve had been able to observe, each failed to see what the other had seen. Finally, in 1851, Lassell announced the discovery of the two inner satellites, now called Ariel and Umbriel, which his observations of October and November 1851 showed to have periods and mean radii in accordance with Kepler's third law. Both Lassell and Struve later observed all four Uranian satellites extensively.

Soon after the 26-inch refractor came into use at the US Naval Observatory in 1873, Newcomb commenced observations of the satellites of Uranus, and these he continued during the oppositions of 1874 and 1875. From them he determined the orbital elements of the satellites and the mass of the planet. His orbital elements were to be used for the computation of ephemerides for many years. In the early 1880s Hall observed the two outer satellites, in an attempt to improve the accuracy of their orbital elements and hence of the mass and the position of the equator of the planet.

Motions of the four satellites are very simple: all can be taken as moving in circular orbits in the plane of the equator of Uranus, which is inclined to the ecliptic by 98°.

Neptune

On 10 October 1846, only 17 days after the discovery of Neptune by J. G. Galle and H. L. d'Arrest, Lassell glimpsed what he suspected to be a satellite of the planet. It was months afterwards before he could confirm his suspicion; he announced his discovery of Triton, as it was later named, on 8 July 1847. At the same time, he had the impression that he was seeing a ring about the planet. For nearly six years he sought to confirm this impression, but finally concluded that the appearance must be due to faults, and in particular to a flexure, in his 2-ft (61-cm) reflector. The satellite, he early noticed, was much brighter in one half of its path than in the other – a phenomenon explicable, as before mentioned, by supposing the periods of axial rotation and orbital revolution to be equal, and the reflectivities on opposite sides of the satellite to be different. By late 1848 W. C. Bond of Harvard had sufficient observations to allow a determination of the mass of the planet. In 1875 Newcomb was able to derive orbital elements for Triton from an analysis of observations made at the US Naval Observatory over the previous two years.

Triton was the first satellite to be discovered whose orbital motion is retrograde, the inclination of its orbit to Neptune's equator being 160°. (In the twentieth century, five additional planetary satellites have been found to have retrograde motions, namely four outer satellites of Jupiter together with Saturn's Phoebe.) The eccentricity of Triton's orbit is almost zero, while the longitude of the ascending node moves at a rate of 61°.5 per year, mainly owing to the oblateness of Neptune; it is not necessary to take into account any other perturbations. The mass of Triton was later determined from observations of the motion of Neptune around the common centre of mass of planet and satellite.

Conclusion

Through the nineteenth century, observations of planetary satellites and analyses of their motions were rewarded by the knowledge they supplied of the planetary masses, a prerequisite for the calculation of planetary perturbations. An additional reward was acquaintance with the variety and beauty of the satellite systems, their mechanical resonances and librations and such special features as the retrograde motion of Triton. In the twentieth century, particularly since space probes have begun to supply new and more accurate data, a lively interest has arisen in the evolution of satellite systems, as depending on tidal interactions of the primaries and satellites; it is suggested that studies of the evolution of the regular satellite systems of Jupiter, Saturn and Uranus may hold the key to the origin of the solar system. In the late twentieth century, satellites of the outer planets have come to play a role in the navigation of space missions, with resulting new demands on the accuracy of satellite ephemerides.

ILLUSTRATIONS: ACKNOWLEDGEMENTS AND SOURCES

14.1 I. Newton, *Principia*, 3rd edn (London, 1726), p. 384.

14.2 *Ibid.*, p. 377.

14.3 From G. W. Leibniz, Tentamen de motuum coelestium causis, *Acta eruditorum*, Feb. 1689, 82–96.

14.7 From *Histoire et mémoires de l'Académie royale des Sciences*, 1736, Plate 9, Fig. 3.

15.1 Collection of Owen Gingerich.

15.2 R. Outhier, *Journal d'un voyage au nord en 1736 et 1737* (Paris, 1744), p. 96.

15.3 P. Bouguer, *Figure de la terre* (Paris, 1749), Planche VII, courtesy of the Institute of Astronomy, Cambridge University.

16.1 I. Newton, *Principia*, 3rd edn (London, 1726), p. 134.

16.2 Courtesy of Académie des Sciences, Paris

17.1 From *Philosophical Transactions*, vol. 45 (1748), p. 16.

17.2 Courtesy of Académie des Sciences, Paris.

18.1 N.-L. de Lacaille, *Astronomiae fundamenta* (Paris, 1757), title-page.

18.2 N.-L. de Lacaille, *Tabulae solares* (Paris, 1758), title-page.

18.3 From Göttingen *Comentarii*, vol. 3 (1754), p. 393, by permission of the Syndics of Cambridge University Library.

19.1 From *Philosophical Transactions*, vol. 24 (1705), p. 1886, by permission of the Syndics of Cambridge University Library.

19.2 From *Philosophical Transactions*, vol. 49 (1755), p. 350.

19.3 From *Hollandsche Maatschappij der Wetenschappen Verhandelingen*, vol. 2 (1755), 288.

19.4 Broadside by B. Martin, courtesy of the British Library.

19.5 From *Mémoires de l'Académie royale des Sciences*, 1760, pp. 385–6, by permission of the Syndics of Cambridge University Library.

19.6 From *Mémoires de l'Académie royale des Sciences*, 1760, plate II, facing p. 465, by permission of the Syndics of Cambridge University Library.

20.1 Courtesy of Académie des Sciences, Paris.

20.2 Courtesy of Öffentliche Bibliothek der Universität Basel.

21.1 Courtesy of Académie des Sciences, Paris.

22.1 Courtesy of Académie des Sciences, Paris.

22.2 P.-S. de Laplace, *Traité de mécanique céleste*, vol. 1 (Paris, 1799), title-page, by permission of the Syndics of Cambridge University Library.

23.1 Redrawn from Harry Woolf, *The Transits of Venus* (Princeton, NJ, 1959), p. 18.

23.2 Redrawn from H. H. Turner, *Astronomical Discovery* (London, 1904), p. 29.

23.3 Simon Newcomb, *Popular Astronomy* (London, 1878), p. 179.

23.4 From *Mémoires de l'Académie royale des Sciences*, 1757, plate III, preceding p. 251, by permission of the Syndics of Cambridge University Library.

23.5 Engraving by J. G. Kaid, 1771, after a portrait from life by W. Pohl, courtesy of I. Bernard Cohen.

23.6 Courtesy of the Royal Greenwich Observatory Archives, Cambridge, and by permission of the Syndics of Cambridge University Library.

24.1 D. Gregory, *The Elements of Physical and Geometrical Astronomy*, 2nd English edn (London, 1726), plate I.

24.2 *Beobachtung über die Natur vom Herrn Karl Bonnet* (Leipzig, 1766), trans. by J. D. Titius, pp. 7–8.

24.3 From the drawing by W. Watson, RAS MS Herschel W.5/5, no. 3, courtesy of the Royal Astronomical Society.

24.4 W. Herschel, *Collected Scientific Papers* (London, 1912), vol. 1, plate A.

24.5 W. Pearson, *An Introduction to Practical Astronomy*, vol. 2 (London, 1829), plate XVIII.

25.1 Lithograph by E. Ritmüller after the painting by C. A. Jensen, courtesy of the Mary Lea Shane Archives of Lick Observatory.

28.1 Courtesy of Académie des Sciences, Paris.

28.2 Engraving by Samuel Cousins from the painting by Thomas Mogford, 1851, courtesy of the Institute of Astronomy, Cambridge University.

28.4 From *Vistas in Astronomy*, vol. 3 (1960), p. 44.

28.6 G. W. Hill, *Collected Mathematical Works*, vol. 1 (Washington, D.C., 1905), frontispiece.

28.7 S. Newcomb, *Reminiscences of an Astronomer* (London, 1903), frontispiece, courtesy of the Institute of Astronomy, Cambridge University.

28.8 From the photograph of Henri Poincaré in his study, Viollet Collection

29.1 *La Connoissance des temps*, vol. 1 (Paris, 1679), title-page, courtesy of the Royal Greenwich Observatory Archives, Cambridge, and by permission of the Syndics of Cambridge University Library.

29.2 *Nautical Almanac*, vol. 1 (London, 1766), title-page courtesy of Institute of Astronomy, Cambridge University.

COMBINED INDEX TO PARTS 2A AND 2B